WITHDRAWN

BIOLOGICAL AND MEDICAL PHYSICS
BIOMEDICAL ENGINEERING

BIOLOGICAL AND MEDICAL PHYSICS BIOMEDICAL ENGINEERING

The fields of biological and medical physics and biomedical engineering are broad, multidisciplinary and dynamic. They lie at the crossroads of frontier research in physics, biology, chemistry, and medicine. The Biological & Medical Physics/Biomedical Engineering Series is intended to be comprehensive, covering a broad range of topics important to the study of the physical, chemical and biological sciences. Its goal is to provide scientists and engineers with textbooks, monographs, and reference works to address the growing need for information.

Continued After Index

M. Zamir

The Physics of Coronary Blood Flow

With 269 Illustrations

With Foreword by Y.C. Fung

M. Zamir
Department of Applied Mathematics
University of Western Ontario
London, Ontario, N6A 5B7
CANADA
zamir@uwo.ca

Library of Congress Cataloging-in-Publication Data
Zamir, M. (Mair)
    The physics of coronary blood flow / M. Zamir.
        p. cm. — (Biological and medical physics, biomedical engineering)
    Includes bibliographical references and index.
        1. Coronary circulation.   2. Hemodynamics.   3. Blood flow.   I. Title.   II. Series.
    QP108.Z36 2005
    612.1′7—dc22   2005042502

ISBN-10: 0-387-25297-5   e-ISBN: 0-387-26019-6   Printed on acid-free paper.
ISBN-13: 978-0387-25297-1

Printed in the United States of America.   (MP)

9 8 7 6 5 4 3 2 1

springeronline.com

To the memory of my father –
how well I know him now.

Foreword

Everybody is interested in his/her coronary blood flow. I am delighted to read Dr Zamir's clear exposition of the dynamics of the coronary blood flow in this book. I have read many of his scientific papers published in professional journals, as well as his book *The Physics of Pulsatile Flow*, published by Springer-Verlag New York in 2000. His writing is a model of clarity. In an unhurried manner, he dwells on points of conceptual difficulty, and takes his reader to look at the difficulties from as many angles as possible. He offers solutions of difficulties. He enhances true understanding, but never dogmatically. As a reader, I am grateful for that. He loves the word *conundrum*. For example, the heading of Section 10.3 of this book is "Coronary Heart Disease and the Conundrum of Coronary Flow Reserve." That section is, of course, the heart of this book: every reader would be interested in this topic. What does conundrum mean? According to Webster's Dictionary, conundrum means "(1) a riddle whose answer involves a pun, (2) anything that puzzles [1590-1600; pseudo-L word of obscure orig.]" That word prepares the reader. Be patient. Listen! Then I became patient, and I got a great deal of enlightenment out of the book.

The book opens with a beautiful introductory Chapter 1, and concludes with a very serious Chapter 10. I read the whole book, but I would like to make a recommendation to any reader: read these two chapters first. Then you can get a clear picture of the whole book right away. Then, again, for every chapter, I would suggest to a reader to read the first and the last sections of that chapter first, in order to get a completely clear picture of that chapter. Dr Zamir's writing is especially clear and persuasive in these sections. Chapters 2-7 are devoted to lumped models. Chapters 8-10 are devoted to unlumped models. The lumped models are extremely clearly described in this book. They are mature. They are black boxes. The unlumped models are clearly less mature. But they are transparent. We can expect many future developments in the unlumped models, especially for the myocytes which are nourished by the coronary blood flow. Thus, we expect attention to microcirculation, venous return, cellular mechanics, effects of stress and strain on and in the cells, heart muscle tissue remodeling, gene expression activities, protein configuration changes, cell membrane behavior, integrins, enzymes, kinases, and their activities. Before long, besides learning about the heart, we have to learn how to engineer the coronary blood vessels for health and longevity. Physics and mathematics will have a lot more to do with biology!

Y.C. Fung
La Jolla, California
November 20, 2004

Series Preface

The fields of biological and medical physics and biomedical engineering are broad, multidisciplinary and dyanmic. They lie at the crossroads of frontier research in physics, biology, chemistry, and medicine. The Biological & Medical Physics/Biomedical Engineering Series is intended to be comprehensive, covering a broad range of topics important to the study of the physical, chemical and biological sciences. Its goal is to provide scientists and engineers with textbooks, monographs, and reference works to address the growing need for information.

Books in the series emphasize established and emergent areas of science including molecular, membrane, and mathematical biophysics; photosynthetic energy harvesting and conversion; information processing; physical principles of genetics; sensory communications; automata networks, neural networks, and cellular automata. Equally important will be coverage of applied aspects of biological and medical physics and biomedical engineering such as molecular electronic components and devices, biosensors, medicine, imaging, physical principles of renewable energy production, advanced prostheses, and environmental control and engineering.

Elias Greenbaum
Oak Ridge, TN

Preface

Coronary blood flow is blood flow to the heart for its own metabolic needs. In the most common form of heart disease there is a disruption in this flow because of obstructive disease in the vessels that carry the flow. The subject of coronary blood flow is therefore associated mostly with the pathophysiology of this disease, rarely with dynamics or physics. Yet, the system responsible for coronary blood flow, namely the "coronary circulation," is a highly sophisticated *dynamical* system in which the dynamics and physics of the flow are as important as the integrity of the conducting vessels. While an obstruction in the conducting vessels is a fairly obvious and clearly visible cause of disruption in coronary blood flow, any discord in the complex dynamics of the system can cause an equally grave, though less conspicuous, disruption in the flow.

This book is devoted specifically to the *dynamics and physics* of coronary blood flow. While it upholds the clinical and pathophysiological issues involved, the book focuses on dynamics and physics, approaching the subject from a strictly biomedical engineering viewpoint. The rationale for this approach is simply that the coronary circulation involves many issues in dynamics and physics, as the book will demonstrate. Also, with this particular focus, the book will complement other books on the subject, that have so far focused largely on clinical and pathophysiological issues.

A study of the dynamics of the coronary circulation requires far more information about the system than is currently available. Whether in terms of anatomical details of the vasculature, system properties such as capacitance and elasticity of the conducting vessels, or the basic and regulatory conditions under which the system operates, the information currently available is highly incomplete. Thus, the scope of this book is limited to dynamical aspects of coronary blood flow, but within these limits it is also constrained to deal necessarily with an incomplete picture of these dynamics. In particular, the book does not include the microcirculation, the venous part of the coronary circulation, Thebesian veins or the lymphatic system. Also, the many-faceted regulatory mechanisms of the coronary circulation are not considered in any systematic or factual way, but only tangentially in how they may affect the dynamics of the system. These omissions reflect the degree of complexity of the coronary circulation and serve as a sober reminder that it may never be possible or practical to deal with this complexity in a single book.

What seems possible at this time is to use known elements or properties of the system in such a way as to construct a meaningful, though incomplete, model of the system. This is the spirit in which the content of this book is presented. The book deals essentially with the dynamics of that part of the coronary circulation extending from the coronary ostia at the base of the aorta to the capillary level of coronary vasculature. It is meaningful to consider this part of the system in isolation because this is where the largest part of the pressure drop driving the flow occurs. While the dynamics of this part of the system may not represent the dynamics of the system as a whole, they demonstrate clearly the role of dynamics in the coronary circulation and illustrate how a disruption in these dynamics can affect coronary blood flow as significantly as can the obstruction of a blood vessel. This is indeed what the present book is about. Other books have in the past focused in a similar way on clinical or pathophysiological aspects of the system, or on the microcirculation. Each of these must clearly be seen as representing an important, though equally incomplete, view of the system.

My foray into the subject of coronary blood flow began in earnest in 1984 when I spent a sabbatical leave in the Department of Pathology at University Hospital in London, ON, and it is fair to say that this book would not have come into being without my ensuing collaboration with Professor Malcolm D. Silver, then department chair and chief of pathology. His passion for the heart and coronary arteries, and the depth of his expertise in cardiovascular pathology in general, was a haven for an engineer/applied mathematician seeking entry into the subject. With his help I came to know the coronary arteries literally "in the flesh" as I attended weekly autopsy review sessions and learned to dissect, cast and measure coronary vasculature. The collaboration was not a hit from the start - he was as puzzled by my preoccupation with branching angles and branch diameters as I was by his preoccupation with shades of pink in myocardial tissue. But a meeting of the minds soon prevailed, and together we embarked on several studies that have since formed the basis of all my subsequent work on the subject. I am deeply indebted to Dr. Silver not only for his continued guidance over the years but also for reading a draft of this book and offering many valuable comments and suggestions.

I am indebted by equal measure to my long-time friend and colleague Dr. Gerry Klassen, formerly professor of medicine, physiology and biophysics at Dalhousie University. His passion for the subject, combined with his love of science and engineering, made him an invaluable "resource" for me for more than two decades. Always ready to explain and ready to help, he made a lasting contribution to my education in the field of coronary blood flow. His enthusiasm for the subject was always a source of inspiration to me. I am grateful to Dr. Klassen for these "hidden" contributions to this book as well as for reading a draft of the book and offering many valuable comments. I received valuable comments also from my friend and colleague Dr. Erik L Ritman, Professor of Physiology and Medicine at the Mayo Clinic College of Medicine, who kindly took time from his ever busy schedule to read the manuscript.

But the mission of this book was to be in Science and Engineering rather than in Pathology or Medicine. For many years I have been inspired by and learned enormously from the books of Professor Y.C. Fung who, in my eyes, is the father of biomechanics and of the notion that engineers should not only "dabble" in the mechanics of the cardiovascular system but should do so deliberately and with no apologies. His books on the subject exemplify this notion and have provided me with the inspiration to write a book in the same spirit. I am grateful to Professor Fung not only for this but for agreeing to read this book and to write the Foreword to it.

It was important for me to have the book scrutinized by experts on both sides of the divide, and I am grateful to three colleagues who were kind enough to plow through the analytical aspects of the book and the tedium of equations and algebra. Professor Guy Kember from the Department of Engineering Mathematics at Dalhousie University, Professor Matt Davison from the Department of Applied Mathematics at the University of Western Ontario, and Dr. Hope Alderson, formerly of the Department of Mathematics at the University of New Brunswick. I am indebted to all three for laboring tirelessly through the manuscript, particularly to Dr. Alderson who did so heroically with a baby in one hand and the manuscript in the other.

I am grateful to my friend and long-time aide, Mira Rasche, for technical help with the manuscript, and to the secretarial brigade in Applied Mathematics - Gayle McKenzie, Pat Malone, and Audrey Kager– for always being there. My deepest thanks go to my wife, Lilian, who may not share my enthusiasm for the subject, yet so willingly plowed through the manuscript and shared the burden in so many different ways. I am grateful to Dr. Elias Greenbaum, Editor-in-Chief of the Biological and Medical Physics, Biomedical Engineering Series, for his encouragement and continued support, and, at Springer NY, to David Packer and Lee Lubarsky for walking the manuscript to production, to MaryAnn Brickner for finally taking the manuscript away from my hands and setting the production wheels in motion, and to Frank McGuckin and Frank Ganz for helping with the nightmare of electronic typsetting.

M.Zamir
London, Ontario
February 7, 2005

Contents

1

Static Design Issues

1.1 The Lone Pump

There are approximately 10^{14} living cells within the human body, each needing to receive food (nutrients, metabolic products) and to dispose of waste products *on an individual basis*. The situation is not unlike that of the (somewhat lower, though equally overwhelming) number of humans on this planet, approximately 6×10^9 in this case, each needing to receive food and to dispose of waste products, again, on an individual basis. The cardiovascular system is responsible for bringing a line of supply and a line of return within reach of every living cell within the human body. These service lines are fully centralized, in the sense that they all originate from one location and all return to it. In the world analogy this would amount to having only one location for the dispatch of food for the entire world population, only one location for the return of all waste products, and a line of supply and one of return within reach of every individual on the planet.

The cardiovascular system achieves this mammoth task within the human body not by *storing* food and waste products in one location, but by having them carried to and from every cell by a *circulating* fluid. The circulating fluid is blood, and the service lines along which it circulates are blood vessels. The central location which all lines of supply originate from and return to is not a massive warehouse but a small, very small, *pump*. The pump keeps the fluid in circulation (Fig.1.1.1). As it circulates, food is added to the fluid and waste products are extracted from it, on a continuing basis, somewhat in the manner of a conveyor belt.

The most important element of the cardiovascular system is therefore not so much the place where nutrients and metabolic products come from but the pump that circulates the fluid which carries these products. That pump is, of course, the heart. Circulation of blood that carries the food is as critical as the food itself because circulation keeps the food coming to the cells. Any disruption in that circulation is a disruption in food supply. Yes, life is not possible without blood, but in truth life is not possible without the *circulation*

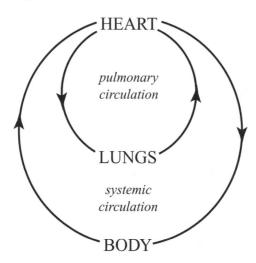

Fig. 1.1.1. Circulation of a fluid medium makes it possible to meet the metabolic needs of several billion cells within the human body by means of only a small pump (HEART). Metabolic substances are carried by the fluid to and from cells in every part of the body (systemic circulation), and the fluid itself is reconditioned by a second circulation to and from the lungs (pulmonary circulation). Circulation of the fluid is therefore as critical as the metabolic goods which it carries. The pump must maintain both circulations at all times and without fail. There is no backup, no contingencies.

of blood. The heart is the pump that drives that circulation. It must pump at all times, which it does by contracting and relaxing in a rhythmic pattern, approximately once every second, more than 86 thousand times every day, and about 2 billion times in a lifetime of 75 years, nonstop. If it fails, the entire body is deprived of its lifeline and fails with it.

Is this a design error? Has nature made a mistake? We suspect not. The human cardiovascular system is highly evolved. Indeed, it has been observed that in the evolution of species the degree of complexity and sophistication of an organism goes hand in hand with the degree of complexity and sophistication of the metabolic system required to support it [119, 117, 118, 201]. We suspect, therefore, that far from there being any design error, the human cardiovascular system with its lone pump is fully tested. Yet questions about the wisdom of a lone pump linger, for when the system fails it does so in a catastrophic manner. There is an element of finality in its failure. There is no backup for the lone pump, no contingencies.

1.2 Heart "Disease"?

While the heart, like other parts of the body, is subject to genetic disorders and infectious diseases [179], only a very small proportion of heart failures in the developed world are caused by such conditions. In all other cases a lack of blood supply to the heart muscle, or a lack of *fuel* which the muscle needs for doing its pumping work, is the principal cause of heart failure [83, 206, 14].

Yet, "heart disease" is the term most widely used in association with heart failure. The term is somewhat misleading because it suggests that the failure is due to a disease *of the heart itself* which, as stated above, occurs in only a very small proportion of cases. The situation is analogous to using the term "malfunction" to describe a car engine failure caused by lack of fuel. Terms that are less commonly used but that are somewhat more accurate are "coronary heart disease" and "coronary artery disease", both at least hinting at a problem in the *lines of blood supply* to the heart.

In this book the term "heart disease" shall be reserved for only the small proportion of cases where the heart is diseased in the true sense of a genetic or infectious disorder. In all other cases the term "heart failure" shall be used instead, to mean *failure of the heart as a pump caused by lack of blood supply.* The lack of blood supply may be the *immediate* cause of heart failure as in the case of a thrombus or an embolus, or it may be the *precipitating* factor as in the case of myocardial infarction.

It is appreciated that this usage of the terms "heart disease" and "heart failure" differs from common usage of these terms in the clinical setting. The intention here is to emphasize the strictly biomedical engineering view of the coronary circulation to be adopted in this book. According to this view, blood supply to the heart is provided by a highly dynamic system which can be disrupted by not only a problem in the lines of supply but also by a disturbance in the dynamics of the system. Indeed, the *dynamics* of blood supply to the heart are the principal subject of this book.

The complication, of course, is that the organ responsible for blood supply to the heart is the heart itself. When the heart fails, it fails to provide blood supply to not only every other part of the body but to itself, too. The situation has the elements of a control system with *positive* feedback. In such a system, a small failure produces a signal which leads to more failure, while in a system with *negative* feedback, a small failure produces a corrective signal which leads to less failure. The second system is regarded as stable and the first unstable, in the sense that after a small departure from equilibrium the second system can recover and return to equilibrium while the first cannot. A small steel ball at the bottom of a big bowl constitute a stable system, but if the bowl is turned upside down and the steel ball is positioned precariously at its peak, the system becomes unstable (Fig.1.2.1).

On the face of it, it would seem that the system of blood supply to the heart is an unstable system, since a small failure of the heart muscle would reduce its ability to pump which in turn would reduce blood supply to every

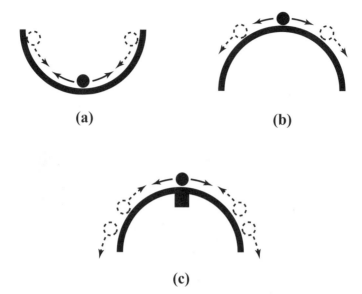

(a)   **(b)**

(c)

Fig. 1.2.1. Stability of coronary blood supply. Blood supply to the heart is provided by the heart itself: Does heart failure, therefore, lead to more failure? Shown here is the analogy of a small steel ball in equilibrium at the bottom of a large bowl (a). A small departure from equilibrium in this case is corrected by a negative signal which acts to move the ball back towards its neutral position. If the bowl is turned upside down as in (b), however, and the ball is positioned precariously at the peak, a small departure from equilibrium produces a positive signal which moves the ball yet further away from the peak. Regulatory mechanisms in coronary blood flow ensure that a small reduction in blood supply to the heart muscle does not lead to further reduction. In the bowl analogy it is as if a small magnet is placed under the bowl's peak as in (c), so that a small departure of the ball from the peak is corrected by the pull of the magnet. But if blood supply is prevented from reaching the heart muscle because of coronary artery disease, this mechanism becomes less effective or inoperative. In the bowl analogy it is as if the magnet is contaminated and can no longer pull back the ball effectively.

part of the body, including the heart muscle itself. But the situation is not in fact as precarious as the steel ball at the top of the upside-down bowl. Regulatory mechanisms in coronary blood flow (autoregulation) ensure that a reduction in the ability of the heart to pump does not translate immediately into a reduction of blood supply to the heart itself [83, 128, 136, 100, 183]. This mechanism provides a local reprieve and allows recovery to occur. In the bowl analogy it is as if a magnet is placed under the peak of the upside-down bowl, so that a small excursion of the steel ball away from the peak is corrected by the magnet pull before the situation becomes beyond recovery (Fig.1.2.1).

This issue does not usually play a key role in heart failure because, in the overwhelming majority of cases, the cause of a reduction in blood supply to the heart muscle is *not* a reduction in the innate ability of the heart muscle to pump but a gradual occlusion or sudden blockage of some of the vessels that carry this supply to the heart, caused in turn by a disease process *within the vessels*. Thus, heart failure (as defined in this book) is not usually mediated by positive feedback but by a *direct* reduction of blood supply to the heart muscle, which is not triggered by failure of the muscle to pump in the first place. The recovery mechanism that compensates for the reduction in blood supply to the heart muscle is thus disrupted because of an inability of blood supply to *reach* the heart. In the bowl analogy it is as if the magnet is contaminated and its ability to bring the steel ball back to the neutral position is diminished. "Heart disease" does not fairly describe the situation in hand, "heart starvation" would be a more accurate term. As stated earlier, the term "coronary heart disease", used less often, is somewhat more accurate in that it hints at the involvement of the coronary arteries, while "coronary artery disease" is yet more accurate because it places the problem precisely where it belongs.

1.3 Origin of Coronary Blood Supply

As a pump, the heart consists of four chambers, two ventricles and two atria, which contract and relax rhythmically. The two ventricles eject blood and the two atria act as receiving chambers for returning blood [133]. Output from the left ventricle is carried by the aorta to every part of the body, then returns to the right atrium, thus producing the so-called "systemic circulation" (Fig.1.1.1). Output from the right ventricle is carried by the pulmonary artery to the lungs where blood is oxygenated and then returned to the left atrium, constituting the "pulmonary circulation" [48].

The two systems must clearly operate in tandem to avoid accumulation (congestion) of fluid proximal to or in any of the four chambers. That is, on average, the systemic and pulmonary circulations must move the same volume of blood per cardiac contraction (Fig.1.1.1). However, the systemic circulation operates at a much higher pressure than the pulmonary, hence the work performed by the two ventricles is not the same. The pumping power produced by the left ventricle and hence its metabolic requirements are considerably higher than those of the right ventricle [135, 141]. Accordingly, the muscular walls of the left ventricle are the main focus of blood supply to the heart. While blood supply must reach every part of the heart for the organ to remain viable, blood supply to the left ventricle dominates coronary blood flow because of its intense requirements.

While the four chambers of the heart contain blood at all times, and while the left ventricle contains oxygenated blood ready for distribution to every part of the body, this blood must leave the heart before it can be tapped to supply the heart itself. The reason for this is that most cardiac cells, like those in other parts of the body, are served by a system of capillaries which are fed, ultimately, by a supply line from the aorta [199]. As it traverses the

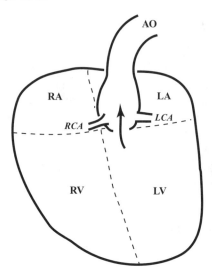

Fig. 1.3.1. Origin of blood supply to the heart. As the aorta (AO) leaves the left ventricle (LV) with oxygenated blood for every part of the body, its first two branches, the so-called left main and right coronary arteries (*RCA, LCA*), are destined to serve the heart itself. The figure is highly schematic, to show functional topology only.

body the aorta gives rise to many branches, destined to different parts of the body or to specific organs [48]. Each organ or territory within the body has its own supply line from the aorta, followed by a vascular tree structure to reach different parts of the organ or territory, and a system of capillaries to reach every cell. The heart is served in the same manner [81].

It would seem appropriate, therefore, that as the aorta leaves the left ventricle, laden with oxygenated blood for every part of the body, its first two branches are destined to serve the heart itself (Fig.1.3.1). They are known as the left main and right coronary arteries [133, 228, 216]. While this may seem as if the heart is being given "first priority", it is more likely the result of physical proximity of the heart to the root of the aorta (Fig.1.3.1).

Not infrequently, more than two coronary branches arise at the root of the aorta to bring blood supply to the heart (Fig.1.3.2), but in most cases only two are prominent and serve the role of the left and right main coronary arteries. Additional branches are usually considerably smaller than the main two and make only a local contribution to coronary blood supply. In rare cases only one supplying artery may arise from the aorta, and in yet others a supplying artery may not arise directly from the aorta but from one of its branches. The range and frequency of these and other variations have been well studied and documented [94, 69, 16, 146, 127, 61, 73, 196, 133, 182, 81, 152].

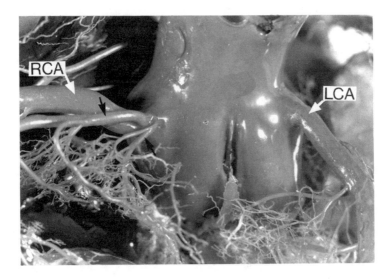

Fig. 1.3.2. Origin of the left main and right coronary arteries (LCA, RCA) and two additional branches (black arrows) as they arise from the aorta. From a cast of a human coronary network [216].

1.4 Coronary Arteries

As the left main and right coronary arteries leave the aorta they circle the heart in the manner of a crown, hence the name "coronary" arteries. The attributes "left" and "right" relate to the fact that the left coronary artery circles the left side of the heart while the right coronary artery circles the right side, though the situation is not accurately so. Furthermore, while the term "coronary" was first applied to the two main supplying arteries, it is now used to include all branches and sub-branches of these vessels as well as the capillary and venous vasculature, which in total comprise the "coronary circulation". Thus, in general terms, the word "coronary" has come to mean any element of blood supply to the heart for its own metabolic needs.

The left main and right coronary arteries encircle the heart along the atrioventricular groove, in the atrioventricular sulcus, formed between the ventricular and atrial chambers of the heart [133]. Their points of origin from the aorta are somewhat above this groove, thus each begins its course by proceeding down towards the groove. In this description, and in what follows, the heart is imagined in an upright position, with the atria at the top and ventricles at the bottom (Fig.1.4.1). If the heart is positioned in this way within the body, its left and right sides coincide approximately with the left and right sides of the body. Wide variations in the layout of the coronary arteries from one heart to another make only approximate descriptions possible. Furthermore, it is important to differentiate between exact anatomical mapping of

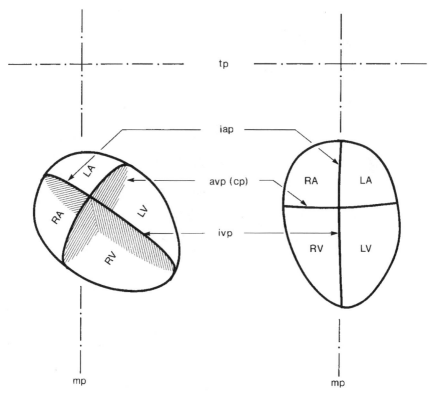

Fig. 1.4.1. (*left*) Approximate anatomical orientation of the heart (in a front view) in relation to the median and transverse planes of the body (mp, tp). (*right*) A "theoretical" upright position in which the heart is turned so that its atrioventricular plane (avp) is approximately parallel to the transverse plane of the body, and the interatrial and interventricular planes (iap, ivp) are approximately parallel to the median plane of the body. This position is convenient for discussion since here the left and right sides of the heart coincide approximately with the left and the right sides of the body. The main coronary arteries circle the heart in the atrioventricular plane, which is hence also known as the coronary plane (cp). From [228].

the coronary arteries in a given heart, and approximate *functional* layout of these vessels in every heart. The first may be important for the purpose of clinical intervention and treatment in a particular heart. The second relates to fluid dynamic design and function of the coronary network in general, and it is the main focus in this book.

As the left main coronary artery reaches the atrioventricular groove, it bifurcates into two major branches [133]. One, known as the left anterior descending artery, heads down along the interventricular groove, which overlies a thick muscular wall between the left and right ventricular chambers (Fig.1.4.2). This wall, known as the interventricular septum, actually func-

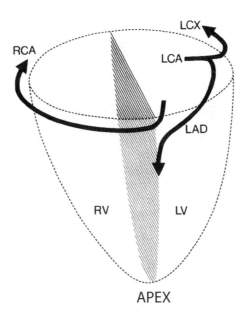

Fig. 1.4.2. The way in which the main coronary arteries circle the heart (in a front view), in relation to the coronary plane (dashed circle) and the interventricular plane (hatched plane). The right coronary artery (RCA) circles the right side of the heart while the left circumflex artery (LCX), which is a main branch of the left main coronary artery (LCA), circles the left side of the heart. The left anterior descending artery (LAD), which is the other main branch of the left main coronary artery, descends along the edge of the interventricular wall toward the apex of the heart. From [216].

tions as part of the *left* ventricle, hence blood supply to it is particularly important. The other branch, known as the left circumflex artery, turns to circle the left side of the heart along the atrioventricular groove. As it does so it gives rise to a number of small branches which head up to serve the right atrial region, and larger branches which head down to serve the lateral and posterior walls of the left ventricle (Fig.1.4.3).

As the right coronary artery reaches the atrioventricular groove, it turns to circle the right side of the heart, coursing along the groove and giving rise to branches heading up to serve the right atrial region and down to serve the anterior and posterior walls of the right ventricle (Fig.1.4.3). In approximately 90% of human hearts, as the right coronary artery reaches the atrioventricular groove at the *back* of the heart, it gives rise to an important branch known as the posterior descending artery [133]. This branch runs down along this groove to serve the posterior interventricular septum, in the same way that the anterior descending artery serves the anterior interventricular septum. Indeed

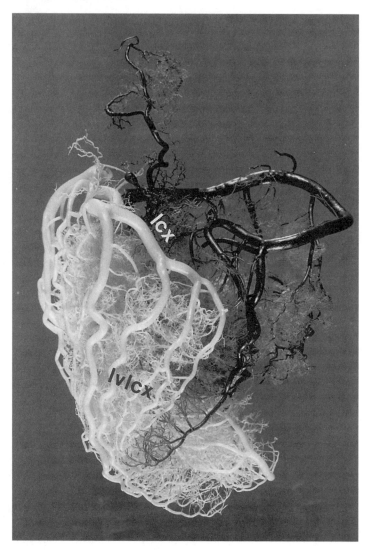

Fig. 1.4.3. Cast of the coronary network of a human heart in which the left main coronary artery and its branches are seen in light grey and the right coronary artery and its branches are seen in black. The heart is here viewed from the back. The left circumflex artery (lcx) is seen circling the left side of the heart and giving rise to a number of large branches (lvlcx) to serve the left ventricular wall. The right coronary artery is seen to do the same on the right side of the heart, giving rise first to some small branches that serve the right ventricular wall, but then a large branch known as the posterior descending artery to serve the interventricular wall from the back. From [228].

Fig. 1.4.4. The left anterior descending artery, which descends along the front edge of the interventricular wall (Fig.1.4.2), is seen here wrapping itself around the apex of the heart (see also Fig.1.4.5). It then begins to *ascend* along the back edge of the interventricular wall, as if to meet the posterior descending artery which is descending toward it along that edge. From [228].

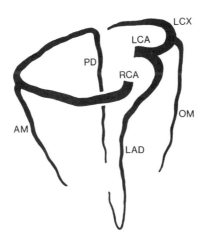

Fig. 1.4.5. The main "skeleton" of the coronary arterial network and the way in which it serves different regions of the heart. The right coronary artery (RCA) and the left circumflex artery (LCX) circle the heart as a belt or a "crown", sending branches up to serve the atria and down to serve the ventricles. Of the latter the most important are the left anterior descending artery (LAD) and the posterior descending artery (PD) which together circle and serve the interventricular wall. Two other main branches, the acute marginal (AM) and obtuse marginal (OM), serve the right and left arterial walls, respectively. This skeleton is supplemented by many other branches of a wide range of sizes and in a highly variable pattern (see Fig.1.4.3).

both arteries head towards each other as they move towards the "apex", the pointed bottom of the heart. In most cases the anterior descending artery then wraps around the apex (Fig.1.4.4) and the two vessels stop short of meeting thereafter (Fig. 1.4.5). In 70% of human hearts the right coronary artery continues to circle the heart, after giving rise to the posterior descending artery, to serve some of the posterior wall of the left ventricle [133].

1.5 Left/Right Dominance

With the heart in an upright position, the right coronary and left circum-flex arteries circle the heart in a horizontal (coronary) plane, and in opposite directions (Fig. 1.4.2). As they reach the back of the heart, the two arteries move toward each other and terminate short of actually meeting (Fig. 1.5.1). The point at which this occurs is a measure of the extent to which the right coronary artery supplies the left side of the heart, which is an important functional aspect of the coronary network usually referred to as left/right dominance [133]. An important anatomical landmark in this subject is the "crux", the point at which the horizontal atrioventricular groove crosses the vertical interventricular groove at the back of the heart (Fig. 1.5.1). At approximately this point the posterior descending artery arises and begins its descent along the interventricular groove to serve the interventricular wall. Whether this artery arises from the right main coronary artery or from the left circumflex artery depends on which of the two arteries reaches the crux.

In only 10% or so of human hearts, the left circumflex artery reaches the crux and gives rise to the posterior descending artery [133]. Such cases are known as "left dominant" since, functionally, blood supply to the left ventricle then depends entirely on the left coronary artery. In approximately 20% of human hearts, known as "balanced" cases [133], the right coronary artery reaches the crux, gives rise to the posterior descending artery and terminates at that point. In the remaining 70% of cases, known as "right dominant", the right coronary artery continues beyond the crux to serve part of the posterior wall of the left ventricle.

Thus, in the large majority of human hearts, blood supply to the left ventricle is shared by the left and right main coronary arteries, not necessarily equally but in parallel, thus providing two sources of supply instead of one. While this has a clear design advantage, the relative extent to which the two arteries share their service to the left ventricle is highly variable. At one extreme the right coronary artery may not serve the left ventricle at all, thus leaving that ventricle, including the interventricular wall, to depend entirely on the left circumflex artery. At the other extreme the right coronary artery may supply the interventricular wall as well as the entire back wall of the left ventricle (Fig. 1.5.2). Since this wide range of variability represents the variable extent to which blood supply to the left ventricle may depend on one or two main lines of supply, the degree of left/right dominance in a given heart

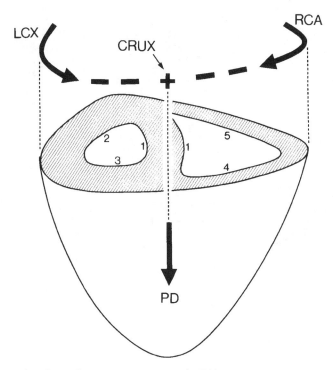

Fig. 1.5.1. As the right coronary artery (RCA) and the left circumflex artery (LCX) reach the back of the heart they move toward each other. Only one of them reaches the "crux" and gives rise to the posterior descending artery (PD). The numbers identify individual walls of the two ventricles: (1) interventricular septum, (2,3) lateral and posterior walls of the left ventricle, respectively, (4,5) posterior and anterior walls of the right ventricle, respectively. It is seen that the interventricular septum, which separates the two ventricles, is in fact an important part of the *left* ventricle. From [216].

may be regarded as an "anatomical risk factor". While an exact numerical measure of that risk factor based on a measure of left/right dominance is not easy to calculate, the connection between the two measures is clear.

More generally, the concept of left/right dominance highlights the wide range of variability in the detailed layout of the coronary arteries and their branches, and hence the need to refer to these vessels by function rather than name. A vessel referred to by the same name in different hearts rarely serves precisely the same fluid dynamic function in every heart. It is as important to assess the role of a given artery in the general scheme of blood supply to a given heart as it is to assess the degree of stenosis in that vessel.

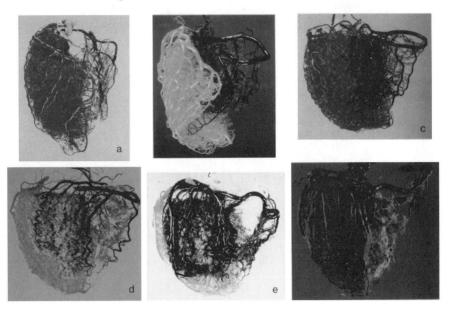

Fig. 1.5.2. Casts of the arterial networks of human hearts in which the distribution of the left main coronary artery is seen in red (a,c,f) or yellow (b,d,e) and the distribution of the right coronary arteries is seen in blue (a,b,c,d,f) or green (e). The hearts, in all cases, are in an upright posterior view. The figure illustrates the range of variability in the way in which the two arteries share the responsibility of cardiac blood supply. At one extreme (a), known as "left-dominant", the left coronary artery gives rise to the posterior descending artery and therefore blood supply to the left ventricle depends on vasculature arising entirely from the left coronary artery. In a "balanced" case (b), the right coronary artery gives rise to the posterior descending artery and thereby contribute to blood supply to the interventricular wall, which is an important part of the left ventricle. In "right-dominant" cases, the right coronary artery continues beyond the crux to supply part or all the posterior wall of the left ventricle (c,d,e,f), thereby taking a greater share of the supply to that ventricle. From [216]. (See color insert.)

1.6 Branching Structure

The question of whether the arterial system of the heart has the structure of an "open tree" or an "interconnected mesh" is one of whether blood supply to any region of the heart can reach it via more than one path. In an open tree structure a main supplying artery, such as the left main or right coronary artery, divides into (usually two) branches, then each of these branches in turn divides into further branches, and so on. There are no "collateral" connections between branches. Flow from the main supplying artery to any peripheral branch can reach it only by progressing along the strict hierarchical structure of the tree, from parent to branch at each junction. That is, the flow can only reach its destination via one distinct path (Fig. 1.6.1). In an interconnected

COLOR PLATE I

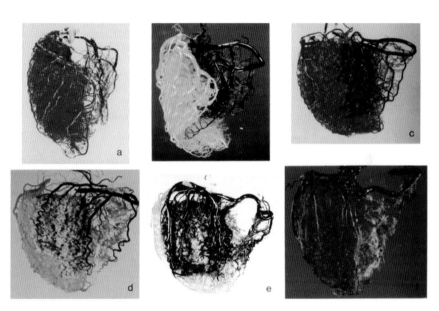

Fig. 1.5.2. Casts of the arterial networks of human hearts in which the distribution of the left main coronary artery is seen in red (a,c,f) or yellow (b,d,e) and the distribution of the right coronary arteries is seen in blue (a,b,c,d,f,) or green (e). The hearts, in all cases, are in an upright posterior view. The figure illustrates the range of variability in the way in which the two arteries share responsibility of cardiac blood supply. At one extreme (a), known as "left dominant," the left coronary artery gives rise to the posterior descending artery and therefore blood supply to the left ventricle depends on vasculature arising entirely from the left coronary artery. In a "balanced" case (b), the right coronary artery gives rise to the posterior descending artery and thereby contribute to blood supply to the interventricular wall, which is an important part of the left ventricle. In "right-dominant" cases, the right coronary artery continues beyond the crux to supply part or all the posterior wall of the left ventricle (c,d,e,f), thereby taking a greater share of the supply to that ventricle. From [216].

COLOR PLATE II

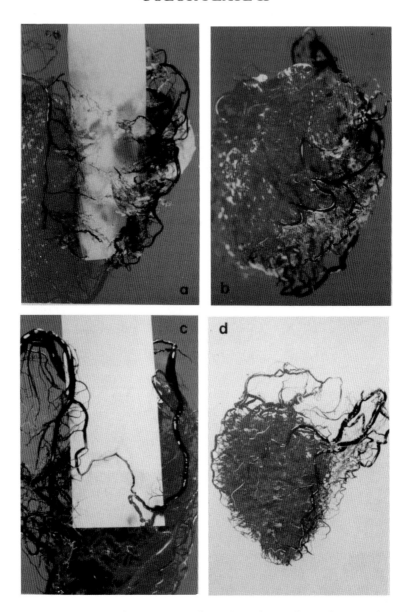

Fig. 1.6.4. Resin casts showing sporadic collateral vasculature between branches of the left and right coronary arteries in four human hearts. The two sides were perfused in different colours to uncover places where mixing of the two colours occured. This mixing could not be mediated by capillary beds since the size of dye particles prevented them from entering these vessels. Thus, the capillary beds actually acted as a *barrier* between the two colours because only clear resin could enter these beds (seen as white fluff in (a,b)). This ensured that any observed mixing occured at higher levels of the tree structure. From [216].

COLOR PLATE III

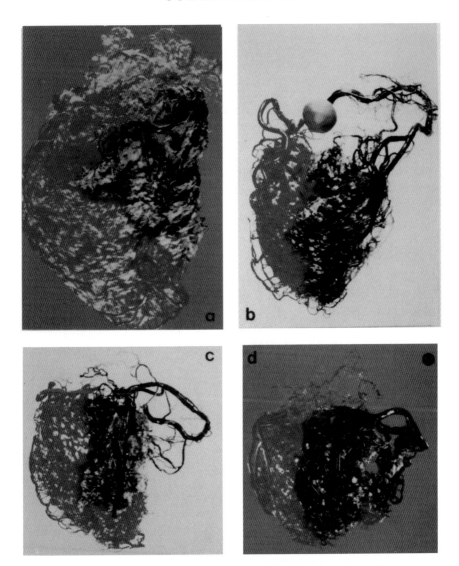

Fig. 1.6.5. Resin casts of four human hearts in which the left main and the right coronary arteries were prefused in different colours. Here it is seen that in the absence of collateral vasculature, the distributions of the left and right coronary arteries are demarcated by clear borders, and no mixing of the two colours is observed. See also caption for Fig. 1.6.4. From [216].

COLOR PLATE IV

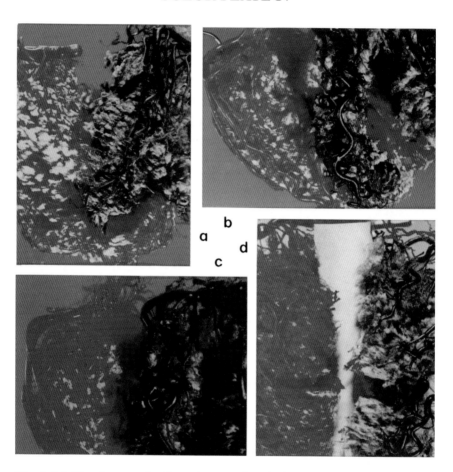

Fig. 1.6.6. The absence of collateral connections is further confirmed by the ability to physically separate the distributions of the left and right coronary arteries. In fact, in (b) and (d) it is seen that the beds of individual vessels on the same side could also be separated from each other. See also caption for Fig. 1.6.4. From [216].

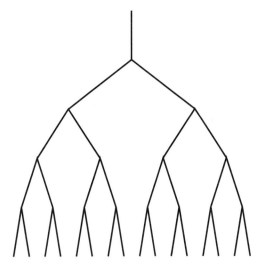

Fig. 1.6.1. In an open tree structure, shown schematically here, flow from the main supplying artery (top) must proceed along the strict branching hierarchy of the tree, that is from parent to branches at each junction. Each destination (bottom) can be reached via only one distinct path. Symmetry and uniformity are not necessary but are used here to emphasize these hierarchical features. Arterial trees are generally nonsymmetrical and highly nonuniform, but their underlying architecture is most often that of an open tree structure.

mesh structure, or in an open tree structure with an overlay of collateral connections, by contrast, there may be several paths to any given destination (Figs. 1.6.2, 3).

This issue is of particular functional and clinical importance. In the presence of multiple paths blood supply can bypass an occluded vessel segment and reach its destination via a different path, an important consideration since occluded or obstructed blood vessels are the cause of most heart failures [3, 17, 209, 83, 81, 14]. But the subject is highly controversial, for several reasons. First, because the presence of collateral vessels in the vascular system of the heart is highly variable in different species. Second, because the presence of collateral vessels in the human heart is found to be highly variable not only from heart to heart but also in the same heart at different times. Thus, the way in which collateral vasculature fits in the hemodynamic design of the coronary circulation in general, and in that of the human heart in particular, is not fully established.

Remarkably, as early as the seventeenth century it was observed that "The vessels which carry blood to the heart .. come together again and here and there communicate by anastomosis" [126]. Two centuries later: "The human heart is potentially able to develop collaterals where they are needed" [93]. And more recently, by various casting and injection techniques, investigators

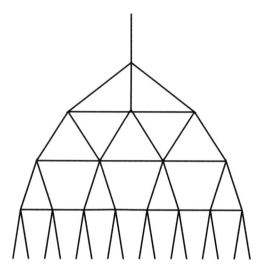

Fig. 1.6.2. In an interconnected mesh structure flow from the main supplying artery (top) can reach its destinations (bottom) via different routes. Early studies of dog coronary vasculature indicated that its underlying architecture has this interconnected structure. Subsequent work on the human heart, on the other hand, showed repeatedly that the underlying architecture of its vasculature is that of an open tree structure. Interconnected mesh structures in the vascular system occur mostly at the capillary level. At higher levels of the system they are rare because an open tree structure is, fluid dynamically, more efficient than an interconnected mesh.

demonstrated clearly the existence of such collateral vasculature, abundant in the dog heart, rarely or scarcely found in the pig, and variable in the human heart [185, 173, 27, 15, 69, 16, 74, 95]. While differences between species suggest that collateral vasculature may be part of the hemodynamic design of the coronary circulation in some species, the matter is far from settled. Much of the focus has been on the situation in the human heart.

An extensive study by Baroldi and Scomazzoni [16] sums up much of our current understanding of collateral vasculature in the human heart and resolves much of the controversy relating to variability. By a systematic survey of human hearts the authors found that both the number and size of collateral vessels were strongly linked to the presence of an ischemic history or conditions within the heart. Very fine collateral vasculature may be found sporadically here and there when ischemic conditions are absent, but in the presence of such conditions collateral vasculature becomes more significant in size and clearer in its mission. The results clearly established the phenomenon of collateral vasculature as a compensatory *mechanism* in coronary heart disease [15, 14], and this is how the matter rests.

But the question of the *effectiveness* of this mechanism is far from settled, and controversy continues to surround this aspect of the phenomenon.

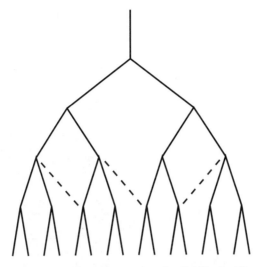

Fig. 1.6.3. When interconnections occur sporadically (dashed lines) in an otherwise open tree structure, they are referred to as "collateral" vasculature. The presence of collateral vasculature has been demonstrated in the coronary circulation of the human heart (see Figs. 1.6.4–6) but controversy continues over its origin, rate of growth, and hemodynamic significance. Collateral vasculature is to be distinguished from *permanent* collateral bridges such as the communicating arteries in the circle of Willis of the human brain. The latter are part of the fluid dynamic design of the cerebral circulation, that is, part of the normal anatomy of the cerebral vasculature. Collateral vasculature in the human heart is not part of the normal anatomy of the coronary vasculature.

The overwhelming majority of heart failures caused by insufficient myocardial blood supply [185, 173, 27, 15, 69, 16, 74, 95] point clearly to a failure of this mechanism in these cases. The difference between the time scale of development and growth of collateral vessels and the time scale of vascular obstruction, whether by a slow disease process or sudden occlusion, is clearly an important factor. The indications are that the mechanism of collateral vasculature cannot deal with sudden occlusion, first because the vasculature would likely not be present before the occlusion occurs, and second because new vasculature cannot develop within the fast time scale of a sudden occlusion. In the case of the much slower pace of stenosis by atherosclerotic disease, the indications are that collateral vasculature *can* develop in time to make a significant contribution, but whether it does or not in every case, and the magnitude of the contribution it makes in each case, is not clear. The subject is decidedly far from settled [172, 44, 180, 77].

The ultimate question, of course, is the evolutionary *origin* of the mechanism of collateral vasculature in the human heart. Is it an integral part of the hemodynamic design of the coronary circulation, or merely a normal angiogenic response to ischemia, not peculiar to the heart? If it is the former,

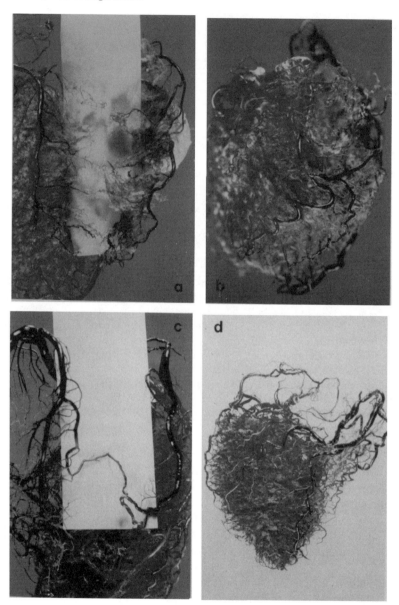

Fig. 1.6.4. Resin casts showing sporadic collateral vasculature between branches of the left and right coronary arteries in four human hearts. The two sides were perfused in different colours to uncover places where mixing of the two colours occured. This mixing could not be mediated by capillary beds since the size of dye particles prevented them from entering these vessels. Thus, the capillary beds actually acted as a *barrier* between the two colours because only clear resin could enter these beds (seen as white fluff in (a,b)). This ensured that any observed mixing occured at higher levels of the tree structure. From [216]. (See color insert.)

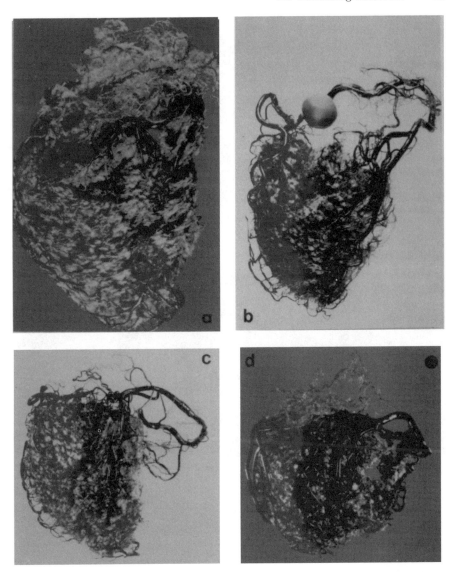

Fig. 1.6.5. Resin casts of four human hearts in which the left main and the right coronary arteries were perfused in different colours. Here it is seen that in the absence of collateral vasculature, the distributions of the left and right coronary arteries are demarcated by clear borders, and no mixing of the two colours is observed. See also caption for Fig. 1.6.4. From [216]. (See color insert.)

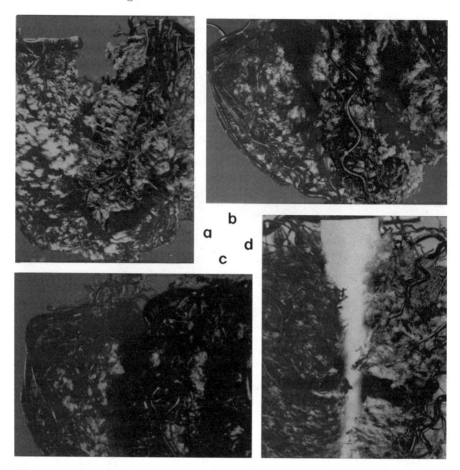

Fig. 1.6.6. The absence of collateral connections is further confirmed by the ability to physically separate the distributions of the left and right coronary arteries. In fact, in (b) and (d) it is seen that the beds of individual vessels on the same side could also be separated from each other. See also caption for Fig. 1.6.4. From [216]. (See color insert.)

collateral vessels would likely have a more permanent, less sporadic, presence within the coronary vasculature. Yet, such permanent collateral bridges as the communicating arteries in the circle of Willis of the brain do not exist in the heart. Studies of the branching architecture of the coronary arteries show repeatedly that the underlying branching pattern is that of an open tree structure [227, 223, 214, 106]. Sporadic connections between branches can be easily demonstrated, particularly in the presence of ischemic disease or history (Fig. 1.6.4). Under normal circumstances, however, the indications are that the distributions of the main coronary arteries are fairly distinct and do not mesh with each other (Figs. 1.6.5, 6). Meshing does occur at the capillary level, of course, but at that level it can no longer serve the function of collateral vasculature since the meshing is highly localized and cannot reach (hemodynamically) from the distribution territory of one major artery to that of the next.

1.7 Underlying Design?

Wide variability in the anatomical layout of the coronary arteries and their major branches calls into question the "naming" of these vessels, because the names imply that they are clearly defined anatomical entities. Yet we have seen that the left or the right coronary artery in one heart is not the same functional entity as the left or the right coronary artery in another. This variability also calls into question the notion of "one-artery" or "two-artery" disease as measures of the severity of coronary heart disease, since, again, these terms suggest that one artery in one heart is the same functional entity as an artery with the same name in another heart, or indeed that two arteries have twice the functional value of one artery. More accurately, wide variability in their size and distribution suggests that the coronary arteries, as represented by their anatomical names, do not represent elements of the underlying *functional design* of the coronary network as a *fluid conveying system*. Elements of that functional design would be represented by features of the coronary network which do not vary from heart to heart.

The situation is not unlike the functional design of the heart itself as a double pump for the maintenance of two circulations. The characteristic four chambers of the heart are essential elements of this design that do not change from heart to heart. On the other hand, the exact size and shape of these chambers, or the size and shape of the heart as a whole, are secondary features that vary considerably from heart to heart. They are not essential elements of the *underlying design* of the heart as a double pump. Thus, by analogy, we ask: are there features of the coronary network that do not change from heart to heart?

A study of human hearts with the purpose of addressing this issue found, in summary, that the coronary network serves the heart by dividing the myocardium into six distinct zones [228]. Each zone is served by two types of vessels: "distributing vessels" that run along the *borders* of these zones, and "delivering vessels" that enter the zones and branch profusely to reach the capillary bed and thereby deliver blood within the zone. Indeed, subsequent studies confirmed that there is a significant difference in the branching patterns of these two types of vessels, consistent with their different roles [214, 218]. Briefly, distributing vessels branch only sporadically and maintain their identity as they circle the zones, while delivering vessels quickly lose their identity as they branch profusely and more uniformly to reach the capillary beds (Figs. 1.7.1–3).

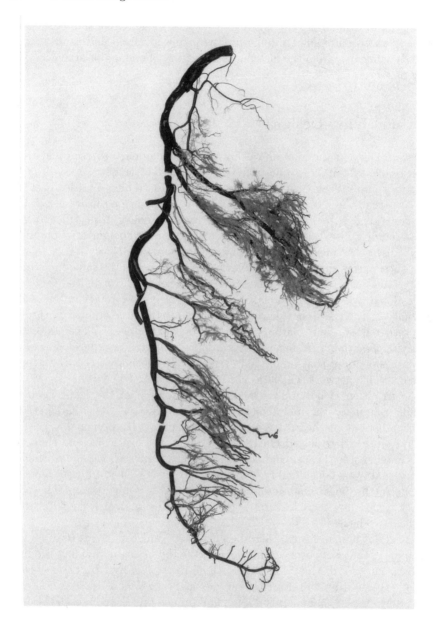

Fig. 1.7.1. A cast of the left anterior descending artery of a human heart and its branches. The artery runs along the anterior interventricular sulcus, as a distributing vessel, while its branches probe into the plane of the interventricular wall, as delivering vessels. This figure illustrates the difference between the branching patterns of the two types of vessels. Distributing vessels branch only sporadically and maintain their identity, while delivering vessels quickly lose their identity as they branch profusely and more uniformly to reach the capillary beds. From [214].

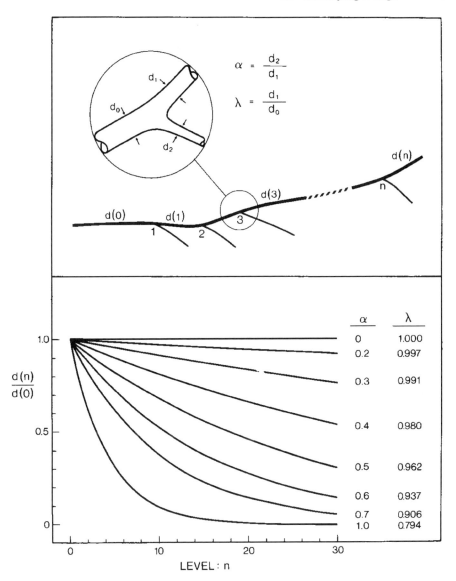

Fig. 1.7.2. A theoretical basis for the difference between the branching patterns of distributing and delivering vessels. According to the cube law of branching, the diameter of an artery that undergoes repeated *symmetrical* or near symmetrical bifurcations diminishes rapidly, following one of the lower curves for which the value of the symmetry ratio α is near unity. The curves describe the diameter of the vessel at different levels n of a tree structure, or after undergoing n bifurcations. By contrast, the diameter of an artery that undergoes *asymmetrical* bifurcations, giving rise to relatively small side branches, diminishes much more slowly, following one of the top curves for which the value of α is near zero. From [214].

Borders defining the six zones of the myocardium in this scheme coincide with certain anatomical landmarks of the heart, namely the "sulci" of the two main dividing walls of the heart: the atrioventricular septum and the interventricular septum. Topographically, the sulci represent the intersections of the two dividing planes with the outer surface of the heart. They appear as faint grooves on the surface of the heart. The distributing vessels run along these grooves, usually covered by a thin layer of fat. The atrioventricular septum, for example, which lies in the coronary plane (Figs. 1.4.1, 2), produces the atrioventricular sulcus on the surface of the heart as a mark of its intersection with it. This sulcus runs as a "belt" around the "waist" of the heart and it is the groove along which the right coronary artery and the left circumflex artery run (Figs. 1.4.1, 2). The analogy used more commonly is that of a "crown" (presumably around the "head" of the heart), hence the term "coronary" as mentioned earlier.

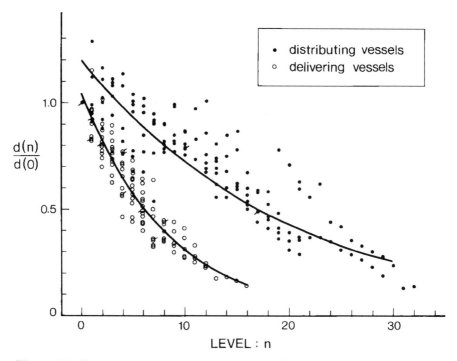

Fig. 1.7.3. Diameter measurements from casts of human coronary arteries. Measurements from distributing vessels are identified by bold circles while those from delivering vessels are identified by empty circles (see Fig. 1.7.2 for notation). The curves represent a statistical fit of the data points. The figure shows clearly that delivering vessels branch more profusely and undergo fewer bifurcations but their diameters diminish rapidly in the process. The diameters of distributing vessels, by contrast, diminish more slowly and the vessels undergo a larger number of bifurcations. From [214].

Another important inner wall of the heart is the interventricular septum, which acts as a divider between the two ventricles but is functionally part of the left ventricle. The intersection of the interventricular plane with the outer surface of the heart produces the interventricular sulcus along which the anterior and posterior descending arteries run (Figs. 1.4.1, 5). By what appears to be a clear design, and in a pattern which does not vary from heart to heart, the two arteries circle the wall as distributing vessels, while branches from them run into the plane of the wall as delivering vessels (Fig. 1.7.1).

Two other borders coincide with the "acute margin" on the right side of the heart, and the "obtuse margin" on the left (Fig. 1.4.5). The names of the six zones and of the borders defining them are shown in Fig. 1.7.4.

The zones and borders are anatomical landmarks which are permanent features that do not change from heart to heart (Fig. 1.7.3). The underlying fluid dynamic design of the coronary network appears to be based on these landmarks. By this functional design, *distributing* arteries run along the dividing borders, giving rise to *delivering* branches that enter the zones and implement blood delivery. *This scheme by which the coronary network serves the heart does not change from heart to heart [228].*

Variability in the size and shape of the zones, and wide variability in the length and size of coronary arteries referred to earlier, do not alter this scheme. A border may be occupied by *different* distributing vessels in different hearts, but the scheme remains constant. The posterior interventricular sulcus, for example, may be occupied by a posterior descending artery arising from the right coronary artery or one arising from the left circumflex. The left posterior atrioventricular sulcus may be occupied fully by the left circumflex artery, fully by the right coronary artery, or partly by both. In these different scenarios there is wide variability in the length and size of the coronary arteries involved, but the underlying design of the coronary network remains the same. In all cases distributing vessels bring blood supply to the borders of zones while delivering vessels enter the zones and implement delivery. Only the *identities* of the vessels vary from heart to heart, hence these identities, as represented by the anatomical names of the vessels, do not represent accurate *functional* elements of the coronary network.

This subject is clearly important in coronary heart disease where one or more arteries may be affected by disease that limits its fluid dynamic function. The foregoing discussion suggests that an accurate clinical assessment of coronary heart disease in a particular heart should be based not on the *anatomical names* of the affected vessels but on the *fluid dynamic role* of each affected vessel in the scheme of blood supply to that heart. What is clearly important is whether the affected vessel is a distributing or a delivering vessel, and what particular position it occupies in that particular heart. Only then is it possible to deduce the effect of the diseased vessels on particular *zones* of the heart, and ultimately on its ability to perform the heart's pumping function.

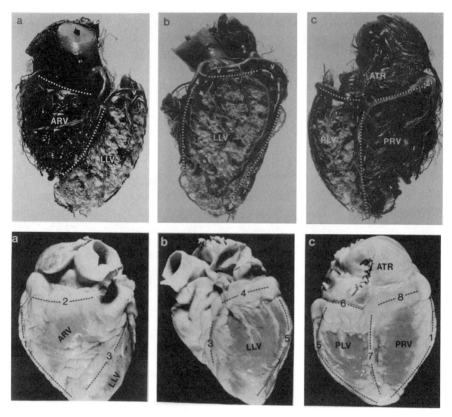

Fig. 1.7.4. Bottom row: a human heart in (a) an upright front view, (b) a left side view, and (c) a back view. Top row: cast of the coronary arteries of the same heart in the same views. Six zones of the myocardium are clearly visible in both sets of panels, defined by the distributing vessels seen in the top panels and by the bands of fat covering the distributing vessels in the lower panels. The borders in both cases are highlighted by dotted lines. A clear demonstration of how delivering vessels enter a zone and lose their identity is seen in panel (b) in the bottom row. Clear demonstrations of the way distributing vessels run along the borders of and thereby define the zones are seen in the top panels. The six zones are: anterior right ventricular (ARV), lateral left ventricular (LLV), posterior left ventricular (PLV), posterior right ventricular (PRV), atrial (ATR), and interventricular septal (IVS), not seen because it is an internal wall. From [228]

1.8 Coronary Flow Reserve

The intensity of the pumping action of the heart varies considerably, depending on the metabolic activity of the rest of the body. The range of demand is fairly wide, extending from a base level when the body is resting to a considerably higher level when the body is at maximal activity. Energy required to support the pumping action of the heart is therefore highly variable, and blood supply to the heart for the purpose of its own metabolic activity must be able to change accordingly over a wide range. The coronary circulation, being the vehicle for that supply, must therefore have the capacity to deliver far more blood flow to the heart than it does under normal resting conditions. This excess capacity is referred to as "coronary flow reserve" [83, 42, 66, 64, 129, 128, 100, 183, 26, 112]. How does the coronary circulation provide this reserve?

The triggers for change in coronary blood flow and the mechanisms by which change is accomplished have been studied widely and are well documented elsewhere [83, 42, 66, 64, 129, 128, 100, 183, 26, 112]. In this section we are concerned with only the ultimate expression of these mechanisms in terms of the fluid dynamic design of the coronary circulation. The first question in that context must clearly be whether there is an element of flow reserve in the size of the main supplying arteries of the heart, namely the left main and right coronary arteries. Do these arteries, by design, have larger diameters than would normally be required for normal blood flow to the heart? In other words, are the diameters of the left and right main coronary arteries better matched to the high or low end of the range of coronary blood flow?

Within the cardiovascular system, it is reasonably well established that the diameters of the main arteries supplying an organ are directly related to the organ's requirements for blood supply. More accurately, the diameter of a supplying artery is generally related to the flow rate which the vessel is destined to convey. An actual relation between diameter and flow rate was proposed many years ago in the form of what is now known as the "cube law" [147, 190, 178, 167, 132]. It proposes that the flow rate through a vessel should optimally be proportional to the cube of the vessel's diameter, or conversely, that the diameter of the vessel should optimally be proportional to the cube root of the flow rate the vessel is destined to convey. Other relations have been explored since then and were shown to have some theoretical or empirical validity, but the cube law continues to have the widest support in terms of biological data [167, 132, 225, 222, 223, 92, 106].

The cube law provides a useful tool for assessing the provisions for blood supply to different organs, with a useful concept in this assessment being that of "bolus speed". A "bolus" within a blood vessel is defined as a cylindrical volume of blood which has the same diameter as the luminal diameter of the vessel and a length equal to that diameter (Fig. 1.8.1). The volume of a bolus is thus proportional to the cube of the diameter of the vessel in which it resides. Thus if the diameters of two vessels are denoted by d_1, d_2 and the

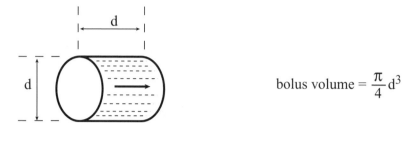

bolus volume = $\dfrac{\pi}{4}d^3$

bolus speed = $q / (\dfrac{\pi}{4}d^3)$

Fig. 1.8.1. The concept of bolus speed. A "bolus" within a blood vessel is defined as a cylindrical volume of fluid which has the same diameter as the lumen diameter of the vessel and a length equal to that diameter. Under the branching "cube law", whereby the diameter of a vessel is optimally proportional to the cube root of the flow rate which the vessel carries, the bolus speed is the same throughout the vascular network.

corresponding volumes of their boluses are denoted by V_1, V_2, then

$$\frac{V_2}{V_1} = \left(\frac{d_2}{d_1}\right)^3 \tag{1.8.1}$$

If the *flow rates* through the two vessels are denoted by q_1, q_2, respectively, then the cube law proposes that the ratio of the diameters of the two vessels is related to the ratio of the two flows by

$$\frac{d_2}{d_1} = \left(\frac{q_2}{q_1}\right)^{1/3} \tag{1.8.2}$$

Combining the two results (Eqs.1.8.1,2) we then have

$$\frac{V_2}{V_1} = \frac{q_2}{q_1} \tag{1.8.3}$$

Now, the volumetric flow rate through a vessel is the product of the volume of a bolus in that vessel and the number of boluses passing through the vessel per unit time. We shall refer to the latter as the "bolus speed". That is, if the bolus speed is denoted by S, then

$$q = VS \tag{1.8.4}$$

and for the two vessels discussed above we then have

$$\frac{q_2}{q_1} = \frac{V_2 S_2}{V_1 S_1} \tag{1.8.5}$$

Results in Eqs.1.8.3,5 lead clearly to

$$S_2 = S_1 \qquad (1.8.6)$$

which is a useful expression of the cube law in terms of bolus speed, that is: *The relation between diameter and flow in blood vessels of different size should optimally be such that the bolus speed is the same in all vessels.* Thus the provisions for blood supply to an organ, on this basis, may be deduced from the diameters of the main supplying arteries to that organ, on the assumption that the bolus speed in these arteries is the same as that in the aorta or in the vascular system in general.

To apply this result we consider a human cardiovascular system in which systemic flow rate through the aorta is $q_a = 5 \; l/min$ and aortic diameter is $a = 2.5 \; cm$. The volume V_a of a bolus in the aorta is then $\pi(2.5)^3/4$ and the corresponding bolus speed S_a is given by

$$S_a = \frac{q_a}{V_a} = \frac{5000/60}{\pi(2.5)^3/4} = 6.79 \; b/s \text{ (boluses per second)} \qquad (1.8.7)$$

This value of the bolus speed can be used as a reference for flow in other parts of the cardiovascular system.

In particular, when we consider a human heart supplied by two main coronary arteries, which for the purpose of discussion are assumed to have equal diameters and carry equal flow rates, if the diameter of each artery is taken as 3.5 mm, and if the same bolus speed is assumed to exist in these arteries as it does in the aorta, then the volume of a bolus in each of the two vessels is $\pi(0.35)^3/4$ and the combined flow rate to the heart would be given by

$$q = 2 \times \frac{\pi(0.35)^3}{4} \times 6.79 = 27.44 \; ml/min \qquad (1.8.8)$$

This coronary flow rate is considerably lower, by almost an order of magnitude, than usual estimates [83, 66, 128, 100, 183, 26, 112]. If the two coronary arteries are each taken to be 4 mm in diameter, then the corresponding flow rate to the heart would be 40.96 ml/min, still considerably lower than estimated. (Under normal conditions it is approximatly 5 percent of systemic blood flow, or 250 ml/min [66]).

These simple calculations indicate clearly that coronary flow reserve in the human heart is not based on having supplying vessels that are "larger than normal". In fact, the reverse appears to be the case. At a base flow rate of 250 ml/min, if the two main coronary arteries were to have the same bolus speed as exists in the aorta, then the diamter of each vessel, using Eq. 1.8.2, would be approximately given by

$$d = 25 \left(\frac{125}{5000} \right)^{1/3} = 7.31 \; mm \qquad (1.8.9)$$

Diameters of the two main supplying coronary arteries in the human heart are usually in the range of $3 - 5\ mm$ [16, 73, 133].

Thus the diameters of the main supplying coronary arteries appear to be matched decidedly to the *lower* end of the flow range. Coronary flow reserve, which makes it possible for blood flow to increase by a factor of 6 or more [83, 42, 66, 64, 129, 128, 100, 183, 26, 112], must therefore be based on a different design paradigm. From a purely fluid dynamic standpoint, flow rate through the coronary vascular tree can only increase by increasing the driving pressure or by decreasing the overall resistance of the tree. Since the driving pressure (for flow to the tree as a whole) can only change by a relatively small amount, it is clear that a large increase in flow can only be achieved by a change of resistance to flow. Design provisions for coronary flow reserve in the human heart must therefore be based on having wide control over the diameters of the *resistance vessels* within the coronary network.

While in *steady* flow the resistance to flow in a tube or a vascular tree is determined predominantly by the tube or vessel radii, in pulsatile flow in general, and in the coronary circulation in particular, the situation is far more complex and is not clearly predetermined. In pulsatile flow the resistance to flow, then more appropriately referred to as impedance, depends on the frequency of pulsation, *and hence on the harmonic composition of the driving pressure wave*, since the harmonic components of the wave propagate at different frequencies. Also, elasticity of the conducting vessels turns the flow in each vessel into a propagating wave, with consequent wave reflections. The relation between pressure and flow in the presence of wave reflections, and hence the "resistance" to flow, are not clearly predetermined as they are in steady flow, nor easy to formulate mathematically [9, 135, 141, 221].

The complex architecture of the coronary network compounds the difficulty, not so much because of its degree of complexity but because of an insufficient amount of architectural data. Also, elasticity of the coronary vessels gives the system the ability to change its volume to an extent and in a manner which are not fully known. This property of the coronary network, generally referred to as its "capacitance", in combination with the pulsatile nature of the flow introduces an element of inflation and deflation which further complicates the relation between pressure and flow. Finally, most of the coronary vasculature is deeply imbedded within the cardiac muscular tissue, and as this tissue contracts and relaxes in the pumping process, the effects on pressure and flow within the vessels are far from known or fully understood. Thus, while the nature of resistance to coronary blood flow is highly complex and not fully understood, it is clear that the design provisions for coronary flow reserve are based entirely on having substantial control over that resistance. Coronary blood flow can increase by a factor of 6 or more not by having supplying vessels that are designed to carry such high flow but by having resistance vessels under strict dynamic control. We may say that coronary flow reserve is not part of the *static* design of the coronary network but rather is a *dynamic* property of the coronary circulation.

1.9 Design Conflict?

Provisions for coronary flow reserve discussed in the previous section are clearly geared, by design, to changes in flow demand associated with normal physiological function. The sudden increase in coronary blood flow required at the onset of vigorous physical exercise, for example, is accomplished by the facility of coronary flow reserve. But the integrity of that facility is seriously compromised in the presence of obstructive coronary artery disease. The reasons for this are somewhat convoluted and have the appearance of a "design conflict" between the availability of coronary flow reserve and the problem of long term fluid dynamic changes associated with obstructive vascular disease.

The reason for this is that as disease gradually obstructs the supplying vessels, thus increasing the resistance to flow and therefore decreasing flow rate through the vessels, the facility for coronary flow reserve counteracts by dilating resistance vessels to restore the overall resistance in the system and thereby restore flow rate. But some of the capacity to further dilate the resistance vessels and further increase the flow rate has now been lost. That is, some of the coronary flow reserve has been "used up". And more is used up, in the same way, as the severity of obstructive disease increases, to a point where the capacity to increase coronary flow rate is completely lost.

Thus, the facility of coronary flow reserve does not appear to be aimed *by design* to deal with the long-term fluid dynamic effects of coronary artery disease. It is indeed well established that coronary flow reserve diminishes in the presence of atherosclerotic coronary artery disease [83, 66, 128, 100, 183, 26, 112]. Are there other facilities or mechanisms in the coronary circulation that are aimed at the long term effects of coronary artery disease?

It is known that in the presence of obstructive coronary artery disease the coronary circulation can develop collateral flow routes aimed at counteracting the fluid dynamic restrictions imposed by the obstruction. While the precise mechanisms for this development are not fully understood or agreed upon, its association with the presence and severity of coronary artery disease is fairly well established [16, 172]. Thus, the development of collateral pathways is a mechanism that appears aimed by design to deal with the long-term fluid dynamic effects of obstructive coronary artery disease.

The most important difference between this mechanism and that of coronary flow reserve is in the *time scale* at which the two mechanisms operate. While the time scale of coronary flow reserve is in the order of seconds or minutes, the time scale of collateral pathways is in the order of weeks, months, or perhaps even years [15, 172, 128].

The grounds for a design conflict are thus apparent: If the faster acting facility of coronary flow reserve is triggered to correct a deficiency in blood flow rate, the slower acting facility of developing collateral pathways is no longer called for and is therefore not implemented. While the trigger for both facilities must ultimately be the presence of ischemic tissue, if the needs of that tissue are met by the facility of flow reserve, the other facility is not triggered.

That is, as coronary flow reserve responds to ischemic events caused by obstructive vascular disease, it *masks* the effects of that disease from the facility of collateral pathways. The effects of the obstruction therefore do not trigger the more permanent measure of developing collateral pathways. That is until coronary flow reserve has been completely depleted.

An analogy which has been used for this conundrum is that of a bank account with a large reserve that acts to mask any deficits between the totals of deposits and debits each month [217]. A monthly deficit is "covered" by the large balance and therefore goes unreported and does not trigger any long-term remedies. But in the process the size of the large reserve has diminished by the amount of the deficit and will continue to diminish if monthly deficits continue [217].

Both in the coronary circulation and in the bank analogy the grounds for a design conflict are clear. A reserve that has the purpose of dealing with short-term (acute) deficits acts inadvertently to mask long-term (chronic) ones, and in both cases a resolution of the conflict can only be achieved by removing the "masking effect" of the reserve. In the bank analogy this would mean to challenge the reserve on a regular basis so as to moniter its size and take remedial action if the reserve is declining. In the coronary circulation this would mean to challenge the coronary flow reserve on a regular basis to create near-ischemic conditions that would trigger collateral pathways development. In the bank analogy the challenge to the reserve may take the form of regularly timed "conditional" spending sprees. In the coronary circulation it may take the form of regularly timed vigorous physical exercise. The benefits to the coronary circulation of regular physical exercise are indeed well known. The conundrum of coronary flow reserve is a context in which these benefits can be explained.

1.10 Summary

There are several billion cells within the human body, which consume food (nutrients, metabolic products) and dispose of waste products *on an individual basis*. The cardiovascular system achieves this mammoth task not by *storing* these products in one location but by having them carried to and from every cell by means of a circulating fluid, namely blood. The most important element of the cardiovascular system is therefore not so much the place where nutrients and metabolic products come from but the *pump* that circulates the fluid within which these products are carried. That pump is of course the heart.

A complication arises because the heart, which is responsible for supplying blood to every part of the body, is also responsible for its own blood supply. This seems to suggest that the system of blood supply to the heart is an unstable positive feedback system in which failure leads to more failure, but this is not the case because of regulatory mechanisms. More commonly, heart failure is caused by coronary artery disease which disrupts cradiac blood supply by obstructing some of the conducting vessels. "Heart disease" is the term most widely used to describe this course of events but the term is somewhat a misnomer because in the overwhelming majority of cases the failure is due only to a lack of blood supply to the heart for its own metabolic needs and hence a lack of the fuel it needs to perform its function as a pump. In this book the term "heart disease" is reserved for only the small proportion of cases where the heart is diseased in the true sense of a genetic or infectious disorder. In all other cases the term "heart failure" is used to mean *failure of the heart as a pump caused by lack of blood supply*. While this usage of these terms differs from their common usage in the clinical setting, the intention here is to emphasize the strictly biomedical engineering view of the coronary circulation to be adopted in this book. According to this view, blood supply to the heart is a highly dynamic system which can be disrupted by not only a problem in the lines of supply but also by a problem in the dynamics of the system. Indeed, the *dynamics* of blood supply to the heart is the principal subject of this book.

Blood supply to the heart comes via two branches of the aorta, known as the left main and right coronary arteries. These two vessels and their branches first circle the heart in the manner of a "crown", hence the name "coronary", then establish branches and sub-branches to every part of the heart. The resulting vasculature, which we refer to collectively as the "coronary network", is likely one of the most compact and complex within the human body.

The overall *functional* picture of the coronary network consists of main arteries circling the heart once in a horizontal plane along the atrioventricular groove and once in a vertical plane along the interventricular groove. This picture is important because it does not vary from heart to heart, therefore representing an *invariant* feature of the coronary network, a feature of its functional design. While there are wide variations in the anatomical details of the coronary arteries, this design feature of the network rarely varies.

As they reach the back of the heart, the left circumflex and the right coronary arteries move toward each other and terminate short of actually meeting. The point at which this occurs is a measure of the extent to which the right coronary artery participates in blood supply to the left side of the heart, an important functional aspect of the coronary network usually referred to as left/right dominance.

The underlying branching pattern of coronary vasculature is that of an open tree structure in which there is a unique pathway to every region of the heart tissue. The question of "collateral" pathways that provide alternate routes is controversial because neither the existence of collateral vessels within the coronary network nor their functional significance are fully agreed upon. It is generally accepted, however, that collateral vasculature is a compensatory mechanism for coronary artery disease, even if the full details of that mechanism are not fully understood.

While there are wide variations in the "details" of coronary vasculature from heart to heart, some underlying functional design can be identified. According to this design the heart appears to be divided into individual zones, each being circled by "distributing" vessels that bring blood supply to that zone and penetrated by branches that act as "delivering" vessels. The most important functional issue in each case is the extent to which a particular zone depends on blood supply from the left main and/or from the right coronary arteries.

Intensity of the pumping action of the heart varies considerably, depending on the metabolic activity of the rest of the body, the range extending from a base level when the body is in a resting state to a considerably higher level when it is at maximal activity. Energy required to support this pumping action is therefore highly variable, and coronary blood flow must be able to change over a wide range. This capacity of the coronary circulation is known as "coronary flow reserve", and it is facilitated *not* by having vessels that are larger than normal but by having control over the resistance to flow within the coronary network.

It would seem that coronary flow reserve is clearly aimed at responding to *immediate* changing demands for blood flow (within seconds) while the mechanism of collateral vasculature is likely aimed at compensating for *gradual* changes caused by coronary artery disease (within weeks or months). A case can be made for a possible "design conflict" between these two mechanisms because the capacity of coronary flow reserve is diminished by coronary artery disease and because that capacity can also be inadvertently "used up" to compensate for the gradual effects of coronary artery disease.

2

Modelling Preliminaries

2.1 Why Modelling?

To solve fluid flow problems and fully determine the dynamics of the flow, including mapping the velocity field and the relation between prevailing pressure and flow fields, is possible only in the most simply constructed cases and mostly in the physical sciences [134, 193]. Fluid flow problems in biology, by contrast, are rarely simply constructed and can rarely be solved directly [34, 120]. The problem of flow in a tube, for example, has the simple "Poiseuille flow" solution when the tube is rigid, its cross-section is perfectly circular, the tube is long enough for flow to fully develop, and the fluid is a smooth "continuum" that has the simple rheological properties of a "Newtonian" fluid in which shear stress is related linearly to the velocity gradients [34, 120]. Barely any of these ideal conditions is met in biological problems involving flow in tubes, most notably the problem of blood flow in arteries, and particularly flow in coronary arteries, which is the subject of this book. Here the fluid is not a smooth continuum but a suspension in plasma of discrete red and other blood cells and, as we saw in the previous chapter, the system does not consist of a single tube but of many millions of tube segments that are joined together in a hierarchical tree structure. The segments are rarely long enough or perfectly circular to support fully developed Poiseuille flow, and the details of flow at their junctions are highly complicated and depend strongly on the exact geometry of each junction [122]. Furthermore, the precise branching structure of the vascular system of the heart cannot be mapped to the last detail to allow a mathematical solution of the flow problem. In fact, it is known that these details vary widely from one heart to another as much as do fingerprints from one individual to the next [228].

The purpose of the vascular system of the heart is to bring blood flow to within reach of every cell of the myocardium. Schematically, the vascular system has the hierarchical form of a tree structure (Fig. 1.6.1), with flow proceeding from the root segment of the tree to the periphery. Pressure at the base of the aorta, where the vascular trees of the left and right coronary arteries have their roots (Fig. 1.3.2), provides the driving force for this flow, but the relationship between this pressure and the ultimate flow at the delivering end of the two trees is everything but simple [98, 97]. Indeed, it is far from clear that pressure at the base of the aorta is the *only* driving force for coronary blood flow, nor is it clear that the resistance to flow, which this driving force must overcome, is limited to that of simple flow in a tube. Other mechanisms may be at play, and while some are known, their exact role in the dynamics of coronary blood flow is as yet not fully understood. Prominent among these are the rhythmic contractions of the myocardium with each pumping cycle and the consequent effect of these contractions on vessels that are totally imbedded within that tissue. It has been demonstrated that one effect of this so-called "tissue pressure effect" is to reduce or even reverse the flow in the main coronary arteries during the contracting (systolic) phase of the pumping cycle [101], but it is possible that this same effect may actually provide a pumping (driving) force for blood flow within the peripheral vessels near the delivering end of the tree. The cyclic compression of coronary vasculature by surrounding tissue also has a "capacitance" effect, namely a cyclic change in the volume of blood contained in the system. This effect plays an important yet unclear role in the dynamics of the coronary circulation, rendering the relation between driving pressure and delivering flow far less tractable [96, 97]. The same is true of the effects of wave reflections from a massive number of vascular junctions within the coronary network and the important yet unclear role which these play in the dynamics of the coronary circulation [219].

Direct measurements of pressure and flow within elements of the coronary network, to establish an empirical relation between them, are fraught with no less difficulty. While some measurements have been made successfully in isolated hearts [98, 97], access is possible only to larger coronary vessels at entry into the coronary network, becoming increasingly difficult with increasing "depth" into the network. Measurements *in vivo* are further hampered by the violent motion of the coronary vessels as the heart contracts and relaxes in its periodic pumping action. Thus, at best some access is possible to one end of the coronary circulation, but this can provide only a limited base for any conclusions because of lack of access to the distal end of the circulation. More precisely, flow measured at entry to the coronary tree does not usually represent flow at exit, because of capacitance and other effects mentioned earlier.

Modelling is thus a necessity rather than a luxury in the study of coronary blood flow. In the absence of adequate access to the system for direct observations or measurements of pressure and flow, the only prospect for a good understanding of the system is by using a model. The accuracy of the model can be improved by testing it against whatever data or observations are available, changing its design so as to produce closer agreement. The obvious and most important advantage of using a model is that its behaviour can be studied easily and more extensively than the actual system which it represents. Indeed a range of such models have been proposed in the past and we examine some of them subsequently, but the emphasis in this book is less on the models themselves than on the elements from which the models are constructed. The reason for this is that a model of the coronary circulation is only useful if it can be tested against some direct measurements. In fact, the model must be tailored to the type of measurements available, and as the nature and availablility of such measurements changes, so must the design and nature of the model to be used.

Our understanding of the dynamics of the coronary circulation is presently at its infancy. Indeed, in the clinical setting a purely *static* view of the system predominates, in which the concern is primarily with whether vessels are fully open or restricted by disease [127, 133, 73]. The reason for this viewpoint is not that the *dynamics* of the coronary circulation are thought unimportant in the clinical setting but that as yet we do not have a clear understanding or a clear model of these dynamics. The purpose of this book is to provide the student, researcher, or indeed clinician, with basic analytical and conceptual tools with which to explore and hopefully improve his or her understanding of the dynamics of the coronary circulation.

2.2 The "Lumped Model" Concept

The relation between pressure and flow in a tube depends on such properties of the tube as its diameter, length, and elasticity. It also depends on the form of the driving pressure, in particular whether the pressure is steady or pulsatile. The relation between pressure and flow in a vascular tree structure consisting of a large number of tube segments depends not only on all such factors in each tube segment but also on events at the junctions between tube segments and on how the properties of individual segments are distributed within the tree structure. The overwhelming complexity of this problem gives rise to the "lumped model" concept. Detailed analysis and results based on this concept are presented in subsequent chapters. Here we discuss only broadly the concept itself as a valid modelling strategy.

Essentially, in a lumped model the complex vascular structure of the coronary network is ignored and the network is replaced by a single tube having properties representative of the network as a whole. It is a variant of the more familiar "black box" concept, in which a complex system is enclosed by an imaginary box and only the relation between input and output from the box is examined to learn something about the characteristics of the system without delving into the complexity that produced these characteristics. In the coronary circulation the lumped model attempts to reproduce a relation between pressure and flow similar to that observed or measured in the physiological system but without going through the overwhelming task of determining how the relation unfolds through the complex structure of the coronary vascular network.

Of particular interest is the relation between pressure and flow at *input* to the system and pressure and flow at *output*. The reason for this is that while some direct measurements of pressure and flow are possible at input to the system, usually at the left or right main coronary arteries, no such measurements are possible at output, that is at the capillary end of the system. The output end of the coronary circulation is of course of particular clinical interest because it represents the ultimate function of the system, namely the delivery of blood to cardiac tissue. But at this end of the system flow is divided into many millions of capillaries in which neither the velocity nor the number of capillaries can be determined with sufficient accuracy to compute total output. A correct model of the system would thus provide a theoretical means of obtaining important information at output which is not available experimentally. However, the "correctness" of the model can ultimately be verified only by testing its results against some measurements from the physiological system. Thus, the modelling process becomes a highly intricate *iterative* process whereby the choice and values of model parameters are guided by a comparison of the results of the model with whatever direct measurements are available [110, 24, 115, 90, 98, 97].

Pressure and flow in the coronary circulation are highly pulsatile because of the pulsatile nature of the input driving pressure and because, in addition to this, much of the coronary vasculature is imbedded in cardiac muscle tissue and is subject to the effects of cyclic contraction of the cardiac muscle, so-called "tissue-pressure effects". Thus, pressure and flow at both ends of the system are time-dependent in the sense that they have cyclic waveforms. The waveforms are not the same at both ends, however. At any point in time within the oscillatory cycle, total inflow into the coronary system is not usually equal to total outflow because of the so-called "capacitance" effect. There is continuous change in the total volume of vascular lumen within the system during the oscillatory cycle. Therefore, some inflow may go towards "inflating" the system and will not contribute to outflow and, conversely, some outflow may be produced by "deflation" of the system rather than by direct inflow. While *average* flow must be the same at both ends of the system, that is, flow averaged over one or more cycles, a relation between average flow and average pressure does not feature the time-dependent characteristics of the system that actually contribute to that relation. Only events *within the oscillatory cycle* exhibit these characteristics, but the nature of these events is lost in the time-averaging process. For these reasons the main focus of lumped models has been on a *time-dependent relation between pressure and flow*, that is on the time course of that relation within the oscillatory cycle.

2.3 Flow in a Tube

At the core of almost every modelling scheme for coronary blood flow and for blood flow in general is the mechanics of flow in a tube. Indeed, the lumped model discussed in the previous section is based on the concept that flow through the complex vasculature of the coronary circulation can be replaced by flow in a single tube with "equivalent" properties. It is important, therefore, to outline the basic properties of flow in a single tube, which we do in this section. The validity of the basic premise of the lumped model concept, namely that flow in a complex system of vessels *can* be considered equivalent to flow in a single tube, can only be discussed in the context of each particular modelling scheme and is therefore deferred to subsequent chapters.

When fluid enters a tube, it does not simply slide along the tube as a bullet, because of a condition of "no-slip" that prevails at the tube wall [13, 34, 174, 71] whereby elements of fluid in contact with the tube wall become arrested there, forming a cylindrical layer of stationary fluid attached to the inner surface of the tube wall. As fluid progresses along the tube, the next layer of fluid adjacent to the first is slowed down by the stationary layer because of the viscosity of the fluid, and similarly, subsequent concentric layers of fluid that are further and further away from the wall are slowed down but to a lesser and lesser extent and are thus able to move more freely, fluid along the axis of the tube able to move the fastest (Fig. 2.3.1).

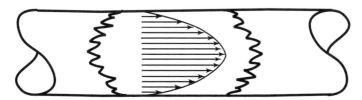

Fig. 2.3.1. Fully developed flow in a tube, commonly referred to as Poiseuille flow, is characterized by a parabolically shaped velocity profile, with zero velocity at the tube wall and maximum velocity along the tube axis.

Ultimately, at some distance downstream from the tube entrance, the flow becomes "fully developed" and is generally referred to as "Hagen-Poiseuille flow" after those who studied it first [168, 192, 174, 135], or more commonly as simply "Poiseuille flow". Flow in this region is characterized by a parabolically shaped "velocity profile" along a diameter of the tube, with zero velocity at the tube wall and maximum velocity at the tube axis, and is given by [221]

$$u = \frac{k}{4\mu}(r^2 - a^2) \tag{2.3.1}$$

where μ is viscosity of the fluid, r is radial coordinate measured from the axis of the tube, a is the tube radius, and k is the pressure gradient driving the flow, which in Poiseuille flow is constant and equal to the pressure difference Δp between any two points along the tube divided by the length of tube l between them, that is [221]

$$k = \frac{dp}{dx} = \frac{\Delta p}{l} \tag{2.3.2}$$

Here p is pressure and x is axial coordinate, positive in the direction of flow. The pressure difference Δp is measured in the direction of flow, that is

$$\Delta p = p_2 - p_1 \tag{2.3.3}$$

where p_1, p_2 are pressures at the upstream and downstream ends of the tube segment, respectively. Since p_1 must be higher than p_2 to produce flow in the positive $x-$direction, Δp is usually referred to as the "pressure drop" along the tube segment.

Eq. 2.3.1 indicates that in Poiseuille flow the flow rate q through the tube is given by

$$q = \int_0^a 2\pi r u \, dr = \frac{-k\pi a^4}{8\mu} \tag{2.3.4}$$

Thus, average flow velocity $\bar{u}$ is given by

$$\bar{u} = \frac{q}{\pi a^2} = \frac{-ka^2}{8\mu} \tag{2.3.5}$$

while maximum velocity $\hat{u}$ occurs on the tube axis where $r = 0$ and from Eq. 2.3.1 is given by

$$\hat{u} = \frac{-ka^2}{4\mu} \tag{2.3.6}$$

The two results show that maximum velocity in Poiseuille flow is twice the average velocity, that is

$$\hat{u} = 2\bar{u} \tag{2.3.7}$$

As described earlier, Poiseuille flow is not established immediately on entry into the tube, but evolves over a length of tube l_e known as the "entry length". Flow in that region of the tube is usually referred to as "developing flow" and an estimate of the entry length is given by [123, 174, 71]

$$l_e = 0.04 N_R d \tag{2.3.8}$$

where d is tube diameter and N_R is the Reynolds number, defined by

$$N_R = \frac{\rho \bar{u} d}{\mu} \tag{2.3.9}$$

where ρ is fluid density.

When the lumped model is used to study flow in the coronary circulation, which means that coronary blood flow is being modelled by an equivalent flow in a single tube, the equivalent flow is invariably considered *fully developed*. This assumption is fairly difficult to deal with because it is at once both necessary and unjustified. The assumption is unjustified because the entry lengths in many millions of tube segments in the coronary circulation will be different and cannot be represented by an "equivalent" entry length in a single tube. Furthermore, the assumption is necessary because the problem of determining the entry length and examining the extent to which flow is fully developed in each of these millions of tube segments is intractable. It is in fact further complicated because flow is entering and leaving tube segments at different stages of development. As a result, the standard entry length analysis leading to the result in Eq. 2.3.8, based on the assumption that flow entering the tube is uniform, no longer applies [31]. The best that can be done is to evaluate the weight of the assumption of fully developed flow in each modelling scheme in context of the particular aspect of coronary circulation being studied.

If flow entering a tube is assumed to have a uniform velocity $\bar{u}$, then a key difference between the developing and fully developed regions of the flow is that in the developing region elements of fluid near the tube axis (where $u = \hat{u}$) are being accelerated to meet the higher velocity there, while elements of fluid near the tube wall (where $u = 0$) are being decelerated because of the condition of no-slip at the tube wall. In the fully developed region, by contrast, fluid elements have reached their ultimate speed and are moving with constant velocity. This difference is compounded when the flow in a tube is *pulsatile*. In that case fluid elements in all regions of the tube are being accelerated and decelerated by the oscillatory driving pressure. Thus, in the entry region of the tube, fluid elements are being accelerated or decelerated *in space* by the entry conditions described above, and accelerated and decelerated *in time* by the oscillatory driving pressure. This makes the length of the entry region time-dependent and more difficult to define [34, 71, 7, 37].

2.4 Fluid Viscosity: Resistance to Flow

Flow in a tube may be resisted in a number of ways. If it is being accelerated, fluid inertia resists the pressure driving the flow. If the tube wall is elastic, its elasticity may oppose the driving pressure as it expands the tube wall. However, in both cases the same effect may also *aid* the flow, as it decelerates in the first instance, and as the tube wall recoils in the second. Thus, when flow in a tube is oscillatory these two forms of resistance do not dissipate energy, except in the second case if the tube wall is not purely elastic but has some viscoelastic properties.

The most important form of resistance to flow in a tube is that due to viscous friction at the interface between fluid and the tube wall. It is important because it is present when flow is steady or oscillatory and it always dissipates energy whether the flow is accelerating or decelerating. Because of this, it is usually referred to simply as "the resistance", and we shall follow this practice in this book. Resistance to flow in a tube arises because of a combination of the no-slip boundary condition at the tube wall and the viscous property of the fluid.

A key property of viscous fluids is that the force required to move adjacent layers of fluid at different velocities, that is, the force required to create shear flow, is an increasing function of the local velocity gradient. For a large class of fluids known as "Newtonian fluids", the force is simply proportional to the velocity gradient, that is

$$\tau = \mu \left(\frac{du}{dr} \right) \tag{2.4.1}$$

where τ is the local shear stress, that is, the local stress required to maintain the shearing motion, and μ is the coefficient of viscosity of the fluid. The velocity gradient du/dr is a measure of the local change in the velocity u of adjacent layers of fluid relative to the distance r between them. In Poiseuille flow this corresponds to the local slope of the parabolic velocity profile shown in Fig. 2.3.1 and given in Eq. 2.3.1.

The linear relation between shear stress and velocity gradient in Eq. 2.4.1 was first derived by Newton, hence the term "Newtonian fluids" has been used for fluids that obey the relation [168, 192]. There is a long-standing question whether blood, because of its corpuscular nature, is or is not a Newtonian fluid [21]. The question is not a very meaningful one because there are blood flow problems in which blood can be treated as a Newtonian fluid and others where it cannot. The question must therefore be directed at the nature of the flow problem being studied rather than at the nature of blood. Many problems relating to the general dynamics of flow in the systemic circulation, with focus on its pulsatile, have been studied successfully on the assumption of a Newtonian behaviour of the fluid, that is, on the assumption that Eq. 2.4.1 is valid [135, 141, 153]. That is not to say that blood *is* a Newtonian fluid, but that any non-Newtonian behaviour of blood does not significantly affect the general dynamics of the systemic circulation as a whole, although it may be important in the study of local flow properties in a single vessel or a single junction. The same is appropriate for a study of the general dynamics of the coronary circulation and we therefore uphold the Newtonian assumption in this book.

An important consequence of the viscous property of fluids is that the velocity difference between adjacent layers of the fluid must be infinitely small so that the velocity gradient remains finite. In other words, change of velocity within the fluid must be smooth. A step change of velocity (Fig. 2.4.1) is not possible because it would produce a locally infinite velocity gradient, and the shear stress required to maintain it would be infinite (Eq. 2.4.1).

It follows from this property that at the interface between a moving fluid and a solid boundary, as at the inner surface of a tube, there can be no finite difference between the fluid velocity tangential to the boundary and the

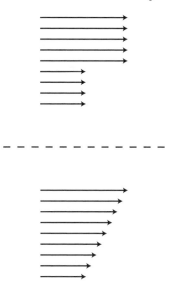

Fig. 2.4.1. An important consequence of the viscous property of fluids is that the velocity difference between adjacent layers of fluid must be infinitely small so that the velocity gradient remains finite. Thus, a step change of velocity (top) is not possible because it would produce a locally infinite velocity gradient and the shear stress required to maintain it would be infinite. Instead, the change of velocity must occur smoothly (bottom) so that the velocity gradient remains finite.

boundary itself. That is, the tangential velocity of fluid elements in contact with the boundary must be zero relative to the boundary, as required by the no-slip boundary condition (Fig. 2.4.2). This does not "prove" the no-slip boundary condition but shows only that the viscous property of fluids is consistent with it. Indeed, the basis of the no-slip boundary condition has been and remains largely empirical [13, 34, 174, 71].

Eq. 2.4.1 applied to Poiseuille flow in a tube, with velocity u as given by Eq. 2.3.1, yields the following result for the shear stress τ_w at the tube wall

$$\tau_w = \mu \left(\frac{du}{dr}\right)_{r=a} = \frac{ka}{2} = \frac{a\Delta p}{2l} \qquad (2.4.2)$$

Since the pressure gradient k or pressure difference Δp are negative in the flow direction, it follows that τ_w is also negative. That is, the shear stress (acting on the fluid) at the tube wall has the effect of opposing the flow. The velocity gradient at the tube wall which is responsible for this shear stress is of course a consequence of the condition of "no-slip" there. It causes fluid in contact with the tube wall to come to rest while fluid along the tube axis charges at maximum velocity. A velocity gradient must therefore exist between the two regions and at the tube wall. Therefore, the condition of no-slip and the viscous property of the fluid *together* produce the shear stress at the tube wall.

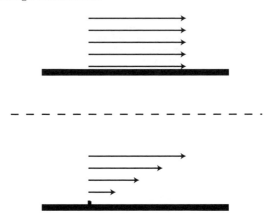

Fig. 2.4.2. The viscous property of fluids requires that at the interface between a moving fluid and a solid boundary, as at the inner surface of a tube, there be no finite difference between the fluid velocity tangential to the boundary and the boundary itself (top). That is, the tangential velocity of fluid elements in contact with the boundary must be zero relative to the boundary itself (bottom), as required by the no-slip boundary condition.

The total resistance to flow R, which results from shear stress acting on the entire surface area of the tube, can be expressed in terms of the flow rate q as

$$R = \frac{\Delta p}{q} \tag{2.4.3}$$

and substituting for the flow rate from Eq. 2.3.4, and using Eq. 2.3.2, this gives

$$R = -\frac{8\mu l}{\pi a^4} \tag{2.4.4}$$

The minus sign indicates that the resistance, which represents the force exerted by the tube wall on the fluid, is opposite to flow direction. The sign is usually omitted because the term "resistance" in fact refers to a force opposing the flow, that is a force in the negative direction when flow represents the positive direction. This is equivalent to modifying the definition of R to

$$R = -\frac{\Delta p}{q} = \frac{8\mu l}{\pi a^4} \tag{2.4.5}$$

It is seen that resistance to flow, which represents the amount of pressure difference required to produce a given amount of flow, depends critically on tube radius, being proportional to the inverse of the radius to the fourth power. Thus, if the tube radius is reduced by a factor of 2, the resistance increases by a factor of 16, that is by $1,600\%$. If the tube radius is *increased* by a factor of 2, the resistance decreases by a factor of 16, that is by approximately 94%.

Writing Eq. 2.4.5 as an equation for the flow rate q, we find the amount of flow that would be produced by a given pressure difference Δp, namely

$$q = -\frac{\pi a^4 \Delta p}{8\mu l} \tag{2.4.6}$$

If in an experiment the amount of flow is found higher than that dictated by Eq. 2.4.6, this could be interpreted as a change in one of the other parameters on the right side of the equation. Indeed, experiments in the past have shown that there is an *apparent* drop in blood viscosity in very small blood vessels, usually referred to as the Fahraeus-Lindqvist effect [63, 45, 221]. The effect is termed "apparent" because it is not based on direct measurement of the viscosity μ but on a measurement of flow for a given pressure drop. Thus, an observed value of q higher than that prescribed by Eq. 2.4.6 was interpreted as a decrease in the viscosity μ because such a decrease would also produce a higher value of q. Another interpretation which has been considered is the possibility of partial slip at the tube wall which would have the effect of requiring a smaller pressure drop for a given amount of flow, or conversely higher flow rate than is prescribed by Eq. 2.4.6, because of lower friction at the tube wall. However, it has been difficult to demonstrate that slip actually occurs in small blood vessels, and this interpretation is still a matter of debate [156, 211, 221]. Similar comments apply to the Fahraeus-Lindqvist effect because of the difficulties involved in actually measuring blood viscosity in small vessels. As a result of these difficulties it has not been possible, so far, to incorporate the concepts of slip or of the Fahraeus-Lindqvist effect into mainstream modelling schemes of the general dynamics of either the systemic or the coronary circulation.

2.5 Fluid Inertia: Inductance

Acceleration in fluid flow may occur in one of two ways: in space or in time. Acceleration in space occurs when the space available to a stream of fluid is decreasing, so the fluid must increase its velocity to go through a reduced amount of space. Flow in a tube with a narrowing, as in a bottle neck, is an example (Fig. 2.5.1). Velocity at the narrowing must be higher than it is elsewhere, since the flow rate through the tube must be everywhere the same by conservation of mass, and since it is assumed here that the flow is *incompressible*, that is fluid density is not changing. Thus, the fluid is in a state of acceleration as it goes through the narrowing. The acceleration is *in space*, that is, in the sense that fluid elements are being accelerated as they progress along the tube.

Another, less obvious, example of acceleration in space occurs at the entrance to a tube. If fluid enters with uniform velocity (Fig. 2.5.2), elements of the fluid along the tube axis must accelerate to meet the maximum velocity

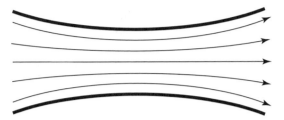

Fig. 2.5.1. Flow in a tube with a narrowing causes fluid elements to accelerate as they approach the narrowing and decelerate as they leave, assuming that the fluid is *incompressible*. Flow velocity is highest at the neck of the narrowing as indicated by the closeness of the streamlines there. Both the acceleration and deceleration are occurring *in space*, in the sense that the change in velocity is occurring as fluid elements progress along the tube.

in Poiseuille flow, while fluid elements near the tube wall are slowed down by the viscous resistance to meet the condition of no-slip at the tube wall. Thus in the entrance region of the tube some fluid is in a state of acceleration and some is in a state of deceleration, in both cases the change is occurring *in space*, that is as the fluid progresses along the tube.

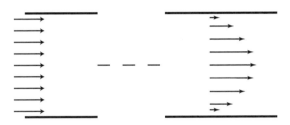

Fig. 2.5.2. Flow in the entrance region of a tube provides another example of acceleration and deceleration *in space*. If fluid enters with uniform veleocity, elements of the fluid along the tube axis must accelerate to meet the maximum velocity in Poiseuille flow, while fluid elements near the tube wall are slowed down by the viscous resistance and condition of no-slip at the tube wall.

One of the most important features of acceleration or deceleration in space is that it occurs in *steady* flow, that is, in a state of flow which does not change in time. In steady flow the velocity field does not change with time, meaning that the velocities at fixed positions within the flow field are constant and acceleration and deceleration occur as fluid elements move from one position to the next. It is in this sense that acceleration and deceleration in steady flow are seen as occurring *in space*.

Acceleration or deceleration *in time*, by contrast, is associated with *unsteady* flow, a state of flow in which the velocity distribution within the flow field changes with time. This situation occurs when the pressure driving the flow is not constant in time, as is the case in pulsatile blood flow where the driving pressure changes in an oscillatory manner (Fig. 2.5.3). In this case acceleration and deceleration are occurring *in time*, in the sense that the velocity at fixed points within the flow field is changing in time.

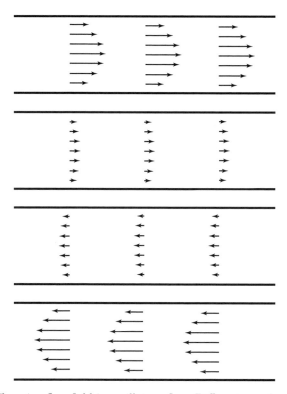

Fig. 2.5.3. Changing flow field in oscillatory flow. Different panels represent different points in time within the oscillatory cycle. Velocity is changing in time at fixed positions in space within the flow field. Acceleration and deceleration are occurring *in time*.

When a mass of fluid is accelerated or decelerated *in time*, the fluid does not respond immediately, because of its inertia. Thus, if the pressure difference Δp driving the flow in a tube changes suddenly to a higher level, it takes the flow rate q some time before it adjusts to a new value appropriate for the new driving pressure difference. This "reluctance" of the fluid to respond immediately is a form of resistance which would appropriately be referred to as "inertance" but is commonly known as *inductance* because of an electrical analogy to be discussed later.

Unlike the *viscous* resistance to flow which is present at constant flow rate, inductance is only present when flow is being accelerated or decelerated, that is, only when there is change in the flow rate. In fact, it is the *rate of change* of flow rate that is being resisted by the fluid, which means that a force is required to bring about such change. In the case of flow in a tube this means that a pressure difference Δp_L would be required specifically for this purpose; the subscript L is there to distinguish this pressure difference from that required to maintain the flow against the viscous resistance. More precisely, the required force is proportional to the rate of change of flow rate, that is

$$\Delta p_L = L\frac{dq}{dt} \qquad (2.5.1)$$

Again, the symbol L is commonly used for the constant of proportionality because of analogy with inductance in electric systems.

The basis of this relation can be found in the mechanics of an isolated mass m, governed by Newton's law of motion, which asserts that the product of mass and acceleration must equal the net force acting on that mass. If the force is denoted by F and the position of the mass is denoted by x, the law can be written as

$$m\frac{d^2x}{dt^2} = F \qquad (2.5.2)$$

where t is time. In general this equation is a vector equation because both F and x are vectors, but for the present purpose it is sufficient to work in only one dimension. In fluid flow the corresponding situation would be that of flow in a tube being accelerated, or decelerated, in one direction, namely along the axis of the tube. If the viscous effect at the tube wall is neglected for now (as it is accounted for separately below), then the body of fluid may be considered to move freely along the tube, as a bolus, in accordance with Newton's law. If the diameter of the tube is d, then the mass of such bolus of length l, being a cylindrical volume of fluid of diameter d and length l, is $\rho l \pi d^2/4$, where ρ is the density of the fluid. If the velocity of the bolus is u and the pressure difference driving it is Δp_L then the law of motion applied to this mass gives

$$\frac{\rho l \pi d^2}{4}\frac{du}{dt} = \Delta p_L \frac{\pi d^2}{4} \qquad (2.5.3)$$

If q is the volumetric flow rate, then $q = u\pi d^2/4$ and the above can be put in the form

$$\Delta p_L = \left(\frac{4\rho l}{\pi d^2}\right)\frac{dq}{dt} \qquad (2.5.4)$$

Comparison of this with Eq. 2.5.1 indicates that the constant L in that equation corresponds to the bracketed term above, that is

$$L = \left(\frac{4\rho l}{\pi d^2}\right) \qquad (2.5.5)$$

Thus, Eq. 2.5.1 and the concept of inductance on which it is based have a basis in simple mechanics.

The total pressure difference Δp required to drive the flow in a tube in the presence of a change in flow rate is the sum of the pressure difference needed to overcome the force of resistance due to inductance, namely Δp_L, and the pressure difference needed to overcome the force of resistance due to viscosity discussed in the previous section, Eq. 2.4.3, now to be denoted by Δp_R, that is

$$\Delta p = \Delta p_R + \Delta p_L \qquad (2.5.6)$$

Substituting for Δp_R from Eq. 2.4.3 and for Δp_L from Eq. 2.5.1, we then have

$$\Delta p = Rq + L\frac{dq}{dt} \qquad (2.5.7)$$

This is a first order ordinary differential equation which has the general solution [116]

$$q(t) = \frac{e^{-t/(L/R)}}{L} \int \Delta p \; e^{t/(L/R)} dt \qquad (2.5.8)$$

If the driving pressure difference is constant, say

$$\Delta p = \Delta p_0 \qquad (2.5.9)$$

Eq. 2.5.8 gives upon integration

$$q(t) = \frac{\Delta p_0}{R} + Ae^{-t/(L/R)} \qquad (2.5.10)$$

where A is a constant of integration. If the flow rate is zero at $t = 0$, we find $A = -\Delta p_0/R$ and the solution finally becomes

$$q(t) = \frac{\Delta p_0}{R}\left(1 - e^{-t/(L/R)}\right) \qquad (2.5.11)$$

As time goes on, the exponential term vanishes, leaving the flow rate at a constant value of $\Delta p_0/R$, which is what it would be against a resistance R and with a driving pressure difference Δp_0 (Eq. 2.4.3). At that value the flow is said to be in *steady state*, while prior to that it is in a *transient state*.

The effect of inertia of the fluid is thus to cause the flow to take a certain amount of time to reach steady state. As the driving pressure difference is applied, the flow increases from zero to its ultimate value, but because of inertia it takes a certain amount of time to reach that value. The higher the inertial effect the longer it takes the flow to reach steady state (Fig. 2.5.4). The ratio L/R has the dimensions of time and is a measure of the time delay caused by the inertial effect. It is usually referred to as the "inertial time constant" and we shall denote it here by t_L, that is we define

$$t_L = \frac{L}{R} \tag{2.5.12}$$

The higher the value of t_L the higher the prevailing inertial effect and the longer is the time required for flow to reach steady state. It is important to note, however, that the approach to steady flow is *asymptotic*, as seen in Fig. 2.5.4, which means that, strictly, the flow takes an infinite amount of time to reach steady state. For practical purposes, however, the flow is sufficiently close to steady state in a finite and usually very short time. The inertial time constant t_L is a measure of that time. More precisely, if in Eq. 2.5.11 we write

$$\overline{q}(t) = \frac{q(t)}{\Delta p_0 / R} \tag{2.5.13}$$

then

$$\overline{q}(t) = 1 - e^{-t/t_L} \tag{2.5.14}$$

and upon differentiation we find

$$\overline{q}'(t) = \frac{1}{t_L} e^{-t/t_L} \tag{2.5.15}$$

$$\overline{q}'(0) = \frac{1}{t_L} \tag{2.5.16}$$

Thus, the reciprocal of t_L represents the initial slope with which the flow curve moves towards its asymptotic value. The higher the inertial effect the higher the value of t_L and hence the lower the initial slope of the the flow curve and the longer it takes flow to reach its asymptotic value. Also, because the asymptotic value of the flow is here set at 1.0, then t_L also represents the time it takes the flow to reach this asymptotic value if, hypothetically, it continued with its initial slope, as illustrated in Fig. 2.5.4

It is important not to confuse transient and steady states here with developing and fully developed flow discussed in Section 2.3. Here, and essentially throughout the lumped model concept, the flow is assumed to be fully developed. Indeed, the relation $\Delta p_R = Rq$ used in Eq. 2.5.7 is based on the results obtained earlier for fully developed flow (Eq. 2.4.3). Steady and transient states here, by contrast, relate to flow development *in time*. Here we start out in a tube where fully developed flow is already established, then the pressure difference driving the flow is changed and we examine how, *in time*, the flow rate q adjusts to this change. Steady state is reached when the flow rate has fully adjusted to the change, while the adjustment period is referred to as the transient state. Thus, broadly speaking, developing and fully developed flow relate to flow development *in space*, as in the entrance region of a tube, while transient and steady states relate to flow development *in time*, as when the pressure difference driving the flow is changed.

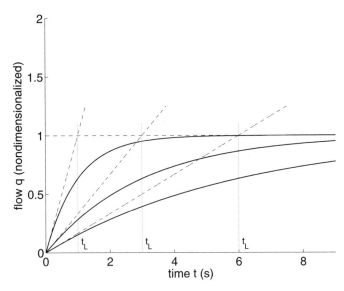

Fig. 2.5.4. If the pressure difference driving the flow in a tube is suddenly increased from 0 to some fixed value Δp_0, the flow increases gradually (solid curves) until it reaches the value $\Delta p_0/R$, which is shown by the dashed line above, normalized to 1.0. At that value the flow is said to be in *steady state*, while prior to that it is in a *transient state*. In steady state the flow rate has the value which it would have against a resistance R and with a driving pressure difference Δp_0 (Eq. 2.4.3), but because of fluid inertia the flow rate takes time to reach this value, the higher the inertia the longer the time. A good measure of the inertia of the fluid is the ratio L/R, which has the dimension of time when L is the inertial constant defined in Eq. 2.5.5 and R is the resistance defined in Eq. 2.4.4. The ratio is usually referred to as the "inertial time constant" and is denoted here by t_L (see Eq. 2.5.12). The three solid curves above, from left to right respectively, correspond to $L/R = t_L = 1.0, 3.0, 6.0$ *seconds*. It is seen clearly how the time it takes the flow curve to reach its ultimate value is directly related to the value of t_L. More specifically, the reciprocal of t_L represents the initial slope with which the flow curve moves towards its asymptotic value as indicated by the sloping dashed lines. The higher the inertial effect the higher the value of t_L and hence the lower the initial slope of the flow curve and the longer it takes the flow to reach its asymptotic value. Also, because the asymptotic value of the flow is here set at 1.0, then t_L also represents the time it takes the flow to reach this asymptotic value if, hypothetically, it continued with its initial slope. In the absence of the inertial effect ($L/R = t_L = 0$), the flow curve would "jump" to the asymptotic value at time $t = 0$ and remain on it thereafter.

If the driving pressure gradient Δp increases linearly with time, say

$$\Delta p = \frac{\Delta p_0}{T} t \tag{2.5.17}$$

where Δp_0 is a constant and T is a fixed time interval, Eq. 2.5.8 gives upon integration (by parts) and simplification

$$q(t) = \frac{\Delta p_0}{TR} \left(t - \frac{L}{R} \right) + Ae^{-t/(L/R)} \tag{2.5.18}$$

where A is a constant of integration. If the flow rate is zero at $t = 0$, we find $A = \Delta p_0 L/(TR^2)$ and the solution becomes

$$q(t) = \frac{\Delta p_0}{TR} \left(t - \frac{L}{R} + \frac{L}{R} e^{-t/(L/R)} \right) \tag{2.5.19}$$

or in nondimensional form

$$\bar{q}(t) = \frac{q(t)}{\Delta p_0/R} = \frac{t}{T} - \frac{t_L}{T} \left(1 - e^{-(t/T)/(t_L/T)} \right) \tag{2.5.20}$$

It is clear from the form of the solution that the appropriate time variable in this case is the fractional time t/T, where T may, for example, be taken as the total interval over which the flow takes place, hence t/T has the range 0 to 1.0. As in the previous case, the effect of inertia is embodied in the value of t_L. Again, since t_L has the dimension of time, it is appropriate in this case to consider values of the inertial time constant t_L/T, as this indeed is the parameter required in the above equation.

Results for $t_L/T = 0.1, 0.3, 0.5$ are shown in Fig. 2.5.5. As the driving pressure difference Δp increases, the flow rate $\bar{q}\,'(t)$ begins to increase, but as in the previous case and because of inertia, it takes a certain amount of time for the flow to reach a value appropriate for the prevailing value of the pressure difference. But since in this case the pressure difference is continually increasing, the flow rate is never able to reach that appropriate value. What the flow rate is able to achieve as time goes on is a state in which its value is a *fixed amount* below what it should be. We may refer to this state as *quasi-steady state* since, strictly, steady state is usually defined as one in which the flow rate is either constant or periodic. In the present case it is continually increasing. Nevertheless, it is possible here to distinguish (Fig. 2.5.5) between an initial period where the flow rate is adjusting to the new pressure difference, which may be referred to as a transient state, and a final period in which the flow rate is still changing but is now changing at a fixed rate, the same rate at which the driving pressure difference is changing. It is in this sense that the latter may be referred to as quasi-steady state.

From Eq. 2.5.20 we see that the quasi-steady state is reached asymptotically, as the exponential term becomes insignificant, and the flow rate reduces to

$$\bar{q}(t) \sim \frac{t}{T} - \frac{t_L}{T} \qquad (2.5.21)$$

Thus, asymptotically, the flow acquires the same form as the driving pressure, namely that of a linearly increasing function with a unit slope (Eq. 2.5.20), but, because of the inertial effect the flow curve is shifted along the time axis by an amount equal to the value of t_L/T as shown in Fig. 2.5.5. This shift represents the time interval by which the flow rate lags behind the prevailing pressure difference. The higher the inertial effect, the higher the value of t_L and the larger this ultimate gap between pressure and flow. Also, this gap between the flow and driving pressure never closes in this case because the driving pressure is continuouly changing. Only in the case of constant driving pressure does the flow ultimately "catch up" with the prevailing pressure and in a sense "overcome" the inertial effect as it reaches steady state. In the case of continuously changing pressure, as in the present case, the inertial effect is present in the transient as well as in the quasi-steady state.

If, finally, the driving pressure difference Δp varies as a *periodic* function of time, say

$$\Delta p = \Delta p_0 \sin \omega t \qquad (2.5.22)$$

where ω is the angular frequency of the oscillation, then Eq. 2.5.8 gives upon integration (by parts again)

$$q(t) = \frac{\Delta p_0 (R \sin \omega t - \omega L \cos \omega t)}{R^2 + \omega^2 L^2} + A e^{-(R/L)t} \qquad (2.5.23)$$

where A is a constant of integration. If the flow rate is zero at time $t = 0$, we find

$$A = \Delta p_0 \omega L / (R^2 + \omega^2 L^2) \qquad (2.5.24)$$

and the solution becomes

$$q(t) = \frac{\Delta p_0}{R^2 + \omega^2 L^2} \left(R \sin \omega t - \omega L \cos \omega t + \omega L e^{-(R/L)t} \right) \qquad (2.5.25)$$

A more useful form of the solution is obtained by combining the two trigonometric terms to give

$$q(t) = \frac{\Delta p_0}{\sqrt{R^2 + \omega^2 L^2}} \left(\sin (\omega t - \theta) - \frac{\omega L}{\sqrt{R^2 + \omega^2 L^2}} e^{-(R/L)t} \right) \qquad (2.5.26)$$

where

$$\theta = \tan^{-1} \left(\frac{\omega L}{R} \right) \qquad (2.5.27)$$

or in nondimensional form

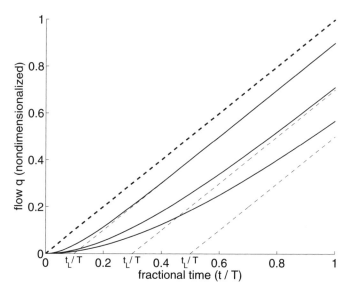

Fig. 2.5.5. If the pressure difference driving the flow in a tube increases linearly from zero, the flow rate begins to increase, but because of inertia it requires a certain amount of time to reach a value appropriate for the prevailing value of the pressure difference. But since in this case the pressure difference is continually increasing, the flow rate is never able to reach that appropriate value. What the flow rate is able to achieve as time goes on is a *quasi-steady state* in which its value is a fixed amount below what it should be. Thus, asymptotically, the flow acquires the same form as the driving pressure, namely that of a linearly increasing function with a unit slope (Eq. 2.5.20), but, because of the inertial effect the flow curve is shifted along the time axis by an amount equal to the value of t_L/T as shown. The three solid curves above, from left to right respectively, correspond to $t_L/T = L/RT = 0.1, 0.3, 0.5$, where T is total time interval over which flow is taking place, here taken as 1.0. The heavy dashed curve represents what the flow rate would be in the absence of inertial effect, that is when the inertial parameter t_L/T is zero. The light dashed curves represent the asymptotes of the flow curves for other values of the inertial parameter, shown at the bottom. It is seen that the higher the value of t_L/T the larger the ultimate gap between pressure and flow and hence the higher the inertial effect.

$$\bar{q}(t) = \frac{q(t)}{\Delta p_0/R} = \frac{1}{\sqrt{1 + \omega^2 t_L^2}} \left(\sin(\omega t - \theta) - \frac{\omega t_L}{\sqrt{1 + \omega^2 t_L^2}} e^{-t/t_L} \right) \quad (2.5.28)$$

$$\theta = \tan^{-1}(\omega t_L) \quad (2.5.29)$$

In this form we see that as the exponential term becomes insignificant, the flow rate becomes the same function of time as the oscillatory pressure difference, but with phase angle shift θ. The size of the shift is higher the higher the inertia of the fluid, that is the higher the value of the inertial time constant

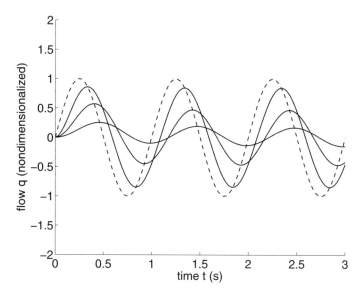

Fig. 2.5.6. If the pressure difference driving flow in a tube changes in an oscillatory manner, the flow rate attempts to follow the same oscillatory pattern, but because of inertia it requires a certain amount of time to reach that pattern. When it does, however, the flow rate lags behind the pressure difference by a fixed phase angle θ and its amplitude is lower than it would be in the absence of inertial effects, which here has the normalized value of 1.0. The three solid curves above, from left to right respectively, correspond to $t_L = L/R = 0.1, 0.3, 1.0$ *seconds*. It is seen that the higher the value of the inertial time constant t_L the larger the phase shift θ and the lower the amplitude of the flow oscillations.

$t_L(= L/R)$. Thus, here we see essentially the same behaviour of the fluid as in the previous case. The flow begins with a transient period in which it attempts to satisfy the prevailing pressure difference, but it never does. Instead, a steady state is reached in which the flow rate oscillates with the same frequency as the pressure difference driving the flow. It is a true "steady state" in this case, by common definition of that term [116]. In this state the flow rate oscillates in tandem with but lags behind the pressure difference by a fixed angle θ. The higher the inertial effect the larger is θ, and in the absence of inertial effects $\theta = 0$ as can be seen from Eq. 2.5.29. Also, from Eq. 2.5.28 we see that the *amplitude* of flow oscillation, which represents the highest flow rate reached at the peak of each cycle, is given by

$$|\bar{q}(t)| = \frac{1}{\sqrt{1 + \omega^2 t_L^2}} \tag{2.5.30}$$

thus the higher the inertial effect, hence the higher the value of t_L, the lower the amplitude of flow oscillation, as seen in Fig. 2.5.6. In the absence of inertial effects the amplitude of flow oscillation would be 1.0.

2.6 Elasticity of the Tube Wall: Capacitance

A tube in which the walls are rigid offers a fixed amount of space within it, hence the volume of fluid filling it must also be fixed, assuming, here and throughout the book, that the fluid is incompressible. By the law of conservation of mass, flow rate q_1 entering the tube must equal flow rate q_2 at exit. There is thus only one flow rate q through the tube, which may vary at different points in time depending on the applied pressure gradient, but at any point in time it must be the same at all points along the tube. Indeed, the relations between pressure gradient and flow considered in previous sections were all of this type, where the flow rate q may be a function of time t but not a function of position x along the tube (Eqs.2.3.4,2.4.6,2.5.11,14,19 and Figs.2.5.3-6). Thus, the analyses and results of previous sections were all based on the implicit assumption that flow is occurring in a rigid tube.

When flow is occurring in a *nonrigid* tube, two new effects come into play. First, the volume of the tube as a whole may change, an effect known as *capacitance*, again by analogy with the effect of a capacitor in an electric circuit. Second, a local change of pressure in an elastic tube *propagates* like a wave crest down the tube at a finite speed known as the wave speed. In a rigid tube, by contrast, a local change of pressure takes effect instantaneously everywhere within the tube. Consequently, the difference between flow of an incompressible fluid in a rigid tube compared with that in an elastic tube can also be expressed by saying that the wave speed is infinite in a rigid tube but is finite in an elastic tube.

While both the effects of capacitance and wave propagation result from elasticity of the tube wall, there is a fundamental difference between them, which provides a basis for dealing with them separately. Under the effect of capacitance there is a change in the *total volume* of the tube or system of tubes. Under the effect of wave propagation there is no change in the total volume of the system– a change of volume occurs only locally, as a local bulge or narrowing, and then propagates down the tube. It is important to emphasize, however, that while this difference makes it possible to separate the two effects on theoretical grounds, it does not necessarily imply that the two effects actually occur separately in practice. Hence, in this and the next section we deal with the effects of capacitance and wave propagation separately, with the understanding that this does not imply that the two effects must or do occur separately.

The key to the capacitance effect on flow in an elastic tube is that it affects the *total volume* of the tube, therefore flow rate at entrance to the tube may no longer be the same as that at exit because some of the flow at entry may go towards inflating the tube while some of the flow at exit may have come from a deflation of the tube. A convenient way of modelling this is to imagine flow going into a rigid tube to which a balloon is attached such that fluid has the option of flowing through the tube as well as inflating the balloon as depicted schematically in Fig. 2.6.1. The choice of a rigid tube is essential in order to eliminate the possibility of local changes in volume that would occur in wave propagation. Thus, the model depicts change in total volume only, consistent with the capacitance effect in isolation.

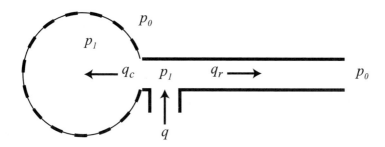

Fig. 2.6.1. Capacitance effect of flow in an elastic tube can be modelled by flow into a rigid tube with a balloon attached at one end. Flow rate q entering the system may go into the balloon or into the tube or both. Pressure p_1 at entry into the system is equal to pressure prevailing inside the balloon. Pressure at exit from the rigid tube is p_0, the same as that outside the balloon.

Initially, we consider the entrance to the balloon to be at entrance to the tube, so that pressure p_1 at entry into the system is equal to pressure prevailing within the balloon. Pressure outside the balloon and at exit from the tube is p_0. Flow through the tube and flow into the balloon are thus *in parallel*, in the sense that they can occur *independently* of each other.

Flow through the tube, to be denoted by q_r, is determined by the viscous resistance R and by the pressure difference Δp, as found previously (Eq. 2.4.3), namely

$$q_r = \frac{\Delta p}{R} \tag{2.6.1}$$

where

$$\Delta p = p_0 - p_1 \tag{2.6.2}$$

For flow into the balloon we note first that the balloon is in an inflated state when pressure inside the balloon is higher than pressure outside it, that is when

$$p_1 > p_0, \quad \Delta p < 0 \tag{2.6.3}$$

If the volume of the balloon in this state is v, then the capacitance C which is a measure of the *compliance* of the balloon is usually defined by the amount of *change* in the pressure difference Δp required to produce a change Δv in the volume of the balloon, that is

$$C = \frac{\Delta v}{\Delta(\Delta p)} \qquad (2.6.4)$$

The notation in the denominator emphasizes that it is not the pressure difference Δp that produces the change in volume but a change in that pressure difference. Also, in this form it is seen that a higher value of C represents a balloon that requires less change in Δp to produce a given change in volume, that is a balloon that is more elastic, or more compliant.

In coronary blood flow and blood flow in general the change in volume Δv is not a useful entity to work with because it is not easily accessible. A more useful entity is the capacitive flow rate q_c representing the amount of flow going into or out of the balloon, which can be related to Δv in the following way. As before, we assume that fluid is incompressible, hence the only way to change the volume of the balloon is to change the amount of fluid within it, that is to have a nonzero flow rate q_c going into or out of the balloon. If a constant flow rate q_c occurs over a time interval Δt, the corresponding change in volume of the balloon will be

$$\Delta v = q_c \Delta t \qquad (2.6.5)$$

Substituting this into Eq. 2.6.4 we then have

$$C = \frac{q_c \Delta t}{\Delta(\Delta p)} \qquad (2.6.6)$$

therefore

$$q_c = C \frac{\Delta(\Delta p)}{\Delta t} \qquad (2.6.7)$$

More generally, if Δp is a continuous function of time, then q_c correspondingly becomes a function of time, given by

$$q_c = C \frac{d(\Delta p)}{dt} \qquad (2.6.8)$$

This result shows clearly, again, that flow rate into the balloon depends not on the pressure difference Δp but on the rate of change of that difference. Also, by noting that total flow rate q into the system must be the sum of flow rates into the balloon and the tube, that is

$$q = q_c + q_r \qquad (2.6.9)$$

we see clearly that, because of the capacitance effect, flow rate q into the system is not necessarily equal to flow rate q_r out of the system.

If the pressure p_0 at exit from the tube and outside the balloon is now fixed, then flow into the system is controlled by only one remaining variable, namely the input pressure p_1. Under these conditions we consider the following three scenarios.

If the input pressure p_1 is constant, that is, if

$$\Delta p = \Delta p_0 \qquad (2.6.10)$$

where Δp_0 is a constant, then Eqs.2.6.1,8 give

$$q_r = \frac{\Delta p_0}{R}$$
$$q_c = 0 \qquad (2.6.11)$$

Thus, in this case flow is entirely through the tube. Flow into the balloon is zero because the rate of change of Δp is zero (although Δp itself is not zero). The volume of the balloon remains unchanged in this case. The balloon comes into play only when Δp is a function of time, which occurs if p_1 is a function of time.

If, for example, p_1 increases linearly with time, then the pressure differences across the tube and across the balloon will also increase linearly with time, say

$$\Delta p = \frac{\Delta p_0}{T} t \qquad (2.6.12)$$

where Δp_0 is a constant as before, t is time, and T is a fixed interval of time over which the change is taking place, which we introduce as in the previous section in order that Δp_0 retains the physical dimensions of pressure, then Eqs.2.6.1,8 now give

$$q_r = \frac{\Delta p_0}{R} \frac{t}{T}$$
$$q_c = C \frac{\Delta p_0}{T} \qquad (2.6.13)$$

There is constant flow into the balloon in this case, because the rate of change of Δp with time is constant. Flow through the tube increases linearly with time as Δp increases with time. To compare the two graphically it is easier to put them in nondimensional forms, namely

$$\bar{q}_r = \frac{q_r}{\Delta p_0/R} = \frac{t}{T}$$
$$\bar{q}_c = \frac{q_c}{\Delta p_0/R} = \frac{RC}{T} \qquad (2.6.14)$$

The product RC is seen to have the physical dimensions of time and is usually referred to as the "capacitive time constant". We shall denote it by t_c, in analogy with the inertial time constant (t_L), and define it by

$$t_c = RC \tag{2.6.15}$$

thus the two flow rates in nondimensional form are finally given by

$$\bar{q}_r = \frac{q_r}{\Delta p_0/R} = \frac{t}{T}$$

$$\bar{q}_c = \frac{q_c}{\Delta p_0/R} = \frac{t_c}{T} \tag{2.6.16}$$

Fig. 2.6.2 compares these flow rates at different values of t_c. We recall that higher values of t_c ($= RC$) are associated with higher compliance, allowing more flow to go into the balloon. Therefore, as seen in the figure, capacitive flow is constant at a value in fact equal to t_c/T, while resistive flow (flow through the tube) increases linearly as t/T.

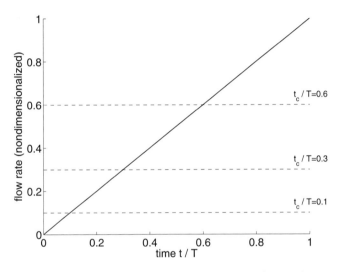

Fig. 2.6.2. Comparison of resistive (solid) and capacitive (dashed) flow rates when the driving pressure Δp is increasing linearly with time over a time interval T and at three different values of the capacitive time constant t_c. In all cases, capacitive flow is constant since it depends on the rate of change of Δp, while resistive flow increases linearly with time since it depends on Δp itself. Higher values of the capacitive constant t_c correspond to higher compliance, thus allowing more flow into the balloon.

Finally, an important scenario to consider is that in which the pressure differences across the tube and across the balloon is oscillatory, say

$$\Delta p = \Delta p_0 \sin \omega t \tag{2.6.17}$$

where Δp_0 is a constant and ω is the angular frequency of oscillation. In this case Eqs.2.6.1,8 give

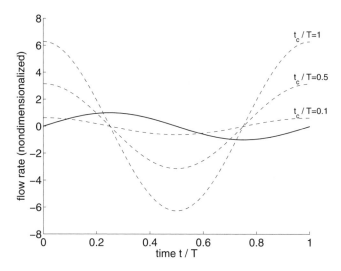

Fig. 2.6.3. Comparison of resistive (solid) and capacitive (dashed) flow rates when the driving pressure Δp is an oscillatory function of time of period T. The resistive flow (solid) has the same form as the driving pressure since inertial effects are not included here and since it is unaffected by the value of the capacitive time constant t_c. The capacitive flow (dashed) in each cycle, on the other hand, is higher with higher values of t_c because of higher compliance of the balloon.

$$q_r = \frac{\Delta p_0}{R} \sin \omega t$$
$$q_c = \Delta p_0 \omega C \cos \omega t \tag{2.6.18}$$

As expected, both q_c and q_r are oscillatory functions of time, with the same frequency as the driving pressure, namely ω. To compare the two it is more appropriate to put them in nondimensional forms, namely

$$\bar{q}_r = \frac{q_r}{\Delta p_0/R} = \sin \omega t$$
$$\bar{q}_c = \frac{q_c}{\Delta p_0/R} = \omega t_c \cos \omega t \tag{2.6.19}$$

The two flows are compared graphically in Fig. 2.6.3, where it is seen that how much of the flow goes into the balloon in each cycle depends on the value of the capacitive time constant t_C. As in the previous case, higher values of the t_c correspond to higher compliance, thus allowing more flow into the balloon. The resistive flow, on the other hand, is unaffected by the value of t_c and has the same form as the driving pressure, noting that inertial effects are not included here.

The results of this section illustrate the important role that capacitance plays in the dynamics of oscillatory flow in a compliant system, and hence its important role in the dynamics of the coronary circulation. While the structure of the coronary vascular system is far more complicated than the simple system in Fig. 2.6.1, the compliance of the system is known to play a role similar to that depicted in Fig. 2.6.1. A key question in the coronary circulation is how much of the oscillatory component of coronary blood flow goes into simply inflating and deflating the volume of the system, and how much goes into forward flow? This question is not properly addressed in the example of Fig. 2.6.1 because the driving pressure used here is a simple harmonic (Eq. 2.6.17) which produces only symmetrical back and forth flow in the rigid tube of Fig. 2.6.1. In coronary blood flow the driving pressure is a more complicated waveform which has a net forward component and some harmonic components. Because the forward and the oscillatory parts of the flow are not entirely separable from each other, capacitance of the system affects both, and much of the work in this subject is aimed at determining the nature and magnitude of this effect [98, 97].

2.7 Elasticity of the Tube Wall: Wave Propagation

As stated in the previous section, a fundamental difference between flow in a rigid tube and flow in an elastic tube is that a local change of pressure in a rigid tube is transmitted instantaneously to every part of the tube while in an elastic tube the change is transmitted with a finite speed. The reason for this is that a local increase in pressure in an elastic tube is able to stretch the tube wall outward, forming a local bulge, and when the change in pressure subsides, the bulge recoils and pushes the excess fluid down the tube [124]. The increase in pressure and the bulge associated with it propagate down the tube like the crest of an advancing wave. This scenario is not possible in a rigid tube because fluid in that case cannot stretch the tube wall, and because, as stated earlier, we assume throughout this discussion that the fluid is incompressible. It is for these two reasons that the local change in pressure in a rigid tube is transmitted instantaneously to every part of the tube. Wave propagation is not possible in a rigid tube.

If a change in pressure occurs at some interior position along an elastic tube, the change will propagate equally in both directions, towards both ends of the tube, as illustrated in Fig. 2.7.1. A scenario of more practical interest, however, is that in which a change in pressure occurs at one end of the tube and propagates in one direction towards the other end, which happens, for example, when a pump is placed at one end of a tube to drive the flow, or simply when there is a change in the pressure difference driving the flow. In this case wave propagation is in only one direction, namely from entrance to exit, as illustrated in Fig. 2.7.2, and this is the case we discuss in what follows under the general heading of wave propagation. However, the possibility exists that a wave propagating in one direction may be totally or partially reflected by an obstacle [221], thus leading to a secondary wave moving in the opposite

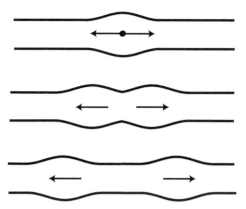

Fig. 2.7.1. A local change in pressure at an interior point in an elastic tube will propagate equally in both directions, towards the two ends of the tube.

direction as illustrated in Fig. 2.7.3. This will be discussed later in the book under the heading of wave reflections. Thus, in this section we consider only a primary wave moving from one end of an elastic tube to the other end.

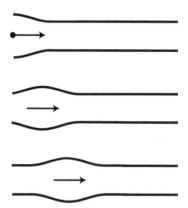

Fig. 2.7.2. A wave propagation scenario of more practical interest is that in which a change in pressure occurs at one end of a tube and propagates to the other end. This occurs, for example, when flow is driven by the stroke of a pump at the tube entrance, or simply when there is a change in the pressure difference driving the flow.

When considering flow in an elastic tube, it is useful to distinguish between *wave motion* and *fluid motion*. If the flow is driven by an increase in pressure at the tube entrance, for example, then wave motion refers to the forward motion of the local swelling or bulge in the tube caused by the increase in pressure, as illustrated in Figs. 2.7.2, 3, much like the motion of the crest of a wave on the surface of a lake. The speed at which the bulge advances along the tube is referred to as the *wave speed*. Fluid motion, on the other hand, refers to the motion of fluid elements within the tube, associated with that wave motion. As the wave crest passes each position along the tube, fluid elements at that location are first swept towards the local bulge in the tube, as illustrated schematically in Fig. 2.7.4, and then as the wave passes and the bulge subsides they are swept back by the decreasing pressure. The situation is again much the same as that experienced by a floating or submerged body swept by the passage of the crest of a wave on the surface of a lake.

Fig. 2.7.3. A wave moving in one direction along an elastic tube may be reflected totally or partially by an obstacle, resulting in a secondary wave moving in the opposite direction.

The wave speed c in an elastic tube depends on the elasticity of the tube, a simple measure of that elasticity being the *Young's modulus* E, sometimes also referred to as the modulus of elasticity. The value of c also depends on the diameter d of the tube and its wall thickness h, and on the density ρ of the fluid. An approximate formula for the speed in terms of these properties is the so called Moen-Korteweg formula [168, 135, 34, 141]

$$c = \sqrt{\frac{Eh}{\rho d}} \tag{2.7.1}$$

The formula is only approximate because it does not take into account some dependence of the wave speed on viscosity of the fluid. Also, the formula is based on the assumption that the wall thickness h is small compared with the tube diameter. Despite these limitations the formula can be used to provide an estimate of the wave speed in the cardiovascular system. This is possible if it is further assumed that an *average* wall-thickness-to-diameter ratio h/d above can be taken for the entire system, which leaves c dependent on E and ρ only. Thus, taking $E = 10^7\ dyne/cm^2$, $\rho = 1\ g/cm^3$, and $h/d = 0.1$, we find $c = 1000\ cm/s$ which, in order of magnitude, is a representative estimate of the wave speed in the cardiovascular system.

If the pressure at the entrance of an elastic tube does not merely rise once but rises and falls in an oscillatory manner, the result is a *train* of wave crests moving in tandem along the tube, the distance between two consecutive crests being referred to as the *wave length* L, as illustrated in Fig. 2.7.5. Fluid motion within the tube then consists of back and forth movements everywhere along the tube as consecutive wave crests pass by. This situation provides a basic working model for flow in the cardiovascular system where the driving

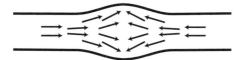

Fig. 2.7.4. As a wave crest passes each position along an elastic tube, fluid elements at that location are first swept towards the local bulge in the tube and then, as the crest passes and the bulge subsides, they are swept back by the decreasing pressure. This *fluid motion* is to be distinguished from the *wave motion*, illustrated in Figs. 2.7.1–3, which is concerned with only the motion of the wave itself. Fluid motion is shown above only schematically in order to illustrate the difference between fluid motion and wave motion, the motion of fluid elements is actually considerably more complicated.

pressure generated by the heart rises and falls in a periodic manner. If the *frequency* of oscillation is f *cycles/s* (Hz), then the wave length is related to the wave speed by

$$L = \frac{c}{f} \tag{2.7.2}$$

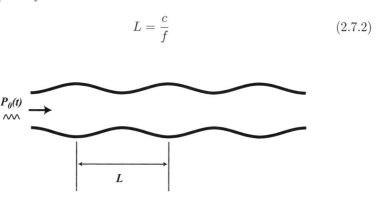

Fig. 2.7.5. If the pressure at the entrance of a tube does not change only once but continuously, in an oscillatory manner, the result is a train of wave crests moving along the tube, or what is commonly referred to as wave propagation. The distance L between two consecutive crests is referred to as the wave length.

   If the frequency of oscillation of the pressure pulse generated by the heart is taken as 1 Hz, then an estimate of the wave length based on the above estimate of the wave speed is $L = 1000$ cm. The wave length is shorter at higher frequency, being only 500 cm at a frequency of 2 Hz. More important, the pressure pulse generated by the heart is actually a *composite wave* consisting of many so called *harmonic components*. Each harmonic component is a perfect *sine* or *cosine* wave but has a different amplitude and different frequency. The frequency of the pressure pulse generated by the heart represents only the so-called *fundamental frequency*, that is, the frequency of the first harmonic component which is referred to as the fundamental harmonic. The frequency of the second harmonic is double the fundamental frequency, and the frequency of the third harmonic is three times the fundamental frequency, and so on. Thus, the higher harmonics have increasingly shorter wave lengths.

Finally, wave speed and wave length are affected by the degree of elasticity of the vessel wall, via the value of Young's modulus E in Eq. 2.7.1. More rigid walls have higher values of E and therefore lead to higher wave speeds and higher wave lengths, which is relevant to blood vessels as they become more rigid, generally with age or more locally because of disease. In the limiting case of a totally rigid tube, E is infinite and hence both the wave speed and wave length become infinite. Wave propagation is therefore not possible in a rigid tube, clearly because a local increase in pressure cannot stretch the tube radially outward and thereby start the propagation process. Nevertheless, it is sometimes convenient to think of wave propagation in a rigid tube as one in which the wave speed is infinite, with a change in pressure at one end reaching all parts of the tube with infinite speed, that is instantaneously, as stated briefly in the previous section. Indeed, if the driving pressure at the entrance of a *rigid* tube changes in an oscillatory manner, the entire body of fluid within the tube oscillates back and forth in unison, which is not to be confused with wave propagation [221].

One of the most important effects of wave propagation in an elastic tube is the possibility of *wave reflections*. Wave reflections arise when a wave meets a change in one of the conditions under which it is propagating, such as the diameter or elasticity of the tube, or more generally any change in the resistance to wave propagation along the tube. It is important to distinguish between the *resistance to flow* in a tube and the *resistance to wave propagation* in that tube. The first represents the opposition to *flow* in a tube caused by the viscous shear at the tube wall, and is usually referred to as "pure resistance" or simply *resistance*. The second represents the opposition to *wave propagation* in a tube caused by a combination of elasticity of the tube wall and inertia of the fluid, and is usually referred to as *reactance*. We have noted earlier, for example, that wave propagation is not possible in a rigid tube. This can now be expressed more accurately by saying that a rigid tube has infinite reactance. More generally, a less elastic tube has higher reactance and offers more resistance to wave propagation than does a more elastic tube. The combined effects of reactance and pure resistance are commonly referred to as "impedance". We shall see later that wave reflections in a tube arise at a point where there is a change of impedance, which may be caused by a change of diameter or elasticity of the tube. Impedance and wave propagation play a central role in the dynamics of coronary blood flow and they are explored more fully in later chapters.

2.8 Mechanical Analogy

The mechanics of flow in a tube or a system of tubes can be identified, by analogy, with the basic mechanics of a solid object in motion under the influence of certain forces and conditions. Indeed, both situations are governed by the same laws of physics, and it should not be surprising that the analytical descriptions of their mechanics are analogous. What is different between the two situations, and what makes the analogy useful, stems from a difference not in the governing laws but in the type of forces and conditions involved and in the corresponding variables used in the two cases.

Thus, in the classical mechanics of a solid object, the familiar mass-damper-spring system is used in which an applied force may be opposed by a spring resistance proportional to the displacement of the object, a damper (or dashpot) resistance proportional to the rate of change of displacement (or velocity), and to an inertial resistance proportional to the second rate of displacement (or acceleration) [139, 76]. While in fluid flow these forces and conditions are not present in the same form, they are present in equivalent forms which obey the same governing laws, hence the basis for the analogy. For example, in fluid flow the capacitance of a tube or a system of tubes plays the role of the spring in the classical mechanics system, the viscous resistance between fluid and the tube wall plays the role of the damper, and the inertia of the fluid plays the role of the inertia of the solid object. These properties have already been discussed in earlier sections, what is required in this section is only to show how they translate into the properties of the classical mechanics system. The translation is not a direct one because the basic variables used in the classical mechanics system, namely mass, displacement and rates of displacement, are not readily available or convenient to work with in the fluid flow system.

Before we carry out this translation it is important to point out that the mechanical analogy has been used extensively in the modelling of coronary blood flow because the classical mechanics of a solid object are familiar and well understood. A model that can be expressed in terms of these mechanics, therefore, has the prospect of unveiling the unknown properties of the coronary circulation in terms which are familiar and well understood. In other words, the analogy is useful because the properties and behaviour of the mechanical system are more familiar and its elements more easily identified than the properties and elements of the fluid system. A potential for error is entailed in this modelling process, however, not because of any inaccuracy in the analogy but because elements of the coronary circulation required for the application of this analogy are not as easily identified as they are in a single tube. Thus, at the core of this modelling process is the fundamental "lumped model" assumption already discussed in Section 2.2, namely that the properties of many millions of tube segments in the coronary circulation can be represented collectively by those of an "equivalent" single tube. While many modelling studies have focused on the likely *values* of these lumped properties [111, 49, 59, 40, 121, 32, 102, 33, 195, 107, 98, 97]– capacitance, resistance, and inertance– the greater potential for error remains in the underlying assumption that these lumped properties actually exist. In other words, the mechanical analogy provides a mechanical model of the coronary circulation only on the assumption that the elements being modelled actually exist in the coronary circulation.

Furthermore, the behaviour of the classical mechanics system depends on a clear relation between the mass, the spring, and the damper. This relation is not known in the coronary circulation and must therefore be *assumed* in any modelling process. The effect of capacitance in the coronary circulation, for example, is produced by a change in the caliber, and hence the volume, of some coronary arteries, resulting in a change in the overall volume of the system [191, 51, 184, 110, 96, 97]. But at the same time this change in diameter also alters the resistance to flow in these vessels. The relation between these two effects is not known. In the classical mechanics system, by contrast, the elements representing capacitance and resistance are entirely separate and have no effect on each other. A related issue is the extent to which the basic elements of capacitance, resistance, and inertance are in series or in parallel in the coronary circulation. In the classical mechanics system this is known *a priori*, but not so in the coronary circulation. Some studies have attempted to deal with these issues by taking more than one lumped element of each type, that is, more than one resistance and more than one capacitance, for example, and by placing them in different combinations of series and parallel arrangements [24, 36, 91, 115].

Despite these difficulties, the mechanical analogy is a useful tool in modelling the coronary circulation because the analogy itself, as it applies to each individual element, is clearly valid. Thus, the relation between the flow rate q and pressure drop Δp in a tube, derived in Section 2.5 (Eq. 2.5.1), namely

$$\Delta p = L\frac{dq}{dt} \tag{2.8.1}$$

where L is the inertance, or inertial constant, of a bolus of fluid within the tube (Eq. 2.5.5), was shown in that section to be equivalent to the basic law of motion

$$F = m\frac{du}{dt} \tag{2.8.2}$$

where m is the mass of a solid object in motion, u is its velocity, and F is the force acting on it. The analogy between the two equations is apparent and the correspondence between the two situations is illustrated in Fig. 2.8.1. The driving pressure difference Δp in the fluid flow system corresponds to the acting force F in the classical mechanics system, while the inertance L corresponds to the mass m, and the flow rate q corresponds to the velocity u. In both cases the underlying law is "force equals mass times acceleration".

Similarly, the viscous resistance to flow in a tube, discussed in Section 2.4, and the resulting relation between the pressure difference Δp and the flow rate q, namely (Eq. 2.4.3)

$$\Delta p = Rq \tag{2.8.3}$$

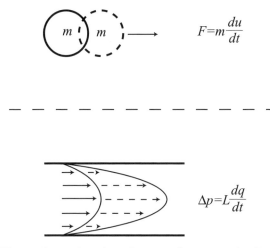

Fig. 2.8.1. The mechanical analogy between flow in a tube (bottom) and the motion of a solid object in classical mechanics (top). The driving pressure difference Δp in the fluid flow system corresponds to the acting force F in the classical mechanics system, while the inertance L corresponds to the mass m, and the flow rate q corresponds to the velocity u. In both cases the underlying law is "force equals mass times acceleration".

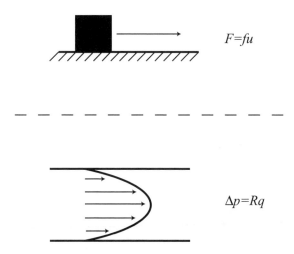

Fig. 2.8.2. Mechanical analogy between the viscous friction at the interface between fluid and tube wall, represented by velocity gradient at the tube wall (bottom), and the friction law in classical mechanics at the interface between two solid objects (top). Here the pressure difference Δp in the tube corresponds to the driving force F in the classical mechanics system, the flow rate q corresponds to the friction velocity u, and the viscous resistance R corresponds to the friction coefficient f.

where R is the resistance to flow due to viscosity (Eq. 2.4.4), is analogous to the classical law of friction at a solid-solid interface

$$F = fu \qquad (2.8.4)$$

where f is the coefficient of friction at the interface, u is the relative velocity between the two surfaces, and F is the driving force. Again, the analogy between the two equations is apparent, and the two situations are illustrated in Fig. 2.8.2. Here the pressure difference Δp corresponds to the driving force F and the flow rate q corresponds to the velocity u, as before, and the viscous resistance R corresponds to the friction coefficient f.

Finally, the capacitance of an elastic tube, discussed in Section 2.6, and the resulting relation between the pressure difference Δp and the change in volume Δv, namely (Eqs. 2.6.4, 6)

$$\Delta(\Delta p) = \frac{1}{C}\Delta v \qquad (2.8.5)$$

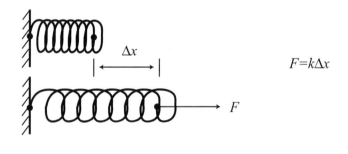

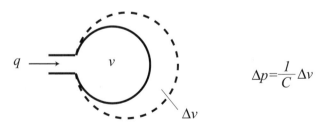

Fig. 2.8.3. Mechanical analogy: between the capacitance effect of flow in an elastic tube, here represented by a balloon, and the stretch of a spring according to Hooke's law. The pressure difference Δp in the flow system corresponds to the applied force F in the spring system, the change in volume Δv of the tube/balloon corresponds to the change in length Δx of the spring, and $1/C$ in the flow system corresponds to the spring constant k, where C is a measure of the compliance of the tube/balloon, as defined by Eq. 2.6.4.

$$= \frac{1}{C} \int q dt; \qquad\qquad \Delta v = \int q dt \qquad (2.8.6)$$

where C is the capacitance of the tube, is analogous to the classical Hooke's law for an elastic spring, namely

$$F = k \Delta x \qquad\qquad\qquad\qquad\qquad (2.8.7)$$

$$= k \int u dt; \qquad\qquad \Delta x = \int u dt \qquad (2.8.8)$$

where k is the spring constant, Δx is the spring extension, and F is the applied force. In the integral terms above, the spring extension is expressed in terms of the velocity u with which the spring is being extended, and the change in volume Δv of the elastic tube/balloon is expressed in terms of the flow rate q. The analogy between the two equations is apparent, with Δp corresponding to the applied force F as before, the change in volume Δv in the flow system corresponding to the change in length Δx of the spring, and $1/C$ in the flow system corresponding to the spring constant k. In the integral terms the flow rate q is seen to correspond to the velocity u in the mechanical system, as in Eqs. 2.8.1, 2. The analogy between the two situations is illustrated in Fig. 2.8.3.

2.9 Electrical Analogy

The dynamics of the coronary circulation can also be modelled, by analogy, in terms of an electric circuit with the basic elements of resistance, capacitance, and inductance. This analogy is subject to the same limitations as the mechanical analogy discussed in the previous section, namely the assumption that these elements can be identified with lumped properties of the coronary circulation. Nevertheless, electrical analogies have been used extensively in the study of the coronary circulation [24, 36, 91, 115] because electric circuits are much easier to manipulate, both analytically and experimentally, and are thus a convenient modelling tool. A model of the coronary circulation based on the electrical analogy can actually be built and tested experimentally. This feature of the electrical model makes it particularly useful in the study of pulsatile flow.

In the electrical analogy the electric potential, or voltage, V corresponds to the pressure difference Δp in the flow system, and the electric current I along a conductor corresponds to the flow rate q along a tube. The basis of the analogy is that the relation between the voltage and current across an inductor L, namely [43]

$$V = L \frac{dI}{dt} \qquad\qquad\qquad (2.9.1)$$

is analogous to the corresponding relation between the pressure difference and flow rate in a tube, as in Eq. 2.8.1, where the inertia of the fluid produces

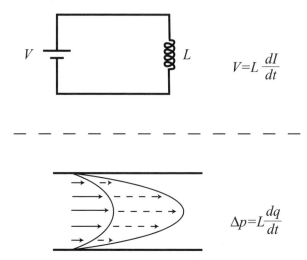

$$V = L\frac{dI}{dt}$$

$$\Delta p = L\frac{dq}{dt}$$

Fig. 2.9.1. Electrical analogy: between flow in a tube and the flow of current in an electric circuit, in the presence of inductance L in both systems. The driving pressure difference Δp in the fluid flow system corresponds to the voltage V in the electrical system, and the flow rate q corresponds to the electric current I. Inductance in the fluid flow system is caused by a change in the flow rate, which is associated with acceleration or deceleration of a mass of fluid, while inductance in the electrical system is due to change in the current, which is associated with acceleration or deceleration of a mass of electrons.

an effect analogous to that of an inductor, as discussed in Section 2.5. The analogy is illustrated in Fig. 2.9.1.

Similarly, the relation between the voltage and current across a resistor R, namely [43]

$$V = RI \qquad (2.9.2)$$

is analogous to the relation between the pressure difference and flow rate in a tube, as in Eq. 2.8.3, where viscous friction between fluid and the tube wall produces an effect analogous to that of a resistor, as discussed in Section 2.4. The analogy is illustrated in Fig. 2.9.2.

Finally, the relation between the voltage across and current into a capacitor namely [43]

$$V = \frac{1}{C}\Delta Q \qquad (2.9.3)$$

$$= \frac{1}{C}\int I\,dt; \qquad \Delta Q = \int I\,dt \qquad (2.9.4)$$

where C is the capacitance and ΔQ the accumulated electric charge on the capacitor, is analogous to the relation between the pressure difference and flow rate into an elastic tube, as in Eqs. 2.8.5,6. Here, because of the elasticity

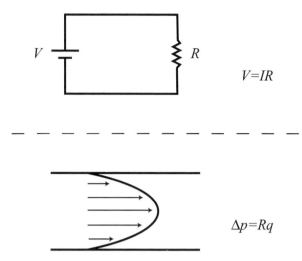

Fig. 2.9.2. Electrical analogy: between flow in a tube and the flow of current in an electric circuit, in the presence of resistance R in both systems. The driving pressure difference Δp in the fluid flow system corresponds to the voltage V in the electrical system, and the flow rate q corresponds to the electric current I. Resistance in the fluid flow system is due to loss of kinetic energy because of viscous friction between fluid and the tube wall, while that in the electrical system it results from a loss of electric energy within the resistor. Interestingly, in both cases the lost energy is converted to heat.

of the tube wall, the accumulated volume of fluid within the tube can change in analogy with a change in the electric charge accumulated on the capacitor. The analogy is illustrated in Fig. 2.9.3.

In summary, the electrical, mechanical, and fluid flow systems have three characteristics in common, namely inductance, resistance, and capacitance, and the dynamics of each system involves two principal variables, namely a driving force and consequent flow or motion. In the electrical system the three elements are an electric inductor, a resistor, and a capacitor, characterized respectively by their intrinsic constants L, R, C. The driving force is the voltage V, the motion is represented by the flow of electric current I, and the governing relations between these variables are:

$$\text{inductance} \quad V = L\frac{dI}{dt} \tag{2.9.5}$$

$$\text{resistance} \quad V = RI \tag{2.9.6}$$

$$\text{capacitance} \quad V = \frac{1}{C}\int I\,dt \tag{2.9.7}$$

In the mechanical system the three elements are a moving object, a damper, and a spring, characterized respectively by the mass m of the moving object, the friction constant f of the damper, and the spring constant k. The driving

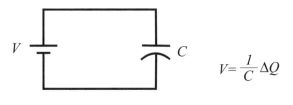

$$V = \frac{1}{C} \Delta Q$$

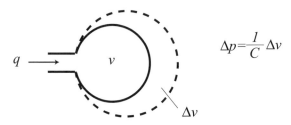

$$\Delta p = \frac{1}{C} \Delta v$$

Fig. 2.9.3. Electrical analogy: between flow in a tube and the flow of current in an electric circuit, in the presence of capacitance C in both systems. The driving pressure difference Δp in the fluid flow system corresponds to the voltage V in the electrical system, and the flow rate q corresponds to the electric current I. Capacitance in the fluid flow system is due to a change in the volume v of fluid within an elastic tube, here represented by an expandable balloon, while in the electrical system it is caused by a change in the total electric charge Q on a capacitor. The change in volume Δv in the fluid flow system is attained by a sustained flow rate into or out of the balloon, while the change in electric charge ΔQ on the capacitor is attained by a sustained current into or out of the capacitor.

force is an applied force F, the motion is represented by the velocity u of the moving object, and the governing relations between these variables are:

$$\textit{inductance} \qquad F = m\frac{du}{dt} \qquad\qquad (2.9.8)$$

$$\textit{resistance} \qquad F = fu \qquad\qquad (2.9.9)$$

$$\textit{capacitance} \qquad F = k\int u\,dt \qquad\qquad (2.9.10)$$

In the fluid flow system, finally, the three elements are the mass of fluid in a tube, viscous resistance between moving fluid and the tube wall, and capacitance produced by the elasticity of the tube wall, characterized respectively by their intrinsic constants L, R, C. The driving force is the pressure difference Δp driving the flow, the motion is represented by the flow rate q, and the governing relations between these variables are:

$$\textit{inductance} \qquad \Delta p = L\frac{dq}{dt} \qquad\qquad (2.9.11)$$

$$resistance \qquad \Delta p = Rq \qquad\qquad (2.9.12)$$

$$capacitance \qquad \Delta p = \frac{1}{C} \int q dt \qquad\qquad (2.9.13)$$

For flow in a tube of length l and radius a, assuming Poiseuille flow throughout, the resistance and inductance constants are respectively given by (Eqs. 2.4.4, and 2.5.5)

$$R = \left(\frac{8\mu l}{\pi a^4}\right); \qquad\qquad L = \left(\frac{\rho l}{\pi a^2}\right) \qquad\qquad (2.9.14)$$

while the capacitance constant C is determined by the elasticity of the tube wall.

2.10 Summary

Modelling of the coronary circulation is necessary because experimental access to the *dynamics* of the system is severely limited. An understanding of the dynamics of coronary blood flow is important because in the absence of such understanding a purely *static* view of the system continues to be used in the clinical setting. In a static view of the system the primary concern is whether vessels are fully open or obstructed by disease. In a dynamic view the concern is more broadly based on all factors that may affects the dynamics of the system, the patency of the conducting vessels being only one such factor.

In a lumped model of the coronary circulation the complex vasculature of the system is essentially replaced by an "equivalent" single tube with "lumped parameters" that are assumed to represent the system as a whole. The model is tested against any measurements that can be obtained from the coronary circulation, and parameter values are adjusted in search of agreement. Despite difficulties associated with this concept, the lumped model has been an invaluable tool in the study of the coronary circulation by establishing some of its basic features.

The mechanics of flow in a tube is at the core of all lumped (as well as unlumped) model analysis. The analysis is usually based on the assumption of *fully developed* flow, ignoring flow in the entrance region of the tube where flow is in a developing phase. This assumption is fairly difficult to deal with because it is necessary, yet not easily justified.

Fluid viscosity together with the condition of no-slip at the tube wall produce "resistance" to *steady* flow in a tube. This resistance increases as the inverse of the tube radius to the fourth power, which means that if the radius of the tube is reduced by a factor of 2, the resistance to flow increases by a factor of 16. This dramatic relationship between vessel radius and resistance to flow figures heavily in clinical practice. It must be remembered, however, that coronary blood flow is not steady but pulsatile, where other forms of resistance exist.

When fluid is accelerated or decelerated, fluid *inertia* gives rise to another form of resistance to flow, commonly referred to as inductance. The immediate effect of inductance is to delay the response of the fluid to a change in the driving pressure difference. The flow rate does not "match" the prevailing pressure difference immediately but with a time delay. In that "transient state" the flow rate is attempting to reach a value appropriate for the prevailing pressure difference, and it ultimately does if the prevailing pressure difference does not change any further. But if the driving pressure difference continues to change, as in oscillatory flow, the flow rate never reaches that appropriate value. It falls short and lags behind, more so at higher values of the inertial constant.

The "capacitance" of a tube or system of tubes arises when the tube wall is elastic (or possibly viscoelastic) and hence the volume of fluid contained within the tube or system of tubes can change. Capacitance is known to play a significant role in the dynamics of coronary blood flow but a definitive model of that role has yet to be formulated.

In addition to giving rise to capacitance, another fundamental consequence of elasticity of the tube wall is that of wave propagation. In an elastic tube, a change of pressure at one end of the tube does not reach the other end instantaneously as it does in a rigid tube. Instead, it stretches the elastic wall of the tube locally at first and then propagates down the tube like the crest of a wave on the surface of a lake. In pulsatile flow this wave propagation is continuous in space (along the tube) and in time. Inductance of the fluid and capacitance of the tube combine to form a new type of resistance, namely "resistance to wave propagation", usually referred to as "reactance", to be distinguished from "pure resistance" caused by viscous shear at the tube wall. Reactance and pure resistance combine to form the "impedance", which plays a key role in "wave reflections", all of which will be discussed more fully in later chapters.

Flow in an elastic tube is governed by the same physical laws and the same equations as the motion of a mass in a mass-damper-spring system. By this so-called "mechanical analogy", inertia of the fluid in the fluid flow system is equivalent to the inertia of the mass in the mechanical system, viscous resistance in the fluid flow system is equivalent to resistance due to damper friction in the mechanical system, and capacitance of the tube in the fluid flow system is equivalent to the stretch of the spring in the mechanical system. The analogy is useful because the elements of the mechanical system are more familiar and their functions can be visualized more clearly than those in the fluid flow system.

Flow in an elastic tube is also analogous to the flow of current in an electric circuit. This so-called "electrical analogy" has been used widely in studying the dynamics of the coronary circulation and forms the basis and "language" of many lumped models of the system. Indeed, the terminology used for elements of resistance, inductance, and capacitance in the fluid flow system has been taken directly from these familiar elements in electrical systems, thus correspondence between the two is fairly clear. A great advantage of the electrical analogy is the availability of well developed mathematical analysis of electric circuits of a wide range of complexity which would be fairly difficult to formulate in terms of either the fluid flow or the mechanical system.

3

Basic Lumped Elements

3.1 Introduction

A lumped model of the coronary circulation must at a minimum include the three basic elements of resistance, inductance, and capacitance, which we shall associate with their respective constants R, L, C, as was done in previous sections, and refer to the combination of the three elements generically as the "RLC system". The reason for this minimum requirement is that the coronary circulation, being a fluid flow system, involves the movement of a fluid that has mass and is therefore associated with inductance when accelerated or decelerated; the moving fluid is viscous and is therefore subject to resistance resulting from the no-slip boundary condition at vessel walls; and the vessels are elastic, thus allowing changes in volume that produce capacitance effects.

What is not known, of course, is whether these elements can indeed be considered as "lumped" in the dynamics of coronary blood flow. Thus, any lumped model analysis must begin with the assumption that this can be done. That is, it must be assumed that the inductance effects of blood flow in the labyrinth of vessels in the coronary circulation can be represented by a *single* inductance parameter L; that the viscous effects at the walls of the same labyrinth of vessels can be represented by a single resistance parameter R; and that the elasticity effects of *all* of these vessels can be represented by a single capacitance parameter C. There are no clear grounds for making these assumptions, nor any *direct* ways of assessing their validity, because of the severe difficulties involved in any attempt to measure these parameters directly from the coronary circulation. A more hopeful prospect of assessing the validity of "lumped" parameters is that of comparing the dynamics of an assumed lumped model with any measurable dynamics of the coronary circulation and thus assessing *indirectly* the validity of the model and of the lumped parameters on which it is based. Indeed, the validity of using the LRC model rests on this prospect rather than on the legitimacy of the lumped model concept.

In using the RLC system it must also be assumed that the dynamics of the coronary circulation are governed by *only* these three parameters. This assumption, again, cannot be supported on any direct grounds, in fact it can be disputed on the indirect ground that there are observed effects in the coronary circulation which cannot be readily represented by the three basic elements. One such effect which has so far proved very difficult to model is the so-called "tissue pressure effect" or the effect of "cardiac muscle contraction". This effect arises because much of the coronary vasculature is actually imbedded within the myocardium, and when the muscle contracts as it does within each cardiac cycle, it exerts considerable pressure on the coronary vessels to the extent of slowing down or even reversing the flow in some vessels, or possibly collapsing the vessels momentarily within each cardiac cycle [84, 20, 68, 187, 41, 101]. Not only are the details of this scenario poorly known, but its modelling has been particularly troublesome [96, 97].

A similar effect in this category is that of wave reflections. Wave reflections are ubiquitous in the coronary circulation because of the pulsatile nature of the flow and because of the large number of vascular junctions where reflections occur [219, 221]. The effect of these on coronary blood flow, or more accurately the way in which these effects are integrated into the design and dynamics of the coronary circulation, is not known.

A related issue is that of the *distinctness* of R, L, C elements within the coronary blood flow system. While it can be argued that resistance, inductance, and capacitance effects must exist in the coronary circulation, that is not to say that they must exist in "pure" form, distinct from each other as they do in an electric circuit. Changes in vessel diameters associated with capacitance effects, for example, clearly have an effect on the resistance to flow because it depends critically on the diameter of the vessel. Thus, the capacitance and resistance effects are not entirely independent from each other, they do not exist in a pure or distinct form.

Furthermore, coronary vessels are not only elastic but muscularized, thus changes in vessel diameters may occur not only *passively* with pressure changes but also *actively* in the course of regulation of coronary blood flow [83, 128]. While such active changes in diameters affect both the resistance to flow and capacitance of the system, these effects cannot usually be included in a primary lumped model of the coronary circulation because they result from neural or humoral regulatory mechanisms that originate outside the primary fluid dynamic system.

It may seem odd to proceed with the RLC system in the face of such major unknowns, but the grounds for proceeding are twofold. First, as argued earlier, these elements must exist within the physiological system in some form, even if not lumped or distinct. Thus, the RLC system, even if not ultimately accurate, provides a reasonable approximation with which to start. Second, the nature of the physiological setting, namely that of a beating heart and deeply imbedded vasculature, is such that the location and arrangement of these elements cannot be determined directly from the physiological system. The lumped model concept is based on the prospect that they can be determined *indirectly*, by modelling.

3.2 RLC System in Series

When the elements of resistance R, inductance L, and capacitance C are in *series*, their effects simply add up against the force driving the flow. Thus, in the electrical system shown in Fig. 3.2.1, the equation governing the flow of current I under the driving voltage V, using the results and notation of Section 2.9, is given by

$$L\frac{dI}{dt} + RI + \frac{1}{C}\int I\,dt = V \tag{3.2.1}$$

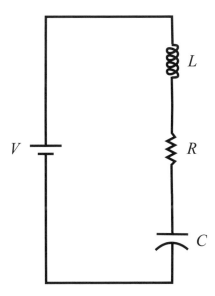

Fig. 3.2.1. The electrical RLC system in series. Flow of electric charge (current) is driven by the voltage V and opposed by an inductor (L), resistor (R), and capacitor (C), in series.

Similarly, in the mechanical analogy shown in Fig. 3.2.2, the equation governing the velocity u of the moving mass under the driving force F, using the results and notation of Section 2.8, is given by

$$L\frac{du}{dt} + Ru + \frac{1}{C}\int u\,dt = F \tag{3.2.2}$$

While in this case the inductance L is equal to the mass m of the moving object (Eq. 2.8.2), the resistance R is equal to the friction coefficient f (Eq. 2.8.4),

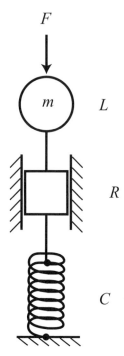

F

m L

R

C

Fig. 3.2.2. The mechanical RLC system in series. Motion (velocity) of mass m is driven by a force F and opposed by the inertia of the mass, thus giving rise to inductance (L), by the friction force of a damper, thus giving rise to resistance (R), and by the force of a spring, thus giving rise to capacitance (C).

and the capacitance C is equal to the reciprocal of the spring constant k (Eqs.2.8.6,8), we continue to use the symbols R, L, C in a generic manner in order to highlight the analogy between the electrical and mechanical systems, and in order to make the discussion and analysis of the governing equation the same for both systems.

Finally, in the fluid flow system shown in Fig. 3.2.3, the equation governing the flow rate q produced by the driving pressure difference Δp, using the results and notation of Sections 2.4,5,6, is given by

$$L\frac{dq}{dt} + Rq + \frac{1}{C}\int qdt = \Delta p \qquad (3.2.3)$$

Again, while the inductance L and resistance R for flow in a tube can be expressed in terms of properties of the fluid and of the tube (Eqs.2.4.5,2.5.5), and for flow into an elastic balloon the capacitance C can be expressed in terms of changes in the volume of the balloon (Eq. 2.6.4), we continue to use the generic symbols R, L, C instead, as discussed earlier.

It is clear that in the fluid flow case of Fig. 3.2.3 the arrangement of the three elements in series is not appropriate as a model of the physiological

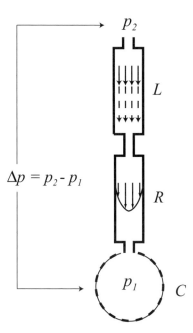

p_2

L

$\Delta p = p_2 - p_1$

R

p_1

C

Fig. 3.2.3. The fluid dynamic RLC system in series. Fluid flow is driven by a pressure difference $\Delta p = p_2 - p_1$ and opposed by the inertia of the fluid, thus giving rise to inductance (L), by the viscous resistance at the tube wall, thus giving rise to resistance (R), and by the elasticity of a balloon, thus giving rise to capacitance (C). Because of the series arrangement, flow is *forced* to go into or out of the balloon and is therefore limited by its capacity. There is no outflow from the system, and continuous flow is clearly not possible.

system because the flow is *forced* into or out of the balloon and is therefore limited by the capacity of the balloon. There is no outflow from the system. While it is possible to modify the model as in Fig. 3.2.4 to produce an outflow, flow through the system, whether inflow or outflow, is still limited by the capacity of the balloon. Continuous flow is not possible. Thus, the RLC system in series is not a good model of the physiological system where continuous flow is an essential feature. Nevertheless, for analytical and pedagogical reasons, this model provides an important starting point and a useful reference.

For the analysis to follow, we use the fluid flow analogue of the RLC system governed by Eq. 3.2.3. It is convenient to reduce the number of physical parameters involved in this equation, so that the solution of the equation does not require the actual values of R, L, C. We do this by dividing the equation by the resistance R to give

$$t_L \frac{dq}{dt} + q + \frac{1}{t_C} \int q dt = \frac{\Delta p}{R} \qquad (3.2.4)$$

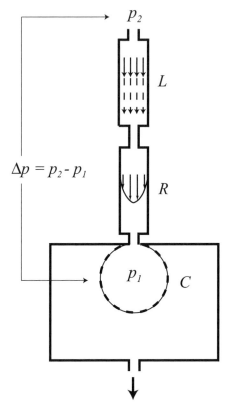

Fig. 3.2.4. Fluid dynamic RLC system in series, as in Fig. 3.2.3, but modified to allow flow out of the system. The balloon is enclosed within a rigid box filled with the same fluid, thus expansion of the balloon forces fluid out of the box as shown by the arrow. Flow is still limited by the capacity of the balloon, however, and continuous flow is not possible.

where t_L $(= L/R)$ and t_C $(= RC)$ are respectively the inertial and capacitive time constants introduced in Sections 2.5,6, which are measures of the inertial and capacitance effects as described in those sections.

Finally, as we found previously, it is more convenient to differentiate this equation once to eliminate the integral term in the equation, giving

$$t_L \frac{d^2q}{dt^2} + \frac{dq}{dt} + \frac{1}{t_C}q = \frac{1}{R}\frac{d(\Delta p)}{dt} \tag{3.2.5}$$

3.3 Free Dynamics of the RLC System in Series

In this section we examine the dynamics of the RLC system in the absence of any external force driving the system and refer to this as a state of "free dynamics". In the governing equation (Eq. 3.2.5) this means that $\Delta p \equiv 0$, thus the term on the right side of the equation is zero and the equation reduces to

$$t_L \frac{d^2q}{dt^2} + \frac{dq}{dt} + \frac{1}{t_C}q = 0 \tag{3.3.1}$$

It is seen from this form of the equation that the inertial and capacitive time constants t_L, t_C play a critical role in the free dynamics of the RLC system, since they are the only parameters present in the equation, thus their values critically affect the solution of this equation. This is particularly meaningful because the values of t_L, t_C actually represent measures of the inertial and capacitance effects in the system as discussed in previous sections. Thus, these time parameters now allow a study of the free dynamics of the system without the need to specify the values of the physical parameters R, L, C individually.

Equation 3.3.1 is a standard second order linear differential equation with constant coefficients [116]. Its solution depends on the nature of the roots of the associated (so called "indicial") equation

$$t_L \alpha^2 + \alpha + \frac{1}{t_C} = 0 \tag{3.3.2}$$

The roots are in general given by

$$\alpha = \frac{-1 \pm \sqrt{1 - (4t_L/t_C)}}{2t_L} \tag{3.3.3}$$

but the solution of the governing equation (Eq. 3.3.1) and hence the dynamics of the system depend critically on whether these roots are real or complex, which in turn depends on the relative values of the time constants t_L, t_C.

If $4t_L < t_C$, then Eq. 3.3.2 has two distinct real roots, given by

$$\alpha_1 = \frac{-1 + \sqrt{1 - (4t_L/t_C)}}{2t_L} \tag{3.3.4}$$

$$\alpha_2 = \frac{-1 - \sqrt{1 - (4t_L/t_C)}}{2t_L} \tag{3.3.5}$$

and the solution of the governing equation is given by

$$q(t) = Ae^{\alpha_1 t} + Be^{\alpha_2 t} \tag{3.3.6}$$

where A, B are arbitrary constants. Given the values of the flow rate at time $t = 0$, namely $q(0)$, and the rate of change of the flow rate at the same time, namely $q'(0)$, we find

$$A = \frac{-\alpha_2 q(0) + q'(0)}{\alpha_1 - \alpha_2} \tag{3.3.7}$$

$$B = \frac{\alpha_1 q(0) - q'(0)}{\alpha_1 - \alpha_2} \tag{3.3.8}$$

If $4t_L > t_C$, then Eq. 3.3.2 has two complex (conjugate) roots, given by

$$\alpha_1 = a + ib \tag{3.3.9}$$
$$\alpha_2 = a - ib \tag{3.3.10}$$

where

$$a = \frac{-1}{2t_L} \tag{3.3.11}$$

$$b = \frac{\sqrt{(4t_L/t_C) - 1}}{2t_L} \tag{3.3.12}$$

$$i = \sqrt{-1} \tag{3.3.13}$$

and the solution of the governing equation is given by

$$q(t) = e^{at}\{A\cos(bt) + B\sin(bt)\} \tag{3.3.14}$$

where A, B are arbitrary constants. Given the values of $q(0)$ and $q'(0)$, again, we find

$$A = q(0)$$
$$B = \frac{-aq(0) + q'(0)}{b} \tag{3.3.15}$$

If $4t_L = t_C$, then Eq. 3.3.2 has two identical real roots, given by

$$\alpha_1 = \alpha_2 = a = \frac{-1}{2t_L} \tag{3.3.16}$$

and the solution of the governing equation is given by

$$q(t) = (A + Bt)e^{at} \tag{3.3.17}$$

where A, B are arbitrary constants. Given the values of $q(0)$ and $q'(0)$, we find

$$A = q(0)$$
$$B = -aq(0) + q'(0) \tag{3.3.18}$$

Examples of these three scenarios are shown in Fig. 3.3.1. It must be remembered that the flow under these scenarios is free from any external driving force, but it is subject to the effects of inertia (L), resistance (R), and capacitance (C). The dynamics of the system are "free" in the sense that the system is under *only* these internal forces.

Unlike the situation discussed in Section 2.6 where the capacitor is *in parallel* with a resistor and hence flow has the option of going into the balloon or into the tube, in the present scenario the capacitor is *in series* and hence flow has no option but to go into the balloon. Clearly, this situation is ultimately

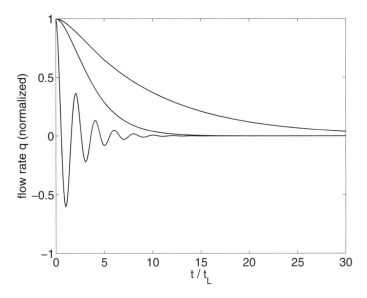

Fig. 3.3.1. Free dynamics of the RLC system in series with $4t_L/t_C = 0.4$ (top curve), $4t_L/t_C = 1$ (middle curve), and $4t_L/t_C = 40$ (bottom curve). The top and bottom curves illustrate two different types of dynamics usually referred to as "overdamped" and "underdamped" respectively. The middle curve represents the singular dynamics, usually referred to as "critically damped", which lies between the two types of behaviour.

limited by the maximum size to which the balloon can stretch, but we ignore this limit for the present and assume that the dynamics of the system are taking place before the limit is reached.

At time $t = 0$ the dynamics are triggered with a pre-existing flow rate $q(0)$ through the system, and for the purpose of illustration we take $q(0) = 1$ and $q'(0) = 0$. This flow will diminish because it has no external driving force and it is opposed by the tube resistance R and by the elasticity of the balloon wall as flow stretches it. The only driving force which keeps some flow going at this phase is an *internal* force due to the inertial effect, namely the momentum which the fluid has by virtue of the pre-existing velocity with which it was started. Since this momentum is finite and is not being renewed by any external force, it is ultimately exhausted and the fluid comes to rest. The flow rate becomes zero.

At this point one of two possible scenarios may unfold: the balloon may recoil and send fluid back, thus reversing the flow direction, or the balloon may simply absorb the increased volume of fluid and come to equilibrium at a new volume. In the mechanical system analogy this is equivalent to the compression (or expansion) of a spring, which may take place either "gently" so that the spring may simply reach equilibrium at a new length, or it may take place more forcefully so that the spring will overshoot its equilibrium length and bounce back.

Which of the two scenarios occurs depends on the rate at which the flow rate diminishes, which in turn depends on the relative values of the inertial and capacitance effects. Recalling that $t_L = L/R$, $t_C = RC$, if the value of the ratio $4t_L/t_C$ is below 1.0, the balloon does not recoil, as seen in the top curve in Fig. 3.3.1. If the value of the ratio is higher than 1.0, the balloon recoils, leading to the oscillations seen in the bottom curve. One particular value of that ratio, namely that corresponding to $4t_L/t_C = 1.0$, is referred to as "critically damped" in the sense that it acts as a separating line between the two types of behaviour, as shown by the middle curve in Fig. 3.4.1.

The free dynamics of a dynamical system do not represent its dynamics under the action of external forces, which is referred to as "forced dynamics", but they do represent the *intrinsic* characteristics of the system. An understanding of these characteristics is important because they ultimately determine how the system responds to external forces.

In the dynamics of the coronary circulation, the above scenarios are relevant to the extent that the elements of inductance, resistance, and capacitance, are known to exist in that system. While it is generally believed that these elements are arranged in parallel rather than in series as they are here, and while the dynamics of the coronary circulation is generally forced rather than free, some of the overall conclusions reached here are relevant nevertheless. More specifically, a change in the relative values of R, L, C, that may occur as a result of disease or surgery, *may change the intrinsic characteristics of the system and hence its dynamic behaviour.* Thus, a narrowing or stiffening of some coronary arteries may change not only the values of R and C and the corresponding resistance and capacitance effects, but it may cause the system to cross over from one type of dynamic behaviour to another. Similarly, a change in the consistency of blood resulting from the administration of certain drugs which may change the density and viscosity of blood, may not only change the values of L and R but may lead to a change in the dynamic behaviour of the system.

3.4 R₁,R₂ in Parallel

When elements of the lumped model are in series, as we have seen in this chapter so far, the flow rate through all elements at any moment in time must be the same because of the law of conservation of mass. When the elements are *in parallel*, however, this is no longer the case, as we saw in Section 2.6 where capacitance and resistance were in parallel. This is because when two elements opposing the flow are in parallel, flow rate through the system is divided in a manner commensurate with the relative opposition presented by each element.

This division of flow is an important property of parallel systems, which is particularly significant in coronary blood flow and in blood flow in general. It allows different flow rates to prevail *simultaneously* within the system, which is not possible when the elements are in series. We shall see in the next chapter that this difference between series and parallel flow systems provides an important tool in the construction of a lumped model of the coronary circulation. In the remainder of this chapter we explore some basic parallel arrangements to highlight their properties, starting with two different resistances in parallel as shown in Fig. 3.4.1. Only elementary properties pertaining particularly to the parallel arrangement are discussed. In each case, the full dynamics of these systems are deferred to subsequent chapters.

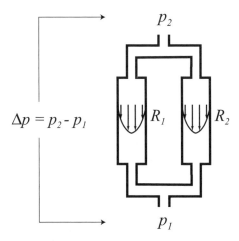

Fig. 3.4.1. Two resistances in parallel, represented schematically by fully developed Poiseuille flow in two tubes in parallel under a driving pressure drop Δp. A parallel system allows different flow rates to prevail *simultaneously* within the system, which is not possible when the elements are in series.

The scenario usually considered in this case is one in which the pressure drop Δp is constant, so any change in one or both resistances will affect only the flow rates. The pressure drops across the two elements will not change because they are both the same under the parallel arrangement and are both equal to the constant value of Δp under this scenario. However, another important scenario to consider is that in which the total flow rate through the system is constant. This condition is likely to arise in the physiological system because of regulatory mechanisms that respond to local oxygen consumption by cardiac tissue, which is related to flow rate rather than to pressure. A change in one or both resistances in this case will affect not only the individual flow rates through the two elements but also the pressure drop across the system, because the pressure drop must change in order to maintain the prescribed constant flow rate through the system. Let us consider these two scenarios in more detail.

Under the scenario of constant pressure drop Δp the individual flow rates through the two tubes are given by (Eq. 2.4.3)

$$q_1 = \frac{\Delta p}{R_1}$$

$$q_2 = \frac{\Delta p}{R_2} \tag{3.4.1}$$

and total flow rate through the system is given by

$$q = q_1 + q_2$$

$$= \left(\frac{R_1 + R_2}{R_1 R_2}\right) \Delta p \tag{3.4.2}$$

From these expressions it is clear that if the resistance in one of the two tubes is changed then the flow rate will change in that tube but not in the other. And since total flow rate is the sum of the two then total flow through the system will change too. To illustrate this more clearly, let us consider R_2 as being fixed and change the ratio R_1/R_2. Using Δp and R_2 as nondimensionalizing constants, the flow rates can then be put in the following nondimensional forms:

$$\bar{q}_1 = \frac{q_1}{\Delta p/R_2}$$

$$= \frac{1}{R_1/R_2} \tag{3.4.3}$$

$$\bar{q}_2 = \frac{q_2}{\Delta p/R_2}$$

$$= 1 \tag{3.4.4}$$

$$\bar{q} = \frac{q}{\Delta p/R_2}$$

$$= 1 + \frac{1}{R_1/R_2} \tag{3.4.5}$$

As the ratio R_1/R_2 changes, these flow rates change as illustrated graphically in Fig. 3.4.2. It is seen that when $R_1/R_2 = 1$, flow through each tube has a normalized value of 1.0 and total flow through the system is 2.0. As the value of R_1/R_2 is decreased from this reference value, flow through R_1 is increased

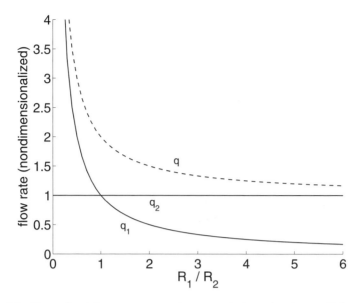

Fig. 3.4.2. Normalized flow rates q_1, q_2 in two resistive tubes in parallel and under a constant driving pressure drop. Total flow rate is q. If R_2 is fixed, then a change in the ratio R_1/R_2 will affect q_1 and q but not q_2.

but flow through R_2 is unchanged, therefore total flow through the system is increased.

Under the scenario of constant flow rate through the system, writing Eq. 3.4.2 as

$$\Delta p = \left(\frac{R_1 R_2}{R_1 + R_2} \right) q \tag{3.4.6}$$

it is clear that a change in any or both of the two resistances will require a commensurate change in Δp in order to maintain the prescribed constant flow rate q. Also, if flow rates are now nondimensionalized in terms of the constant flow rate, we have

$$\overline{q}_1 = \frac{q_1}{q}$$

$$= \frac{1}{1 + R_1/R_2} \tag{3.4.7}$$

$$\overline{q}_2 = \frac{q_2}{q}$$

$$= \frac{R_1/R_2}{1 + R_1/R_2} \tag{3.4.8}$$

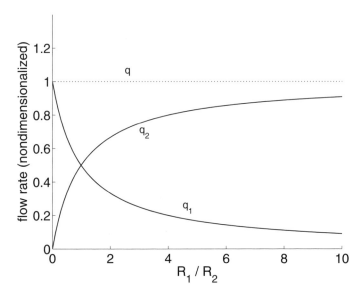

Fig. 3.4.3. Normalized flow rates q_1, q_2 in two resistive tubes in parallel and under a condition of constant flow rate q through the system. If R_2 is fixed, then a change in the ratio R_1/R_2 will affect both q_1 and q_2 in such a way that total flow through the system $(q = q_1 + q_2)$ remains unchanged.

$$\bar{q} = \frac{q}{q}$$

$$= 1 \tag{3.4.9}$$

As before, if R_2 is fixed and the ratio R_1/R_2 is changed, then the change will affect flow in both tubes in such a way that total flow rate remains constant as prescribed under the present scenario. The results are shown graphically in Fig. 3.4.3.

3.5 R,L in Parallel

Consider now an element of resistance R and an element of inductance L in parallel under a driving pressure drop Δp as shown in Fig. 3.5.1. Schematically, the two elements in that figure are represented by fully developed Poiseuille flow in one tube, in which the viscous resistance to flow is R, and a hypothetically "ideal" flow in which the viscous resistance is absent but fluid is being accelerated from rest, hence the inertial effect is present. We shall refer to these briefly as the "resistive tube" and "inductive tube", respectively.

The pressure drop across the first tube will act to overcome the viscous resistance while across the second tube it will act to accelerate the flow, as

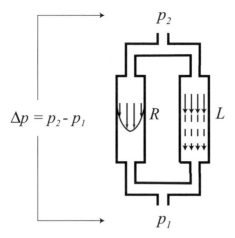

Fig. 3.5.1. Resistance R and reactance L in parallel, under a pressure drop Δp. Resistance is represented here by fully developed flow in a tube in which the only opposition to flow is the viscous resistance at the tube wall but with no inertial effect since the flow is fully developed. Reactance is represented schematically by idealized flow in a tube in which the viscous effect is absent but the fluid is being accelerated from rest, hence the inertial effect is present.

discussed in more detail in Sections 2.4,5. If the flow rates in the two tubes are denoted by q_R and q_L, respectively, then these are related to Δp as determined in Sections 2.4,5, namely

$$\Delta p = q_R R = L \frac{dq_L}{dt} \tag{3.5.1}$$

Thus

$$q_R = \frac{\Delta p}{R} \tag{3.5.2}$$

$$q_L = \frac{1}{L} \int \Delta p \, dt \tag{3.5.3}$$

and the total flow rate is given by

$$q = q_R + q_L$$

$$= \frac{\Delta p}{R} + \frac{1}{L} \int \Delta p \, dt \tag{3.5.4}$$

or

$$\frac{dq}{dt} = \frac{1}{R} \frac{d(\Delta p)}{dt} + \frac{\Delta p}{L} \tag{3.5.5}$$

If the driving pressure drop Δp is assumed constant, Eq. 3.5.5 reduces to

$$\frac{dq}{dt} = \frac{\Delta p}{L} \qquad (3.5.6)$$

and its solution is

$$q(t) = \frac{\Delta p}{L}t + A \qquad (3.5.7)$$

where A is a constant. Also, with Δp constant, Eqs.3.5.2,3 give

$$q_R(t) = \frac{\Delta p}{R}$$

$$q_L(t) = \frac{\Delta p}{L}t + B \qquad (3.5.8)$$

where B is a constant. Since at time $t = 0$ it is assumed that fluid in the reactance tube is at rest, that is $q_L(0) = 0$, then $B = 0$ and

$$q_L(t) = \frac{\Delta p}{L}t \qquad (3.5.9)$$

The total flow rate through the parallel system is thus given by

$$q(t) = q_R(t) + q_L(t)$$

$$= \frac{\Delta p}{R} + \frac{\Delta p}{L}t \qquad (3.5.10)$$

Comparing Eq. 3.5.7 and Eq. 3.5.10 we see that

$$A = \frac{\Delta p}{R} \qquad (3.5.11)$$

and the two equations become identical, as they should. Thus, under the scenario of constant pressure drop, the individual and total flow rates, in nondimensional form, are given by

$$\bar{q}_R(t) = \frac{q_R(t)}{\Delta p/R} = 1$$

$$\bar{q}_L(t) = \frac{q_L(t)}{\Delta p/R} = \frac{t}{t_L}$$

$$\bar{q}(t) = \frac{q(t)}{\Delta p/R} = 1 + \frac{t}{t_L} \qquad (3.5.12)$$

where, as in previous sections, $t_L = L/R$. These are shown graphically in Fig. 3.5.2 from which we see that flow rate through the resistive tube is constant while that through the inductive tube increases linearly and indefinitely with time. The reason for the latter is that in the inductive tube the only

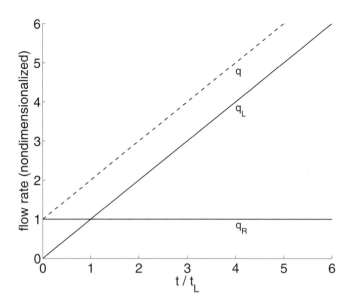

Fig. 3.5.2. Flow rates q_R, q_L in a resistive tube and an inductive tube, respectively, in parallel and under a condition of constant pressure drop. Total flow rate is q. In the inductive tube the only opposition to flow is fluid inertia, hence the fluid continues to accelerate under the constant driving force, a scenario which is unrealistic physiologically under normal conditions but may arise under pathological conditions involving a breach in the vascular system through which blood can escape.

opposition to flow is the inertia of the fluid, and in the presence of a constant driving force the fluid continues to accelerate.

As discussed in the previous section, another scenario of physiological interest is that in which the parallel system in Fig. 3.5.1 is under a condition of constant flow rate rather than constant pressure drop. Thus, setting q constant in Eq. 3.5.5, the equation reduces to

$$\frac{1}{R}\frac{d(\Delta p)}{dt} + \frac{\Delta p}{L} = 0 \qquad (3.5.13)$$

with the solution

$$\Delta p(t) = \Delta p(0)e^{-t/t_L} \qquad (3.5.14)$$

where $\Delta p(0)$ is the value of Δp at time $t = 0$, which highlights the fact that under the present scenario of constant flow rate the pressure drop driving the flow is *not* constant. Using this result for the individual flows in Eqs.3.5.2,3, we obtain

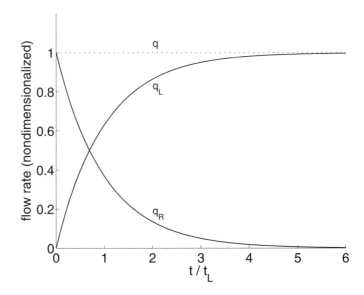

Fig. 3.5.3. Flow rates q_R, q_L in a resistive tube and an inductive tube, respectively, in parallel and under a condition of constant flow rate q through the system. In the inductive tube the flow rate is increased from an initial value of zero to the value of total flow rate through the parallel system. In the resistive tube there is viscous resistance at any flow rate, thus the flow rate there, starting from total flow rate at time $t = 0$ (when the inductive flow is zero) drops continuously as time goes on, as more flow, and ultimately all flow, is diverted via the inductive tube.

$$q_R(t) = \frac{\Delta p(0)}{R} e^{-t/t_L}$$

$$q_L(t) = \frac{\Delta p(0)}{R} \left(1 - e^{-t/t_L} \right) \tag{3.5.15}$$

having assumed again that at time $t = 0$ fluid in the inductive tube is at rest, that is $q_L(0) = 0$. We note that

$$q(t) = q_R(t) + q_L(t)$$

$$= \frac{\Delta p(0)}{R} e^{-t/t_L} + \frac{\Delta p(0)}{R} \left(1 - e^{-t/t_L} \right)$$

$$= \frac{\Delta p(0)}{R} \tag{3.5.16}$$

which indicates that total flow rate through the system is constant at all times as it should be (by design) under the present scenario.

Writing these results in nondimensional form, we have

$$\bar{q}_R(t) = \frac{q_R(t)}{\Delta p(0)/R}$$

$$= e^{-t/t_L} \tag{3.5.17}$$

$$\bar{q}_L(t) = \frac{q_L(t)}{\Delta p(0)/R}$$

$$= 1 - e^{-t/t_L} \tag{3.5.18}$$

$$\bar{q}(t) = \frac{q(t)}{\Delta p(0)/R}$$

$$= 1 \tag{3.5.19}$$

which are shown graphically in Fig. 3.5.3. We see that as time goes on, under this scenario resistive flow diminishes to zero while inductive flow reaches a constant value equal to total flow. In other words, ultimately the entire flow rate through the parallel system is diverted via the inductive tube. The reason for this, of course, is that in the inductive tube there is no opposition to flow at a constant rate, while in the resistive tube there is viscous resistance at any nonzero flow rate.

3.6 R,C in Parallel

We consider next an element of resistance R and an element of capacitance C in parallel under a driving pressure drop Δp as shown in Fig. 3.6.1, but only briefly, since this configuration was discussed more fully in Section 2.6. Again, the two elements in that figure are represented schematically by fully developed Poiseuille flow in one tube, in which there is viscous resistance to flow, and an expandable "balloon" in which the viscous resistance is absent but a capacitance effect is present, as discussed at great length in Section 2.6. The pressure drop across the resistance tube will act to overcome the viscous resistance at the tube wall, while across the balloon (p_2 inside and p_1 outside) it will act to overcome the capacitance effect of the balloon. More accurately, Δp will act to keep the balloon inflated, and depending on whether and how Δp is changing, it may drive flow into or out of the balloon, as discussed in detail in Section 2.6.

If the flow rate going through the resistance tube is denoted by q_R and referred to as "resistive flow", and that going into the balloon is denoted by q_C and referred to as "capacitive flow", it is important to note that capacitive flow is not a "through-flow" but a flow "into" (or out) of the balloon. The extent of this flow is therefore clearly limited by the capacity of the balloon, that is by the maximum volume to which the balloon can be stretched. Here,

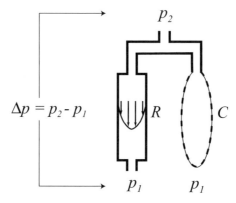

Fig. 3.6.1. Resistance R and capacitance C in parallel, under a pressure drop Δp. Resistance is represented here by a fully developed flow in a tube in which the only opposition to flow is the viscous resistance at the tube wall but with no inertial effect since the flow is fully developed. Capacitance is represented by an expandable "balloon" in which the viscous resistance is absent but a capacitance effect is present. The pressure drop across the resistance tube acts to overcome the viscous resistance at the tube wall, while across the balloon (p_2 inside and p_1 outside) it acts to keep it inflated and, depending on whether and how Δp is changing, it may drive flow into or out of the balloon, as discussed in Section 2.6.

as before and in subsequent analysis, we always assume that the balloon has not reached this limit and that flow is occuring within the elastic range of the balloon. This assumption is fairly consistent with physiological reality where the part of blood flow that goes into stretching blood vessels normally does so within their elastic limits. There have been suggestions that vessel walls may in fact be not purely elastic but viscoelastic [204, 52], but this issue is not important for the purpose of the present section.

Using the results established in Section 2.6, the two parallel flow rates are related to Δp by

$$\Delta p = q_R R = \frac{1}{C} \int q_C \, dt \qquad (3.6.1)$$

or

$$q_R = \frac{\Delta p}{R} \qquad (3.6.2)$$

$$q_C = C \frac{d(\Delta p)}{dt} \qquad (3.6.3)$$

and total flow is given by

$$q = q_R + q_C$$

$$= \frac{\Delta p}{R} + C \frac{d(\Delta p)}{dt} \qquad (3.6.4)$$

Under a scenario of constant pressure drop Δp, we have

$$q_C = 0$$
$$q_R = q = \frac{\Delta p}{R} \tag{3.6.5}$$

Capacitive flow is zero because Δp is constant, and resistive flow is equal to total flow through the system.

Under the alternate scenario of constant total flow rate q into the system, the pressure drop Δp is no longer constant, as it must adjust in such a way as to maintain the prescribed constant flow rate q. This behaviour of Δp is governed by the following differential equation, from Eq. 3.6.4

$$C\frac{d(\Delta p)}{dt} + \frac{\Delta p}{R} = q \tag{3.6.6}$$

which has the following solution, noting that q here is a constant

$$\Delta p(t) = qR + Ae^{-t/t_C} \tag{3.6.7}$$

where A is a constant and, as before, $t_C = RC$. Using this result for the individual flows in Eqs.3.6.2,3, we obtain

$$q_R(t) = q + \frac{A}{R}e^{-t/t_C} \tag{3.6.8}$$

$$q_C(t) = -\frac{A}{R}e^{-t/t_C} \tag{3.6.9}$$

To examine the interplay between capacitive and resistive flow rates we begin with the two being equal and follow their subsequent time course. In other words we set

$$q_R(0) = q_C(0) \tag{3.6.10}$$

which gives

$$A = -\frac{Rq}{2} \tag{3.6.11}$$

Using this value of A in Eqs.3.6.8,9, the flow rates can now be put in the following nondimensional forms

$$\bar{q}_R(t) = \frac{q_R}{q} = 1 - \frac{1}{2}e^{-t/t_C} \tag{3.6.12}$$

$$\bar{q}_C(t) = \frac{q_C}{q} = \frac{1}{2}e^{-t/t_C} \tag{3.6.13}$$

$$\bar{q} = \frac{q}{q} = 1 \tag{3.6.14}$$

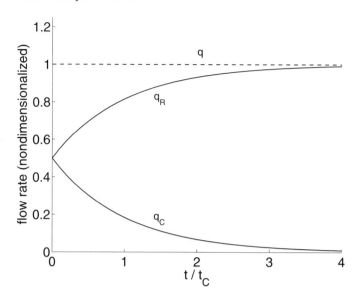

Fig. 3.6.2. Flow rates q_R, q_C in a resistive tube and a capacitive chamber (balloon), respectively, in parallel and under a condition of constant flow rate q through the system, normalized at a value of 1.0. At time $t = 0$ resistive and capacitive flows are set at normalized values of 0.5 each. Subsequently, capacitive flow diminishes from this value to an ultimate value of zero, while resistive flow increases gradually to ultimately encompass total flow into the system. Recalling that resistive flow is driven by the pressure drop while capacitive flow is driven by the *derivative* of the pressure drop, these changes in flow rates are accompanied by corresponding changes in the pressure drop and its derivative as described in the text.

recalling again that under the present scenario q is constant. The results are shown in Fig. 3.6.2, where we see that as time goes on, capacitive flow diminishes while resistive flow increases gradually to encompass total flow into the system. The reason for this can be seen from the behaviour of the pressure drop Δp as time goes on. From Eqs.3.6.7,11 we have

$$\Delta p(t) = qR \left(1 - \frac{1}{2} e^{-t/t_C} \right) \tag{3.6.15}$$

and by differentiation

$$\Delta p'(t) = \frac{1}{2} \frac{qR}{t_C} e^{-t/t_C} \tag{3.6.16}$$

We recall from Section 2.6 that capacitive flow is driven not by the pressure drop Δp but by the rate of change of the pressure drop, namely $\Delta p'(t)$. Accordingly, from Eq. 3.6.16 we see that $\Delta p'(t)$ has its maximum value at time $t = 0$ and diminishes continuously thereafter to an ultimate value of zero. Consequently, capacitive flow has its maximum value at time $t = 0$ and then diminishes gradually to an ultimate value of zero. Since total flow into the system is constant under the present scenario, resistive flow begins with its lowest value at time $t = 0$ and then increases gradually to encompass total flow into the system, as illustrated in Fig. 3.6.2. Interestingly, the corresponding change in the pressure drop is such that it actually increases from its initial value to the value required to drive the ultimate resistive flow, but the *derivative* of the pressure drop (which drives the capacitive flow) decreases continuously from its initial value to an ultimate value of zero.

3.7 RLC System in Parallel Under Constant Pressure

Finally, we conclude this chapter, which started out by considering the three elements of resistance R, inductance L, and capacitance C, in *series*, by considering now these three elements in *parallel* under a driving pressure drop Δp as shown in Fig. 3.7.1. As before, the pressure drop will act to overcome the viscous effect in the resistive tube, the inertial effect in the inductive tube, and the capacitance effect in the balloon (capacitor).

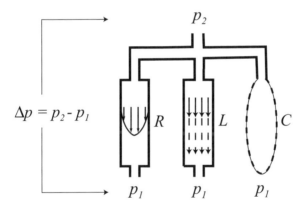

Fig. 3.7.1. Resistance R, inductance L, and capacitance C, in parallel under a pressure drop Δp, other details as in Fig. 3.5.1 and Fig. 3.6.1.

If, as in Sections 3.5,6, resistive, inductive, and capacitive flow rates are denoted by q_R, q_L, q_C, respectively, then, as before, these are related to the pressure drop by

$$\Delta p = q_R R \tag{3.7.1}$$

$$= L\frac{dq_L}{dt} \tag{3.7.2}$$

$$= \frac{1}{C}\int q_C\,dt \tag{3.7.3}$$

or

$$q_R = \frac{\Delta p}{R} \tag{3.7.4}$$

$$q_L = \frac{1}{L} \int \Delta p\, dt \tag{3.7.5}$$

$$q_C = C \frac{d(\Delta p)}{dt} \tag{3.7.6}$$

Total flow rate into the parallel system is given by

$$q = q_R + q_L + q_C \tag{3.7.7}$$

$$= \frac{\Delta p}{R} + \frac{1}{L} \int \Delta p\, dt + C \frac{d(\Delta p)}{dt} \tag{3.7.8}$$

and, after differentiation, we have the following differential equation governing the dynamics of the system

$$\frac{dq}{dt} = \frac{1}{R} \frac{d(\Delta p)}{dt} + \frac{\Delta p}{L} + C \frac{d^2(\Delta p)}{dt^2} \tag{3.7.9}$$

For the purpose of obtaining solutions of this equation, it is convenient to put it in the following form, using the inertial and capacitive time constants introduced in earlier sections

$$t_C \frac{d^2(\Delta p)}{dt^2} + \frac{d(\Delta p)}{dt} + \frac{1}{t_L} \Delta p = R \frac{dq}{dt} \tag{3.7.10}$$

where, as before

$$t_L = L/R \tag{3.7.11}$$
$$t_C = CR \tag{3.7.12}$$

Under the scenario of constant pressure drop ($\Delta p = $ constant), Eq. 3.7.10 reduces to

$$\frac{dq}{dt} = \frac{1}{L} \Delta p \tag{3.7.13}$$

with the solution

$$q(t) = \frac{\Delta p}{L} t + A \tag{3.7.14}$$

where A is a constant. Also, from Eqs. 3.7.4-6, with Δp constant, we have

$$q_R(t) = \frac{\Delta p}{R} \tag{3.7.15}$$

$$q_L(t) = \frac{\Delta p}{L} t + B \tag{3.7.16}$$

$$q_C(t) = 0 \tag{3.7.17}$$

where B is a constant. If at time $t = 0$ it is assumed that fluid in the inductance tube is at rest, that is $q_L(0) = 0$, then $B = 0$ and

$$q_L(t) = \frac{\Delta p}{L} t \tag{3.7.18}$$

Total flow rate through the parallel system is thus given by

$$
\begin{aligned}
q(t) &= q_R(t) + q_L(t) + q_C(t) \\
&= \frac{\Delta p}{R} + \frac{\Delta p}{L} t
\end{aligned}
\tag{3.7.19}
$$

and comparing this with Eq. 3.7.14 we see that $A = \Delta p/R$ and the two equations become identical, as they should. Thus, under the scenario of constant pressure drop, the individual and total flow rates, in nondimensional form, are given by

$$\bar{q}_R(t) = \frac{q_R(t)}{\Delta p/R} = 1 \tag{3.7.20}$$

$$\bar{q}_L(t) = \frac{q_L(t)}{\Delta p/R} = \frac{t}{t_L} \tag{3.7.21}$$

$$\bar{q}_C(t) = 0 \tag{3.7.22}$$

$$\bar{q}(t) = \frac{q(t)}{\Delta p/R} = 1 + \frac{t}{t_L} \tag{3.7.23}$$

These results are identical with those obtained in Section 3.5, Eq. 3.5.12, and shown in Fig. 3.5.2. Thus, under a scenario of constant pressure drop, the LRC system in parallel is the same as the LR system in parallel. The reason for this, of course, is that capacitive flow is driven not by Δp but by changes in Δp.

3.8 RLC System in Parallel Under Constant Flow

With a constant flow rate, that is a constant total flow rate into the parallel system, setting q constant in Eq. 3.7.10, the equation reduces to

$$t_C \frac{d^2(\Delta p)}{dt^2} + \frac{d(\Delta p)}{dt} + \frac{1}{t_L} \Delta p = 0 \tag{3.8.1}$$

This is a standard second order linear homogeneous differential equation with constant coefficients [116]. Its solution depends on the roots of the associated (so-called "indicial") equation

$$t_C \alpha^2 + \alpha + \frac{1}{t_L} = 0 \tag{3.8.2}$$

The roots are in general given by

$$\alpha = \frac{-1 \pm \sqrt{1 - (4t_C/t_L)}}{2t_C} \tag{3.8.3}$$

but the solution of the governing equation (Eq. 3.8.1) and hence the dynamics of the system depend critically on whether these roots are real or complex, which in turn depends on the relative values of the inertial and capacitive time constants t_L, t_C.

If $4t_C < t_L$, then Eq. 3.8.2 has two distinct real roots, given by

$$\alpha_1 = \frac{-1 + \sqrt{1 - (4t_C/t_L)}}{2t_C} \tag{3.8.4}$$

$$\alpha_2 = \frac{-1 - \sqrt{1 - (4t_C/t_L)}}{2t_C} \tag{3.8.5}$$

and the solution of the governing equation (Eq. 3.8.1) is given by

$$\Delta p(t) = Ae^{\alpha_1 t} + Be^{\alpha_2 t} \tag{3.8.6}$$

where A, B are arbitrary constants. Using this result for Δp in Eqs. 3.7.4–6, we find

$$\begin{aligned} q_R(t) &= \frac{\Delta p(t)}{R} \\ &= \frac{A}{R}e^{\alpha_1 t} + \frac{B}{R}e^{\alpha_2 t} \end{aligned} \tag{3.8.7}$$

$$\begin{aligned} q_L(t) &= \frac{1}{L}\int \Delta p(t)dt \\ &= \frac{A}{L\alpha_1}e^{\alpha_1 t} + \frac{B}{L\alpha_2}e^{\alpha_2 t} + K \end{aligned} \tag{3.8.8}$$

$$\begin{aligned} q_C(t) &= C\frac{d(\Delta p(t))}{dt} \\ &= AC\alpha_1 e^{\alpha_1 t} + BC\alpha_2 e^{\alpha_2 t} \end{aligned} \tag{3.8.9}$$

where K is a constant of integration. Using the condition of constant total flow rate under the present scenario, namely

$$q_R(t) + q_L(t) + q_C(t) = q \quad \text{(constant)} \tag{3.8.10}$$

we find, after some algebra,

$$K = q \tag{3.8.11}$$

The flow rates in Eqs. 3.8.7-9 can now be put in the following nondimensional form:

$$\overline{q}_R(t) = \frac{q_R(t)}{q} = \overline{A}e^{\alpha_1 t} + \overline{B}e^{\alpha_2 t} \tag{3.8.12}$$

$$\overline{q}_L(t) = \frac{q_L(t)}{q} = \frac{\overline{A}}{t_L \alpha_1}e^{\alpha_1 t} + \frac{\overline{B}}{t_L \alpha_2}e^{\alpha_2 t} + 1 \tag{3.8.13}$$

$$\overline{q}_C(t) = \frac{q_C(t)}{q} = \overline{A}t_C \alpha_1 e^{\alpha_1 t} + \overline{B}t_C \alpha_2 e^{\alpha_2 t} \tag{3.8.14}$$

where

$$\overline{A} = \frac{A}{Rq} \tag{3.8.15}$$

$$\overline{B} = \frac{B}{Rq} \tag{3.8.16}$$

The pressure drop can also be put in nondimensional form by writing

$$\overline{\Delta p}(t) = \frac{\Delta p}{Rq} = \overline{A}e^{\alpha_1 t} + \overline{B}e^{\alpha_2 t} \tag{3.8.17}$$

from which we note that, in their nondimensional form, the pressure drop $\overline{\Delta p}$ and the resistive flow $\overline{q}_R$ are the same function of time (Eqs. 3.8.12, 17).

The constants $\overline{A}, \overline{B}$ can be determined in terms of prescribed values for the initial flow rates by setting $t = 0$ in Eqs. 3.8.12–14, to get

$$\overline{q}_R(0) = \overline{A} + \overline{B} \tag{3.8.18}$$

$$\overline{q}_L(0) = \frac{\overline{A}}{t_L \alpha_1} + \frac{\overline{B}}{t_L \alpha_2} + 1 \tag{3.8.19}$$

$$\overline{q}_C(0) = \overline{A}t_C \alpha_1 + \overline{B}t_C \alpha_2 \tag{3.8.20}$$

Since there are three initial flow rates and only two unknown constants, only two of the flow rates can be prescribed. There are a number of different combinations in which this can be done. From a practical standpoint we may assume that at time $t = 0$ the entire inflow q is going through the resistive tube while the *sum* of the inductive and capacitive flow rates is zero, that is

$$\overline{q}_R(0) = 1 \tag{3.8.21}$$

$$\overline{q}_L(0) + \overline{q}_C(0) = 0 \tag{3.8.22}$$

From a strictly mathematical standpoint, this is equivalent to setting

$$\overline{q}_R(0) = 1 \tag{3.8.23}$$

$$\overline{q}_L(0) = x \tag{3.8.24}$$

$$\overline{q}_C(0) = -x \tag{3.8.25}$$

where x is as yet unknown. The second and third of these conditions can now be used to find $\overline{A}, \overline{B}$ in terms of x, and the first equation can then be used to find the value or values of x that would satisfy that equation. When this is done, we find in fact that all values of x satisfy this condition. From a practical standpoint, the choice $x = 0$ is appropriate since it can be achieved by having a pre-existing constant flow rate q through the resistive tube at time $t < 0$ with the entrances to the inductive and capacitive tubes closed, then at time $t = 0$ these entrances are opened. And if the choice $x = 0$ is to be made, then we may find $\overline{A}, \overline{B}$ by simply setting the initial conditions

$$\overline{q}_L(0) = 0 \tag{3.8.26}$$
$$\overline{q}_C(0) = 0 \tag{3.8.27}$$

which give, after some algebra

$$\overline{A} = \frac{\alpha_2}{\alpha_2 - \alpha_1} \tag{3.8.28}$$

$$\overline{B} = \frac{-\alpha_1}{\alpha_2 - \alpha_1} \tag{3.8.29}$$

We note that these values satisfy the condition $\overline{A} + \overline{B} = 1$ in Eq. 3.8.18, as well as the condition $\overline{q}(0) = \overline{q}_R(0) + \overline{q}_L(0) + \overline{q}_C(0) = 1$ required under the present scenario of constant flow rate into the parallel system. With these values of $\overline{A}, \overline{B}$, the solution is now complete, and the nondimensional flow rates in Eqs. 3.8.12-14 can be plotted as functions of time. Values of the time constants t_L, t_C are required in order to complete the process. Results using $t_L = 1.0$, $t_C = 0.1$, are shown in Fig. 3.8.1. It is seen that at time $t = 0$ the inductive and capacitive flow rates have the prescribed nondimensional values $q_L = q_C = 0$, while the resistive flow rate, as a consequence, has the value 1.0. Thus, initially all flow is going through the resistive tube, but fairly soon thereafter this flow diminishes while the inductive flow grows to encompass total flow into the system. Capacitive flow is very small throughout this process, and it too becomes zero as time goes on.

If $4t_C > t_L$, then Eq. 3.8.2 has two complex (conjugate) roots, given by

$$\alpha_1 = a + ib \tag{3.8.30}$$
$$\alpha_2 = a - ib \tag{3.8.31}$$

where

$$a = -1/2t_C \tag{3.8.32}$$

$$b = \frac{\sqrt{(4t_C/t_L) - 1}}{2t_C} \tag{3.8.33}$$

$$i = \sqrt{-1} \tag{3.8.34}$$

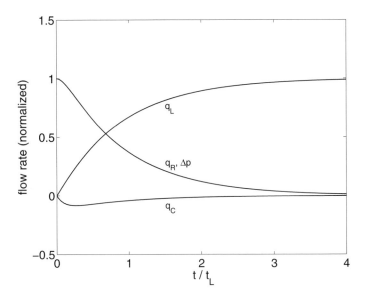

Fig. 3.8.1. Flow rates q_R, q_L, q_C in a resistive tube, an inductive tube, and a capacitive chamber (capacitor), in parallel under a condition of constant flow rate q into the system, normalized at a value of 1.0. At time $t = 0$ inductive and capacitive flows are set at zero, and by consequence resistive flow has an initial value of 1.0, that is, the resistive tube intially carries the entire flow rate into the system. Subsequently, however, the situation reverses as inductive flow grows rapidly to encompass the entire flow into the system. Since under the present scenario total flow into the system is fixed at a normalized value of 1.0, capacitive and resistive flow rates diminish to zero as a consequence. The way in which the flow rates approach their ultimate values depends on the values of the inertial and capacitive time constants t_L, t_C. Results in this figure are based on $t_L = 1.0$, $t_C = 0.1$.

and the solution of the governing equation (Eq. 3.8.1) is given by

$$\Delta p(t) = e^{at}\{A\cos(bt) + B\sin(bt)\} \tag{3.8.35}$$

where A, B are arbitrary constants. Using this result for Δp in Eqs. 3.8.7–9, we find, after a considerable amount of algebra,

$$q_R(t) = \frac{\Delta p}{R}$$
$$= \frac{e^{at}}{R}\{A\cos(bt) + B\sin(bt)\} \tag{3.8.36}$$

$$q_L(t) = \frac{1}{L}\int \Delta p(t)dt$$
$$= \frac{e^{at}}{2R}\{A(-\cos(bt) + 2bt_C\sin(bt))$$

$$+B(-\sin(bt) - 2bt_C\cos(bt))\} + K \qquad (3.8.37)$$

$$q_C(t) = C\frac{d(\Delta p)}{dt}$$

$$= \frac{e^{at}}{2R}\{A(-\cos(bt) - 2bt_C\sin(bt))$$
$$+B(-\sin(bt) + 2bt_C\cos(bt))\} \qquad (3.8.38)$$

where K is a constant of integration. As before, using the condition of constant flow rate under the present scenario, namely $q_R(t) + q_L(t) + q_C(t) = q$ (constant), we find, after some algebra,

$$K = q \qquad (3.8.39)$$

The flow rates in Eqs. 3.8.36–38 can now be put in the following nondimensional form

$$\bar{q}_R(t) = e^{at}(\overline{A}\cos(bt) + \overline{B}\sin(bt)) \qquad (3.8.40)$$

$$\bar{q}_L(t) = \frac{e^{at}}{2}\{\overline{A}(-\cos(bt) + 2bt_C\sin(bt))$$
$$+\overline{B}(-\sin(bt) - 2bt_C\cos(bt))\} + 1 \qquad (3.8.41)$$

$$\bar{q}_C(t) = \frac{e^{at}}{2}\{\overline{A}(-\cos(bt) - 2bt_C\sin(bt))$$
$$+\overline{B}(-\sin(bt) + 2bt_C\cos(bt))\} \qquad (3.8.42)$$

where

$$\overline{A} = \frac{A}{Rq} \qquad (3.8.43)$$

$$\overline{B} = \frac{B}{Rq} \qquad (3.8.44)$$

At time $t = 0$ these give

$$\bar{q}_R(0) = \overline{A} \qquad (3.8.45)$$

$$\bar{q}_L(0) = -\frac{\overline{A}}{2} - \overline{B}bt_C + 1 \qquad (3.8.46)$$

$$\bar{q}_C(0) = -\frac{\overline{A}}{2} + \overline{B}bt_C \qquad (3.8.47)$$

and setting the initial conditions as before, namely

$$\bar{q}_L(0) = 0 \qquad (3.8.48)$$
$$\bar{q}_C(0) = 0 \qquad (3.8.49)$$

we find

$$\overline{A} = 1 \tag{3.8.50}$$
$$\overline{B} = 1/(2bt_C) \tag{3.8.51}$$

We note that the condition of constant flow rate into the system required under the present scenario, which in nondimensional form reads

$$\overline{q}_R(t) + \overline{q}_L(t) + \overline{q}_C(t) = 1 \tag{3.8.52}$$

is satified at time $t = 0$ and at all other times.

As before, only values of the time constants are required now to complete the solution. Results using $t_L = 1.0, t_C = 2.5$ are shown in Fig. 3.8.2. At time $t = 0$ the inductive and capacitive flow rates are zero as prescribed, while the resistive flow rate, as a consequence, has the value 1.0. As in the case of $t_L = 1.0, t_C = 0.1$ (Fig. 3.8.1), initially all flow is going through the resistive tube, but fairly soon thereafter this flow diminishes while the inductive flow grows to encompass total flow into the system. The difference between the two cases, however, is that in the present case the process is accompanied by oscillations which are usually associated with "underdamping", while the behaviour observed in the previous case (Fig. 3.8.1) is associated with "over-damping". Since the inertial time constant is the same in both cases, namely $t_L = 1.0$, this different behaviour is due entirely to the difference in the values of the capacitive time constant t_C ($= CR$). Assuming, for the purpose of comparison, that the value of C is the same in the two cases, then the value of R in the present case where $t_C = 2.5$ is 25 times larger than the value of R in the previous case where $t_C = 0.1$. Thus, the underdamped dynamics observed in Fig. 3.8.2 is here associated with a *higher* value of the resistance R, which is the reverse of what occurs when the R, L, C elements are in *series*. The reason for this is that in this section we are dealing with a *parallel LRC* system and under a condition of *constant flow rate* into the system. Under these circumstances, a lower value of the resistance R diverts more flow into the resistance tube which has a more stabilizing effect on the dynamics of the system, while higher values of R have the opposite effect. By contrast, when the LRC system is in series, lower values of R lead to an increase of flow to *all* components of the system, which has a destabilizing effect, while higher values of R have a damping and hence a stabilizing effect.

If $4t_C = t_L$, finally, then Eq. 3.8.2 has two identical real roots given by

$$\alpha_1 = \alpha_2 = -1/2t_C = a \tag{3.8.53}$$

and the solution of the governing equation (Eq. 3.8.1) is given by

$$\Delta p(t) = (A + Bt)e^{at} \tag{3.8.54}$$

where A, B are arbitrary constants and

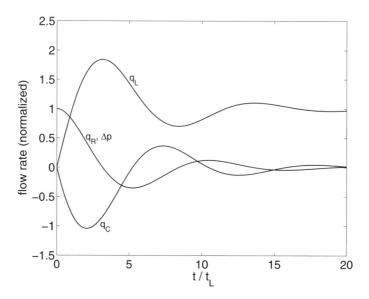

Fig. 3.8.2. Flow rates q_R, q_L, q_C in an LRC system in parallel under a condition of constant flow, as in Fig. 3.8.1, but results here are based on $t_L = 1.0, t_C = 2.5$ (compared with $t_L = 1.0, t_C = 2.5$ in Fig. 3.8.1). While the ultimate outcome is the same in both cases, the oscillations in this figure are indicative of what is usually referred to as an "underdamped" system, while their absence in Fig. 3.8.1 is indicative of an "overdamped" system. Since the higher value of t_C in the present case is associated with a higher value of R, these results show that in the parallel LRC system higher values of the resistance have a destabilizing effect, in contrast with the LRC system in *series* where the reverse is true (see text).

Using this result for Δp in Eqs. 3.8.7–9, we find

$$q_R(t) = \frac{\Delta p(t)}{R}$$
$$= \left(\frac{A}{R} + \frac{B}{R}t\right)e^{at} \tag{3.8.55}$$

$$q_L(t) = \frac{1}{L}\int \Delta p(t)dt$$
$$= -\left(\frac{2t_C}{Rt_L}\right)\{A + B(t + 2t_C)\}e^{at} + K \tag{3.8.56}$$

$$q_C(t) = C\frac{d(\Delta p(t))}{dt}$$
$$= \frac{-1}{2R}\{A - B(2t_C - t)\}e^{at} \tag{3.8.57}$$

where K is a constant of integration. As before, using the condition of constant flow rate under the present scenario, namely $q_R(t) + q_L(t) + q_C(t) = q$ (constant), we find, after some algebra,

$$K = q \tag{3.8.58}$$

The flow rates in Eqs. 3.8.55–57 can now be put in the following nondimensional form

$$\bar{q}_R(t) = (\overline{A} + \overline{B}t)e^{at} \tag{3.8.59}$$

$$\bar{q}_L(t) = \left(\frac{-2t_C}{t_L}\right)\{\overline{A} + \overline{B}(t + 2t_C)\}e^{at} + 1 \tag{3.8.60}$$

$$\bar{q}_C(t) = \left(\frac{-1}{2}\right)\{\overline{A} - \overline{B}(2t_C - t)\}e^{at} \tag{3.8.61}$$

where, as before

$$\overline{A} = \frac{A}{Rq} \tag{3.8.62}$$

$$\overline{B} = \frac{B}{Rq} \tag{3.8.63}$$

At time $t = 0$ these give

$$\bar{q}_R(0) = \overline{A} \tag{3.8.64}$$

$$\bar{q}_L(0) = \left(\frac{-2t_C}{t_L}\right)\{\overline{A} + \overline{B}(2t_C)\} + 1 \tag{3.8.65}$$

$$\bar{q}_C(0) = \left(\frac{-1}{2}\right)\{\overline{A} - \overline{B}(2t_C)\} \tag{3.8.66}$$

and setting the initial conditions as before, namely

$$\bar{q}_L(0) = 0 \tag{3.8.67}$$
$$\bar{q}_C(0) = 0 \tag{3.8.68}$$

we find

$$\overline{A} = 1 \tag{3.8.69}$$
$$\overline{B} = 1/(2t_C) \tag{3.8.70}$$

We note that the condition of a constant flow rate into the system required under the present scenario, namely $\bar{q}_R(t) + \bar{q}_L(t) + \bar{q}_C(t) = 1$, is satified at time $t = 0$ and at all other times. As before, only values of the inertial and

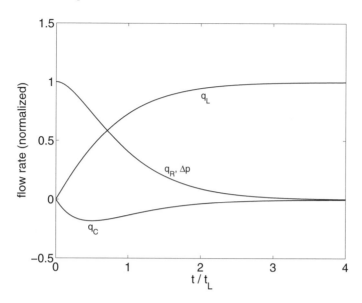

Fig. 3.8.3. Flow rates q_R, q_L, q_C in an LRC system in parallel under a condition of constant flow, as in Fig. 3.8.1,2, but results here are based on $t_L = 1.0, t_C = 0.25$ (hence $4t_C = t_L$). While the ultimate outcome is the same in all three cases, the present case represents a singular case of "critical damping" which occurs with only this particular ratio of the time constants (see text).

capacitive time constants are required now to complete the solution. Results using $t_L = 1.0, t_C = 0.25$ (hence $4t_C = t_L$) are shown in Fig. 3.8.3. Dynamics of the system observed in this figure indicate what is usually referred to as "critical damping", which is a singular condition between the overdamped and the underdamped conditions. With $t_L = 1.0$, only a single value of t_C produces critically damped dynamics, while a range of values can produce overdamped or underdamped dynamics.

3.9 Summary

A lumped model of the coronary circulation must at a minimum include the three basic elements of resistance, inductance, and capacitance, because that circulation, being a fluid flow system, involves the movement of a fluid that has mass and is therefore associated with inductance when accelerated or decelerated; the moving fluid is viscous and is therefore subject to resistance resulting from the no-slip boundary condition at vessel walls; and the vessels are elastic, thus allowing changes in volume and hence producing capacitance effects. While the location and arrangement of these elements cannot be determined directly from the physiological system, the lumped model concept is based on the prospect that they can be determined *indirectly*, by modelling.

The *RLC* system *in series*, though clearly not an appropriate model of the physiological system, provides an important starting point and a useful reference.

Free dynamics of the *RLC* system represent basic *intrinsic* characteristics of the system which ultimately determine how it responds to external forces. The results are relevant to the dynamics of the coronary circulation to the extent that they demonstrate clearly how any change in the relative values of R, L, C, that may occur as a result of disease or clinical intervention, *may change the intrinsic characteristics of the system and hence its dynamic behaviour.*

Flow in two *parallel* resistive tubes may be examined under two different conditions: constant pressure drop across the two tubes, or constant total flow rate through the system. The second of these is of particular interest in the physiological system because of regulatory mechanisms that respond to local oxygen consumption by cardiac tissue, which is related to flow rate rather than pressure. Under constant pressure, a change in the resistance of one tube will change the flow rate in that tube but not in the other. Under constant flow rate, a change in the resistance of one tube will affect the flow rate in both tubes.

Flow through a resistive and an inductive tube *in parallel* produce results that are physiologically unrealistic under the scenario of constant pressure drop as well as that of constant flow rate. Inductive flow rate in one case becomes infinite and in the other it becomes equal to total flow rate through the system. The reason for this, of course, is that under the present parallel arrangement the inductive tube is *in isolation* in the sense that flow has the option of taking the inductive route without passing through the resistance. The situation is somewhat artificial because while the inductive effect due to fluid inertia certainly exists in the physiological system, it does not exist in isolation. We have seen in earlier sections that when fluid in a tube is accelerated from rest, the inertial effect is normally tied *in series* with resistance due to the viscous effect at the tube wall. Thus, one is led to conclude that in the *normal* physiological setting inductive flow is always in series with some other elements of the system, not in parallel. However, under *pathological* conditions, specifically an injury, a breach in the vascular system through which blood can escape, a parallel inductive route is created as depicted in the present model.

For flow through a resistor and a capacitor in parallel, under the constant pressure drop scenario there is no interaction between the resistive and capacitive flows, but under the scenario of constant flow rate the pressure drop driving the two parallel flows changes as time goes on and the two flows change with it. Since total flow rate must remain constant as prescribed under this scenario, changes in the two parallel flows must occur in a complimentary way, a change in one affecting the other. From a prescribed condition of the two parallel flows being equal to each other initially, as time goes on the system moves towards an ultimate condition of zero capacitive flow and resistive flow equal to total flow. Thus, while inductive flow in parallel grows, as was found in the previous section, capacitive flow in parallel diminishes ultimately to zero.

Under a condition of constant pressure drop the RLC system in parallel behaves as if the capacitor C does not exist, that is, it behaves as an RL system in parallel under constant pressure drop.

Under a condition of constant flow rate, we note first the important difference between the dynamics of the series and parallel LRC systems, namely that in the series system the resistance R has a damping effect while in the parallel system it has a destabilizing effect. In the series system, overdamped dynamics occur at higher values of the resistance R, and underdamped dynamics occur at lower values of R. In the parallel system the reverse is true.

Second, in the parallel LRC system, under constant flow conditions, the inductive flow ultimately encompasses the entire flow into the system, thus reducing resistive and capacitive flow to zero. This occurs at all three dynamic modes of the system, namely the overdamped, underdamped, and the critically damped modes.

Finally, while inductance is an important factor in the normal dynamics of the coronary circulation and in the cardiovascular system in general, as it occurs whenever fluid is being accelerated or decelerated, it is usually present in *series* with the resistance and capacitance rather than in parallel. Parallel reactance routes are not normally present in the coronary circulation or in the cardiovascular system in general, since they would ultimately "steal" the entire flow away from other elements of the system. Under abnormal circumstances, however, as in the presence of a breach in the vascular system through which blood can escape freely, the new route for blood flow will act precisely as a inductive route. Fluid will be accelerated freely through the breach until it ultimately steals the entire flow rate available, away from all other routes within the system, precisely as seen in Figs.3.8.1-3. This leads to failure of the organ or organism if the breach is not stopped. While the ultimate cause of this failure is *metabolic*, in the sense that tissue is being deprived from its metabolic needs, it is important to note that the primary cause of the failure is related to the *dynamics* of the system.

4

Forced Dynamics of the RLC System

4.1 Introduction

In the previous chapter we examined some basic dynamics of the R, L, C elements, first arranged in series under a condition of zero driving pressure drop ("free" dynamics), and then in parallel under a condition of either constant flow rate into the system or constant driving pressure drop. While these scenarios are somewhat artificial, they illustrate some intrinsic characteristics of the RLC system and highlight important differences between the dynamics of the series and parallel arrangement of the system.

The dynamics of the coronary circulation are neither "free" nor driven by constant pressure or constant flow rate. They are "forced" dynamics, driven by the same pressure wave that drives blood flow in the systemic circulation, namely the pulsatile pressure wave generated by the pumping action of the left ventricle, the so-called "cardiac pressure wave". In order to bring the dynamics of the RLC system of the previous chapter closer to the dynamics of the coronary circulation, therefore, it is necessary to give the driving pressure drop Δp the form of the cardiac pressure wave.

In free dynamics, the behaviour of the RLC system is determined by the characteristics of the system only, namely the series or parallel arrangement of the R, L, C elements relative to each other, and their relative values. In *forced* dynamics, the behaviour of the system is determined by these same factors but also by the form of the driving pressure, that is, by the form of Δp as a function of time. We shall see later that the cardiac pressure waveform is a "composite" function of time consisting of an assortment of simple sine and cosine waves which are referred to as its "harmonics", as discussed in Chapter 5. In the present chapter, as a first step, therefore, we examine the dynamics of the RLC system under a driving pressure drop that consists of a single harmonic, a simple sine or cosine function, to focus on the response of the system to a simple periodic driving force. In Chapter 6 we consider the dynamics of the system under the force of the full composite form of the cardiac pressure wave.

4.2 The Particular Solution

We begin with the RLC system in series illustrated schematically in Fig. 3.2.3 and governed by Eq. 3.2.4. Taking the driving pressure drop in that equation as a simple cosine function, namely

$$\Delta p = \Delta p_0 \cos \omega t \qquad (4.2.1)$$

where Δp_0 is a constant and ω is the frequency of oscillation of the cosine function. With this form of Δp Eq. 3.2.4 becomes

$$t_L \frac{dq}{dt} + q + \frac{1}{t_C} \int q dt = \frac{\Delta p_0 \cos \omega t}{R} \qquad (4.2.2)$$

As before, for the purpose of solving this equation it is convenient to differentiate it once in order to eliminate the integral term, to get

$$t_L \frac{d^2 q}{dt^2} + \frac{dq}{dt} + \frac{q}{t_C} = \frac{-\Delta p_0 \omega \sin \omega t}{R} \qquad (4.2.3)$$

The solution of this equation consists of two parts, usually referred to as the "homogeneous" part, to be denoted by $q_h(t)$, and the "particular" part, to be denoted by $q_p(t)$. The total solution is given by the sum of these two parts, that is,

$$q(t) = q_h(t) + q_p(t) \qquad (4.2.4)$$

The *homogeneous* part of the solution is, by definition, the general solution of the homogeneous form of Eq. 4.2.3, that is, it satisfies

$$t_L \frac{d^2 q_h}{dt^2} + \frac{dq_h}{dt} + \frac{q_h}{t_C} = 0 \qquad (4.2.5)$$

This equation is identical with Eq. 3.3.1 whose solutions were obtained fully in Section 3.3. Thus $q_h(t)$ represents the flow rate obtained in the free dynamics scenarios considered in Section 3.3, and the solutions of Eq. 4.2.5 can be taken directly from that section.

The *particular* part of the solution, or the particular solution, represents any *particular* solution (not containing arbitrary constants) of Eq. 4.2.3, that is any particular solution of

$$t_L \frac{d^2 q_p}{dt^2} + \frac{dq_p}{dt} + \frac{q_p}{t_C} = \frac{-\Delta p_0 \omega \sin \omega t}{R} \qquad (4.2.6)$$

In the theory of differential equations [116] it has been found that the form of a particular solution of this equation depends on the functional form of the term on the right hand side of the equation. Specifically, when the term on the right-hand side is a simple sine or cosine function, as it is in this case, a particular solution is of the form

$$q_p(t) = K_c \cos \omega t + K_s \sin \omega t \tag{4.2.7}$$

where K_c, K_s are constants to be determined by the equation itself, as shown below, and are not to be confused with the *arbitrary constants* A, B in the *general* solution of Eq. 4.2.5 which are determined by the initial flow conditions as will be shown in the next section, and as was done in Section 3.3. The constants K_c, K_s here are determined simply by substituting for $q_p(t)$ and its derivatives from Eq. 4.2.7 into Eq. 4.2.6 to get

$$L(-K_c\omega^2 \cos \omega t - K_s\omega^2 \sin \omega t) + R(-K_c\omega \sin \omega t + B\omega \cos \omega t)$$
$$+ \frac{1}{C}(K_c \cos \omega t + K_s \sin \omega t) = -\Delta p_0 \sin \omega t \tag{4.2.8}$$

Equating terms in $\sin \omega t$ and $\cos \omega t$ on both sides of this equation, we find

$$-LK_s\omega^2 - RK_c\omega + \frac{1}{C}K_s = -\Delta p_0 \omega \tag{4.2.9}$$

$$-LK_c\omega^2 + RK_s\omega + \frac{1}{C}K_c = 0 \tag{4.2.10}$$

These are two equations in the two unknown constants K_c, K_s. Their solution is standard, though involving some tedious algegra, giving

$$K_c = \frac{\Delta p_0 R}{R^2 + \left(\omega L - \frac{1}{\omega C}\right)^2} \tag{4.2.11}$$

$$K_s = \frac{\Delta p_0 \left(\omega L - \frac{1}{\omega C}\right)}{R^2 + \left(\omega L - \frac{1}{\omega C}\right)^2} \tag{4.2.12}$$

Inserting these values of the constants in Eq. 4.2.7, we obtain, finally, the particular solution as

$$q_p(t) = \frac{\Delta p_0}{R^2 + \left(\omega L - \frac{1}{\omega C}\right)^2} \left\{ R \cos \omega t + \left(\omega L - \frac{1}{\omega C}\right) \sin \omega t \right\} \tag{4.2.13}$$

4.3 Using the Complex Exponential Function

Results obtained in the previous section can be obtained more elegantly and more easily by considering a driving pressure drop Δp in the form of a complex function rather than a simple sine or cosine function. Thus, instead of the cosine function in Eq. 4.2.1, we now write

$$\Delta p = \Delta p_0 e^{i\omega t} \tag{4.3.1}$$

where Δp_0 is a constant as before, and the other element on the right is the well known complex exponential function

$$e^{i\omega t} \equiv \cos \omega t + i \sin \omega t \tag{4.3.2}$$

where $i = \sqrt{-1}$. The great usefulness of this function, as we shall see below, stems from it having two different yet equivalent forms, namely those on the two sides of Eq. 4.3.2. The two forms are not merely equal for some values of t as an equality sign would imply, but are actually *equivalent* for all values of t as indicated by the equivalence operator in that equation.

Since the real part of the complex exponential function in Eq. 4.3.2 is a simple cosine while the imaginary part is a simple sine, then the real and imaginary parts of the pressure drop, to be denoted respectively by $\Re\{\Delta p\}$ and $\Im\{\Delta p\}$, are correspondingly given by

$$\Re\{\Delta p\} = \Delta p_0 \cos \omega t \tag{4.3.3}$$
$$\Im\{\Delta p\} = \Delta p_0 \sin \omega t \tag{4.3.4}$$

Thus, if the complex form of Δp in Eq. 4.3.1 is used in Eq. 3.2.4, namely

$$t_L \frac{dq}{dt} + q + \frac{1}{t_C} \int q dt = \frac{\Delta p_0 e^{i\omega t}}{R} \tag{4.3.5}$$

and differentiating this equation as in the previous section in order to eliminate the integral sign, we then have

$$t_L \frac{d^2 q}{dt^2} + \frac{dq}{dt} + \frac{q}{t_C} = \frac{i\omega \Delta p_0 e^{i\omega t}}{R} \tag{4.3.6}$$

As for Eq. 4.2.3 in the previous section, the solution of Eq. 4.3.6 above consists of two parts: a *homogeneous* part $q_h(t)$ which is the general solution of

$$t_L \frac{d^2 q_h}{dt^2} + \frac{dq_h}{dt} + \frac{q_h}{t_C} = 0 \tag{4.3.7}$$

and a particular part $q_p(t)$ which is a particular solution of

$$t_L \frac{d^2 q_p}{dt^2} + \frac{dq_p}{dt} + \frac{q_p}{t_C} = \frac{i\omega \Delta p_0 e^{i\omega t}}{R} \tag{4.3.8}$$

The total solution of Eq. 4.3.6 is as before the sum of these two parts, namely

$$q(t) = q_h(t) + q_p(t) \tag{4.3.9}$$

However, because of the complex term on the right-hand side of Eq. 4.3.6, $q(t)$ is now a *complex* function. It has a real part and an imaginary part which we shall denote by $\Re\{q(t)\}$ and $\Im\{q(t)\}$, respectively. The great advantage of using the complex exponential function here stems from the fact that the general solution of Eq. 3.2.4 using the complex exponential form of the pressure drop, in other words the solution of Eq. 4.3.6, now yields both the real and imaginary parts of $q(t)$. Furthermore, the real part of $q(t)$ will correspond to the general solution of Eq. 3.2.4 using the real part of Δp, namely $\Delta p_0 \cos \omega t$, while the imaginary part of $q(t)$ corresponds to the general solution of Eq. 3.2.4 using the imaginary part of Δp, namely $\Delta p_0 \sin \omega t$.

We note further that the two parts of $q(t)$ in Eq. 4.3.9 are in themselves complex functions, with real and imaginary parts. The particular part of $q(t)$, namely $q_p(t)$, is complex because its governing equation (Eq. 4.3.6) contains the complex pressure drop term on the right hand side. And the homogeneous part of $q(t)$, namely $q_h(t)$, is also complex, even though its governing equation (Eq. 4.3.7) does not involve the complex pressure drop. The reason for this is that the *arbitrary constants* A, B in the general solution of Eq. 4.3.7 become complex when the initial flow conditions are implemented, as we shall see in the next section.

4.4 Overdamped Forced Dynamics

To implement results from the last two sections we consider now the full solution of Eq. 3.2.4, including the homogeneous and particular parts of the solution, that is

$$q(t) = q_h(t) + q_p(t) \tag{4.4.1}$$

As discussed earlier, the homogeneous part of the solution, which represents the free dynamics of the system, has already been obtained in Section 3.3. However, the form of that solution was found to be different in each of the three scenarios considered in that section, namely the overdamped, underdamped, and critically damped scenarios. In this section we consider the first of these scenarios, in which the solution takes the form (Eq. 3.3.6)

$$q_h(t) = Ae^{\alpha_1 t} + Be^{\alpha_2 t} \tag{4.4.2}$$

where A, B are arbitrary constants and α_1, α_2 are roots of the indicial equation, given by (Eqs.3.3.4,5)

$$\alpha_1 = \frac{-1 + \sqrt{1 - (4t_L/t_C)}}{2t_L} \tag{4.4.3}$$

$$\alpha_2 = \frac{-1 - \sqrt{1 - (4t_L/t_C)}}{2t_L} \tag{4.4.4}$$

The *particular* part of the solution was obtained in Section 4.2, Eq. 4.2.13, for a pressure drop in the form of a simple cosine function, but for the purpose of illustration we shall rederive it here using the complex exponential function. Our starting point is Eq. 4.3.8 governing the particular part of the solution when the pressure drop is in the form of a complex exponential function, that is

$$t_L \frac{d^2 q_p}{dt^2} + \frac{dq_p}{dt} + \frac{q_p}{t_C} = \frac{i\omega \Delta p_0 e^{i\omega t}}{R} \tag{4.4.5}$$

It is known that the particular solution of this equation, because of the exponential term on the right, has the form [116]

$$q_p(t) = K e^{i\omega t} \tag{4.4.6}$$

where K is a constant to be determined as shown below and is not to be confused with the arbitrary constants A, B in the homogeneous part of the solution (Eq. 4.4.2). The constant K is determined by following the method used in Section 4.2 to find the constants K_c, K_s. Eq. 4.4.6 is differentiated twice to find the first two derivatives of $q_p(t)$, then substituting these in Eq. 4.4.5 gives

$$-L\omega^2 K e^{i\omega t} + Ri\omega K e^{i\omega t} + \frac{K}{C} e^{i\omega t} = i\omega \Delta p_0 e^{i\omega t} \tag{4.4.7}$$

This is a simple equation for K from which we readily find

$$K = \frac{\Delta p_0}{R^2 + \left(\omega L - \frac{1}{\omega C}\right)^2} \left\{ R - i \left(\omega L - \frac{1}{\omega C}\right) \right\} \tag{4.4.8}$$

which can in fact be simplified to

$$K = \frac{\Delta p_0}{R + i \left(\omega L - \frac{1}{\omega C}\right)} \tag{4.4.9}$$

To show that this yields the result obtained in Section 4.2, using Eqs.4.4.6,8 we find

$$q_p(t) = K e^{i\omega t}$$
$$= K(\cos \omega t + i \sin \omega t) \tag{4.4.10}$$
$$= \frac{\Delta p_0}{R^2 + \left(\omega L - \frac{1}{\omega C}\right)^2} \left\{ R \cos \omega t + \left(\omega L - \frac{1}{\omega C}\right) \sin \omega t \right\}$$
$$+ \frac{i \Delta p_0}{R^2 + \left(\omega L - \frac{1}{\omega C}\right)^2} \left\{ R \sin \omega t - \left(\omega L - \frac{1}{\omega C}\right) \cos \omega t \right\} \tag{4.4.11}$$

The first term in Eq. 4.4.11 represents the real part of $q_p(t)$ which in turn represents the particular solution for the real part of the complex exponential form of the pressure drop, namely

$$\Re\{\Delta p\} = \Re\{\Delta p_0 e^{i\omega t}\} \tag{4.4.12}$$
$$= \Re\{\Delta p_0(\cos \omega t + i \sin \omega t)\} \tag{4.4.13}$$
$$= \Delta p_0 \cos \omega t \tag{4.4.14}$$

which is the pressure drop used in Section 4.2 (Eq. 4.2.1). Therefore the *real part* of $q_p(t)$ in Eq. 4.4.11 should be identical with the particular solution obtained in Section 4.2, Eq. 4.2.13. We observe from Eqs.4.2.13,4.4.11 that the two are indeed identical. The *imaginary part* of $q_p(t)$ in Eq. 4.4.11 above then, similarly, represents a particular solution corresponding to a pressure drop of the form $\Delta p_0 \sin \omega t$.

Having found the constant K, the two parts of the solution can now be put together, using Eqs.4.4.1,2,6, namely

$$q(t) = Ae^{\alpha_1 t} + Be^{\alpha_2 t} + Ke^{i\omega t} \qquad (4.4.15)$$

and proceed to find the arbitrary constants A, B in terms of initial flow conditions. Differentiating Eq. 4.4.15 twice and evaluating at time $t = 0$, we find

$$q(0) = A + B + K \qquad (4.4.16)$$
$$q'(0) = A\alpha_1 + B\alpha_2 + i\omega K \qquad (4.4.17)$$

These are two equations for the unknown constants A, B from which we readily find

$$A = \frac{-\alpha_2 q(0) + q'(0) + K(\alpha_2 - i\omega)}{\alpha_1 - \alpha_2} \qquad (4.4.18)$$

$$B = \frac{\alpha_1 q(0) - q'(0) - K(\alpha_1 - i\omega)}{\alpha_1 - \alpha_2} \qquad (4.4.19)$$

We note again that the amount of tedious algebra in this process has been reduced considerably by using the complex exponential function.

The solution is now complete, the flow rate being given by Eq. 4.4.15 and the arbitrary constants are given by Eqs.4.4.9,18,19. A plot of $q(t)$ for a value of $(4t_L/t_C) < 1$ where overdamping conditions prevail is shown in Fig. 4.4.1. Typically, the flow rate starts from a prescribed normalized value of 1.0 then gradually enters a phase of regular oscillations about a zero mean, consistent with the driving oscillatory pressure drop. These two phases in the dynamics of the RLC system will be discussed at great length in subsequent sections. Here we note only that the pattern of dynamics observed in Fig. 4.4.1 is analogous to the pattern observed in Fig. 3.3.1 for the overdamped case in free dynamics where the flow also starts from a prescribed normalized value of 1.0, then gradually diminishes to zero. The difference between the two cases is only in the forced oscillations imposed in the present case.

Thus, the study of free dynamics of the RLC system in Section 3.3, though at first seemed somewhat artificial as a model of the physiological system, is now seen as an important integral part of an overall study of the system. We now see that the properties of the RLC system observed in free dynamics under the three different damping conditions are in fact intrinsic properties of the system that are equally relevant in forced dynamics.

Finally, we see now that in the general solution of the forced dynamics problem, broadly speaking, the homogeneous part of the solution, namely

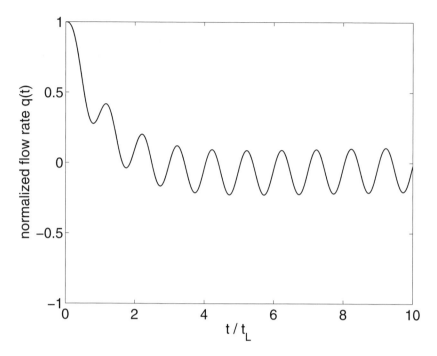

Fig. 4.4.1. Flow rate $q(t)$ in an RLC system in series and in forced dynamics, with an oscillatory driving pressure drop. System parameters are such that $(4t_L/t_C) = 0.4$, therefore producing overdamped conditions.

$q_h(t)$, represents the free dynamics of the system while the particular part of the solution, namely $q_p(t)$, represents the forced part of the dynamics. However, the two are not entirely separate from each other because the forced dynamics constant K appears in the final expressions for the free dynamics constants A, B (Eqs.4.4.18,19). We also note that the constants A, B in the solution for $q_h(t)$ are *complex* even though the equation governing $q_h(t)$ (Eq. 4.3.7) does not involve the complex exponential expression for the pressure drop Δp. In fact, it does not involve the pressure drop at all. The constants become complex in the process of implementing the initial flow conditions.

4.5 Underdamped Forced Dynamics

Here, again, the flow rate consists of two parts:

$$q(t) = q_h(t) + q_p(t) \tag{4.5.1}$$

The *particular* part is the same as in the previous section (Eq. 4.4.9), namely

$$q_p(t) = Ke^{i\omega t} \tag{4.5.2}$$

$$K = \frac{\Delta p_0}{R + i\left(\omega L - \frac{1}{\omega C}\right)} \tag{4.5.3}$$

The *homogeneous* part of the solution, $q_h(t)$, comes from the general solution of the free dynamics equation under the underdamped scenario and is given by (Eq. 3.3.14)

$$q_h(t) = e^{at}(A\cos bt + B\sin bt) \tag{4.5.4}$$

$$a = \frac{-1}{2t_L} \tag{4.5.5}$$

$$b = \frac{\sqrt{(4t_L/t_C) - 1}}{2t_L} \tag{4.5.6}$$

where A,B, are arbitrary constants.

The complete solution is then given by

$$q(t) = e^{at}(A\cos bt + B\sin bt) + Ke^{i\omega t} \tag{4.5.7}$$

As in the previous section, to find the arbitrary constants A, B, differentiating this expression and evaluating at time $t = 0$ gives

$$q(0) = A + K \tag{4.5.8}$$
$$q'(0) = Aa + Bb + i\omega K \tag{4.5.9}$$

from which we readily find

$$A = q(0) - K \tag{4.5.10}$$

$$B = \frac{q'(0) - aA - i\omega K}{b} \tag{4.5.11}$$

Again, we note the reduced amount of algebra involved in this process because of the use of the complex exponential function.

Results are shown in Fig. 4.5.1 for a value of $(4t_L/t_C) > 1.0$ where underdamped conditions prevail. The flow rate starts from a prescribed normalized value of 1.0, undergoes some large irregular oscillations, then gradually enters a phase of regular oscillations about a zero mean, consistent with the driving oscillatory pressure drop. This pattern is analogous to the pattern observed in Fig. 3.3.1 for the underdamped case in free dynamics where the flow also starts from a prescribed normalized value of 1.0, undergoes some large irregular oscillations, then gradually diminishes to zero.

We note that both in free and in forced dynamics, underdamped conditions occur when

$$\frac{4t_L}{t_C} = \left(\frac{4L}{R^2C}\right) > 1.0 \tag{4.5.12}$$

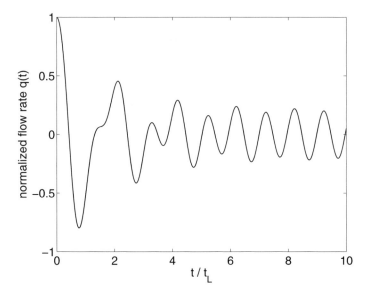

Fig. 4.5.1. Flow rate $q(t)$ in an RLC system in series and in forced dynamics, with an oscillatory driving pressure drop. System parameters are such that $(4t_L/t_C) = 40$, therefore producing underdamped conditions.

compared with the overdamped case considered in the previous section where the quantity in brackets is *less than* 1.0. Thus, all else being equal, underdamped conditions correspond to *lower* values of C which in turn corresponds to an elastic balloon which is *less* elastic. Thus, the initial oscillations observed in the underdamped case result from the recoiling of a stiffer balloon. A balloon that is more elastic would instead absorb the filling without recoiling, as observed in the overdamped case.

4.6 Critically Damped Forced Dynamics

Between the overdamped and underdamped conditions, which occur over a range of values of $4t_L/t_C$ below and above 1.0, there is a singular condition corresponding to a single value of this ratio, namely

$$\frac{4t_L}{t_C} = 1.0 \qquad (4.6.1)$$

whereby the flow rate in the RLC system in series under conditions of forced dynamics moves "most directly" from its prescribed initial value to the state of forced oscillations being imposed on it. This is the scenario of critically damped forced dynamics, analogous to that of critical damping encountered in free dynamics.

As in the previous two sections, the flow rate solution again consists of two parts:

$$q(t) = q_h(t) + q_p(t) \qquad (4.6.2)$$

where the *particular part* of the solution, $q_p(t)$, is the same as before (Eq. 4.4.9), namely

$$q_p(t) = Ke^{i\omega t} \qquad (4.6.3)$$

$$K = \frac{\Delta p_0}{R + i\left(\omega L - \frac{1}{\omega C}\right)} \qquad (4.6.4)$$

The homogeneous part of the solution, $q_h(t)$, comes from the general solution of the free dynamics equation under the critically damped scenario and is given by (Eq. 3.3.17,16)

$$q_h(t) = (A + Bt)e^{at} \qquad (4.6.5)$$

$$a = \frac{-1}{2t_L} \qquad (4.6.6)$$

where A,B, are arbitrary constants.

The complete solution is given by

$$q(t) = (A + Bt)e^{at} + Ke^{i\omega t} \qquad (4.6.7)$$

and, to find the constants A, B, differentiating this expression and evaluating at time $t = 0$ gives

$$q(0) = A + K \qquad (4.6.8)$$
$$q'(0) = B + Aa + i\omega K \qquad (4.6.9)$$

from which we readily find

$$A = q(0) - K \qquad (4.6.10)$$
$$B = q'(0) - Aa - i\omega K \qquad (4.6.11)$$

Results are shown in Fig. 4.6.1 for the value of $(4t_L/t_C) = 1.0$ where critically damped conditions occur. The flow rate starts from a prescribed normalized value of 1.0, then gradually enters a phase of regular oscillations about a zero mean, consistent with the driving oscillatory pressure drop. This pattern is analogous to the pattern observed in Fig. 3.3.1 for the critically damped case in free dynamics where the flow also starts from a prescribed normalized value of 1.0, then gradually diminishes to zero. Both in free and in forced dynamics, the difference between the overdamped and critically damped dynamics is one of degree only. In the critically damped case the flow rate moves towards its ultimate oscillations more "expediently" than it does in the overdamped case.

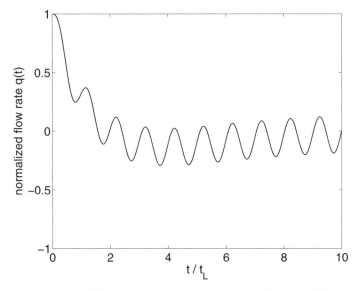

Fig. 4.6.1. Flow rate $q(t)$ in an RLC system in series and in forced dynamics, with an oscillatory driving pressure drop. System parameters are such that $(4t_L/t_C) = 1.0$, therefore producing critically damped conditions.

4.7 Transient and Steady States

Both in free and in forced dynamics the RLC system undergoes two distinct phases (Figs. 3.3.1, 4.4.1, 4.5.1, 4.6.1). In the first, which is widely referred to as the "transient state", the system is adjusting to the imposed initial flow conditions and the flow rate continues to change accordingly. In the second, referred to as "steady state", the adjustment is complete and no further change in the pattern of flow rate takes place.

We saw that in free dynamics the steady state is a state of zero flow because there is no imposed pressure drop to drive the flow. Any flow within the system is there because of the prescribed initial flow conditions. The transient state takes the system from these initial conditions to the steady state. Thus, the steady state may be viewed as that appropriate for the system under the particular driving force being imposed on it externally. In the absence of such forces, as in the case of free dynamics, the appropriate state is that of zero flow.

Similarly, in forced dynamics the steady state of the system is that appropriate for the applied external driving force. When the latter is oscillatory, as we saw in previous sections, the steady state of the system is that in which the flow rate oscillates in tandem with the externally imposed pressure drop. Again, the transient state takes the system from whatever initial conditions are prescribed to this steady state.

In analytical terms, the *steady* state of the *RLC* system is a *particular* solution of the governing equation (Eq. 3.2.5)

$$t_L \frac{d^2q}{dt^2} + \frac{dq}{dt} + \frac{1}{t_C}q = \frac{1}{R}\frac{d(\Delta p)}{dt} \qquad (4.7.1)$$

while the *transient* state of the system is a *general* solution of the reduced equation

$$t_L \frac{d^2q}{dt^2} + \frac{dq}{dt} + \frac{1}{t_C}q = 0 \qquad (4.7.2)$$

In *free dynamics* the pressure drop is zero, therefore the term on the right-hand side of Eq. 4.7.1 is zero and the equation becomes identical with Eq. 4.7.2. Therefore, in this case the steady state is a particular solution of Eq. 4.7.2 and the transient state is a general solution of the same equation. The general solution was obtained in Section 3.3 under the three different scenarios of overdamped, underdamped, and critically damped conditions. A particular solution of Eq. 4.7.2 is clearly $q = 0$, which is consistent with results in Section 3.3 indicating that the steady state of the system is that of zero flow under all three damping scenarios.

In *forced dynamics* the steady state of the *RLC* system depends on the form of the driving pressure drop Δp since this determines the form of the particular solution of Eq. 4.7.1. We saw that when the driving pressure drop is an oscillatory function of the form $\Delta p(t) = \Delta p_0 e^{i\omega t}$ where Δp_0 is a constant, the particular solution is of the same form, namely $q(t) = K e^{i\omega t}$ where K is a constant. Thus, the steady state of the system is one in which the flow rate oscillates with the same frequency as the pressure drop. It is appropriate to refer to this as "steady state" even though the flow rate is a function of time. The term "steady" here is not to be confused with "constant", it merely implies that the pattern of flow rate as a function of time is no longer changing.

The most important property of the transient state is that it occupies only a relatively short time from the onset of the initial flow conditions, then leaving the system in steady state thereafter. We saw that both in free and in forced dynamics, and under all three damping scenarios, the transient part of the flow rate, namely $q_h(t)$, vanishes soon after the initial flow onset, leaving the system with only the steady part of the flow, namely with $q_p(t)$. These results are illustrated graphically in Figs.4.7.1-4. Because of this, many studies find it appropriate to ignore the transient state dynamics of the *RLC* system and focus on the steady state dynamics only (see Chapter 7).

The relevance of this discussion to coronary blood flow is that practically all lumped model studies of the coronary circulation are based on only the steady state dynamics of the system.

One might attempt to examine the validity of this practice by noting from Fig. 4.7.1 that in free dynamics the *RLC* system comes very close to steady state when $t/t_L \approx 15$. In forced dynamics, results in Figs.4.7.2-4 indicate that

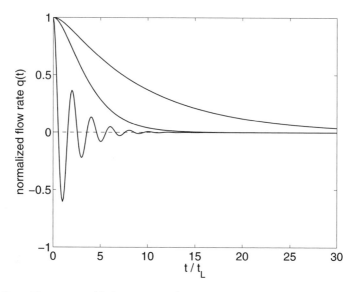

Fig. 4.7.1. Flow rate $q(t)$ (solid curves) in an RLC system in series and in free dynamics, under conditions of overdamping, underdamping, and critical damping. Dashed line represents steady state flow rate.

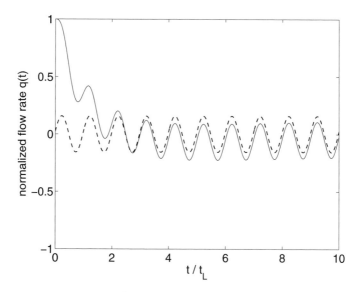

Fig. 4.7.2. Flow rate $q(t)$ (solid curve) in an RLC system in series and in forced overdamped dynamics, with an oscillatory driving pressure drop. Dashed line represents steady state flow rate.

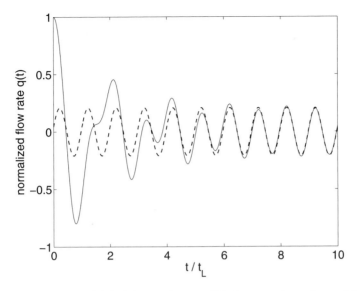

Fig. 4.7.3. Flow rate $q(t)$ (solid curve) in an RLC system in series and in forced underdamped dynamics, with an oscillatory driving pressure drop. Dashed line represents steady state flow rate.

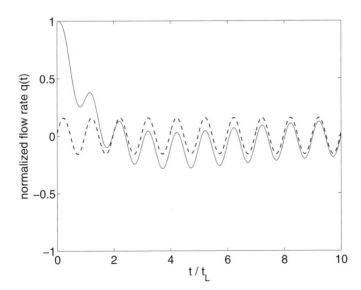

Fig. 4.7.4. Flow rate $q(t)$ (solid curve) in an RLC system in series and in forced critically damped dynamics, with an oscillatory driving pressure drop. Dashed line represents steady state flow rate.

the corresponding time is $t/t_L \approx 5$. To determine these times in seconds requires an estimate of the inertial time constant t_L for the coronary circulation which, of course, is not known. Indeed, one of the main aims of lumped model studies is to *determine* such properties of the system.

Nevertheless, we may attempt to estimate t_L by recalling from sections 2.4,5 (Eqs.2.4.5,2.5.5) that for *flow in a single tube* the resistance to flow and the inertial constant are given by

$$R = \frac{8\mu l}{\pi a^4} \tag{4.7.3}$$

$$L = \frac{4\rho l}{\pi d^2} \tag{4.7.4}$$

where μ, ρ are respectively the viscosity and density of the fluid, and a, d are respectively the radius and diameter of the tube. From this, and from the definition of t_L we then have

$$t_L = \frac{L}{R} = \frac{\rho a^2}{8\mu} \quad seconds \tag{4.7.5}$$

Taking $\rho = 1.0 \ g/cm^3$ and $\mu = 0.04 \ g/(cm \ s)$, this becomes

$$t_L = \frac{a^2}{0.32} \quad seconds \tag{4.7.6}$$

where a is in centimeters.

Now, in a lumped model of the coronary circulation the characteristic parameters of the numerous tubes in the system are "lumped" together, which is equivalent to considering the system as a single tube that has these lumped properties. The problem, of course, is that these lumped properties are not known. Indeed, whether by theory or by experiment, the ultimate goal of lumped model studies is to determine not only the values of such properties but the legitimacy of the lumped parameter concept on which they are based.

Thus, from Eq. 4.7.6 we can only conclude that for a tube of 1cm radius, the equation gives a value of t_L of approximately 3 *seconds*, while for a tube of 1mm radius, the corresponding value is 0.03 *seconds*. The transient state of an RLC system with these values of t_L would then be approximately 15 t_L in free dynamics and 5 t_L in forced dynamics. The highest of these values is 45 *seconds*, the lowest is 0.15 *seconds*. Whether the true transient time of the coronary circulation lies within or indeed outside these estimates is not known. In the forced oscillatory dynamics of the coronary circulation t_L would represent the time it takes the system to recover from a disturbance and return to steady state, thus knowing the actual value of this time constant would be of considerable clinical importance.

4.8 The Concept of Reactance

The concepts of reactance and impedance arise in the dynamics of the RLC system in *steady state*. It is important to emphasize this point because these concepts are so widely used that it is usually only implied, but rarely stated, that their use is limited to steady state dynamics only, not to the transient state. Thus, in introducing these concepts here, and using them in subsequent sections, it must be clear from the outset that we are now dealing with only the *particular* solution of the governing equation (Eq. 3.2.5)

$$t_L \frac{d^2q}{dt^2} + \frac{dq}{dt} + \frac{1}{t_C} q = \frac{1}{R} \frac{d(\Delta p)}{dt} \tag{4.8.1}$$

and for this reason we no longer use the subscripts for the homogeneous and particular parts of the solution discussed in the previous section. It is implied here, and whenever these concepts are used, that $q(t)$ now represents only the particular part of the solution, which, as discussed in the previous section, corresponds to the steady state dynamics of the RLC system.

Furthermore, a meaningful definition of reactance and impedance is only possible when Δp above, and consequently $q(t)$, are simple oscillatory functions such as the trigonometric sine and cosine functions. We thus begin with the problem solved in Section 4.2, where the driving pressure drop is given by

$$\Delta p = \Delta p_0 \cos \omega t \tag{4.8.2}$$

Steady state solution of Eq. 4.8.1 with this form of the pressure drop was obtained in Section 4.2 (Eq. 4.2.13), which we now write in the form

$$q(t) = \Delta p_0 \left\{ \frac{R \cos \omega t + S \sin \omega t}{R^2 + S^2} \right\} \tag{4.8.3}$$

where

$$S = \omega L - \frac{1}{\omega C} \tag{4.8.4}$$

Using standard trigonometric identities, Eq. 4.8.3 can be simplified further to

$$q(t) = \frac{\Delta p_0}{\sqrt{R^2 + S^2}} \cos(\omega t - \theta) \tag{4.8.5}$$

where

$$\tan \theta = \frac{S}{R} \tag{4.8.6}$$

To discuss the nature and effect of the quantity S now, we begin by noting that when $S = 0$, Eqs. 4.8.5,6 give

$$q(t) = \frac{\Delta p_0}{R} \cos \omega t \qquad (4.8.7)$$

$$= \frac{\Delta p}{R} \qquad (4.8.8)$$

which we recognize as the simple expression for the flow rate through a resistance R when the driving pressure drop across it is Δp. Thus, in the dynamics of the RLC system the quantity S as defined in Eq. 4.8.4 embodies the combined effects of inductance L and capacitance C such that when $S = 0$ the system behaves as a simple resistance R. When $S \neq 0$, it is clear from a comparison of Eq. 4.8.5 with Eq. 4.8.7 that S acts as an added form of resistance to flow, resulting from the presence of inductance and capacitance. It is also noted from the presence of ω in the expression for S that this form of resistance occurs only in oscillatory flow. By analogy with the same phenomenon in the flow of alternating current in an electric circuit, S is generally referred to as the "reactance". It is a form of resistance to flow, but it differs from R in that it only occurs in oscillatory flow. Also, unlike the viscous resistance R, the reactance S does not actually dissipate flow energy, it merely stores it and releases it within each oscillatory cycle [221].

From Eq. 4.8.4 we note that there are two ways in which the reactance S can be zero. First, when the capacitance and inductance effects are simply absent, that is

$$L = 0 \qquad (4.8.9)$$

$$\frac{1}{C} = 0 \qquad (4.8.10)$$

which together lead to $S = 0$. Second, when the values of ω, L, C are such that

$$C = \frac{1}{\omega^2 L} \qquad (4.8.11)$$

which again leads to $S = 0$. While the first of these circumstances is trivial, the second has clear physiological significance because it deals with the critical balance between the effects of capacitance and inductance. To pursue this further, consider, in Eq. 4.8.4, that the frequency ω and inductance L are fixed so that the value of S now depends on the capacitance C only. It is convenient in this discussion to use the term "compliance" for the capacitance C. It is a more expressive term than "capacitance" because at higher values of the capacitance C a balloon is more elastic, more compliant, while at lower values of the capacitance C it is less elastic, less compliant.

Starting at the extreme where a balloon is rigid, compliance (C) is zero, and its reciprocal is infinite. The value of S from Eq. 4.8.4 is then infinite and negative, and the corresponding value of the phase angle θ from Eq. 4.8.6 is $-\pi/2$. Eq. 4.8.5 then indicates that the flow is *leading* the pressure drop by $\pi/2$. As the compliance gradually increases from this extreme value, the values of S and θ remain negative at first but continue to increase, until at some point they both become zero. The value of C at which this point is reached is given in Eq. 4.8.11, and we shall denote this value by C_0, that is

$$C_0 = \frac{1}{\omega^2 L} \tag{4.8.12}$$

At this critical value of C the reactance is zero, the phase angle between the flow and pressure drop is zero, and the *amplitude* of the flow rate, which from Eq. 4.8.5 is given by

$$|q(t)| = \frac{\Delta p_0}{\sqrt{R^2 + S^2}} \tag{4.8.13}$$

clearly has its highest value because $S = 0$. Thus,

$$\text{at } C = C_0: \qquad S = 0 \tag{4.8.14}$$
$$\theta = 0 \tag{4.8.15}$$
$$|q(t)| = \frac{\Delta p_0}{R} \tag{4.8.16}$$

As compliance continues to increase beyond this point, the value of S becomes positive, the phase angle θ becomes positive, which means that flow is now *lagging* the pressure drop, and the amplitude of the flow rate begins to decrease again from its maximum value.

To illustrate these results graphically, it is convenient to use a normalized form of the flow rate, namely

$$\bar{q}(t) = \frac{q(t)}{\Delta p_0 / R} \tag{4.8.17}$$

$$= \frac{1}{\sqrt{1 + (S/R)^2}} \cos(\omega t - \theta) \tag{4.8.18}$$

Also, instead of using the actual capacitance C, it is more convenient to use the capacitive time constant

$$t_C = CR \tag{4.8.19}$$

Since R is assumed to be constant, t_C is a direct measure of C. Also, using the inertial time constant

$$t_L = L/R \tag{4.8.20}$$

the reactance S can be expressed in terms of these time constants as

$$\frac{S}{R} = \omega t_L - \frac{1}{\omega t_C} \tag{4.8.21}$$

Fig. 4.8.1 shows the variation of the reactance as t_C increases from zero (where capacitance and compliance are zero) to large values (where capacitance and compliance are large). In that sequence the value of S changes from large negative to positive, thus passing zero at one particular value of t_C which we shall denote by t_{C0}, and which from Eq. 4.8.12 is given by

$$t_{C0} = \frac{1}{\omega^2 t_L} \qquad (4.8.22)$$

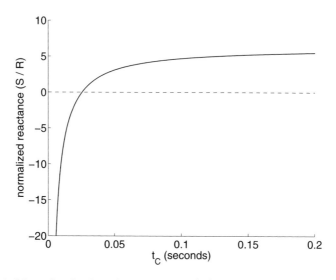

Fig. 4.8.1. Normalized value of the reactance (S/R), as a function of the capacitive time constant t_C. Of particular significance is the point at which reactance becomes zero, which occurs at $t_C = 1/4\pi^2 \approx 0.0253$.

If the frequency of oscillation in cycles per second (Hz) is denoted by f, then the *angular* frequency ω is given by

$$\omega = 2\pi f \quad \text{radians} \qquad (4.8.23)$$

As in previous sections, the inertial time constant t_L may be used as the normalizing unit of time, which is equivalent to taking

$$t_L = 1.0 \quad \text{seconds} \qquad (4.8.24)$$

With these values of ω and t_L, Eq. 4.8.22 gives

$$t_{C0} = \frac{1}{(2\pi)^2} \approx 0.0253 \quad \text{seconds} \qquad (4.8.25)$$

as seen in Fig. 4.8.1. At higher values of t_C, as capacitance effects become more significant, the normalized reactance S/R approaches the constant value, from

Eq. 4.8.21, $2\pi \approx 6.283$, again as seen in that figure. Corresponding values of the phase angle θ are shown in Fig. 4.8.2, where it is seen that the phase angle is zero at the same critical value of t_C where the reactance is zero, namely at $t_C \approx 0.0253$. At higher values of t_C (higher capacitance and compliance) the angle is positive, which means that flow is lagging behind the pressure drop, while at lower values of t_C the reverse is true.

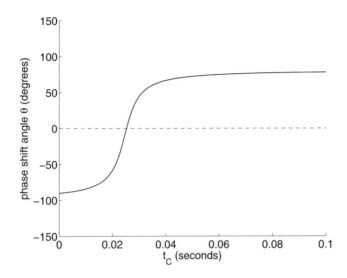

Fig. 4.8.2. Phase angle θ between flow rate and pressure drop, as a function of the capacitive time constant t_C. The angle becomes zero and changes sign at the same critical value of t_C where reactance is zero (Fig. 4.8.1), namely $t_C = 1/4\pi^2 \approx 0.0253$.

This change in phase shift is illustrated in Figs.4.8.3-5 where values of t_C near the critical value are taken. It is remarkable that only a small departure from the critical value of t_C is needed to produce a significant change in phase angle. A similar change occurs in the *amplitude* of the flow wave, which can be put in the normalized form

$$|\bar{q}(t)| = \frac{1}{\sqrt{1 + (S/R)^2}} \qquad (4.8.26)$$

In this form the amplitude has the normalized value of 1.0 when $S = 0$, which occurs at the critical value of t_C. Figs.4.8.3-6 show clearly that this is the *maximum* value of the flow amplitude. At all other values of t_C the flow wave is affected both in phase and amplitude.

These results have remarkable implications regarding the phyiological system. While the RLC system *in series* does not appropriately model the coronary circulation (RLC in *parallel* will be considered in Chapter 6), it nevertheless points to the existence of conditions under which the system may

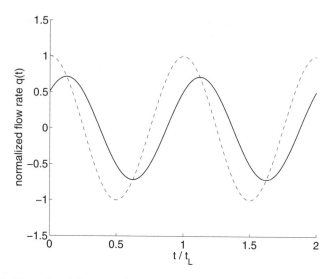

Fig. 4.8.3. Normalized flow rate (solid curve) compared with pressure drop (dashed curve) within the oscillatory cycle, and with $t_C = 0.03$ seconds, which is just above critical value of $t_C = 0.0253$ at which the two curves would be identical. Flow rate lags behind pressure drop and flow amplitude is below maximum.

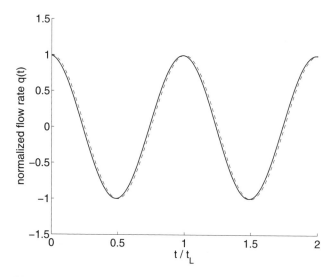

Fig. 4.8.4. Normalized flow rate (solid curve) compared with pressure drop (dashed curve) within the oscillatory cycle, and with $t_C = 0.025$ seconds, which is very close to critical value (at the critical value the two curves would be indistinguishable). Flow rate is in phase with pressure drop and flow amplitude is maximum.

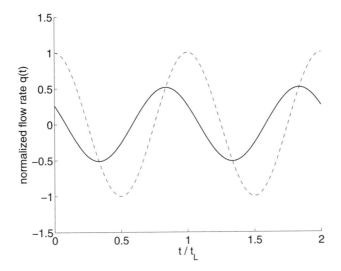

Fig. 4.8.5. Normalized flow rate (solid curve) compared with pressure drop (dashed curve) within the oscillatory cycle, and with $t_C = 0.02$ seconds, which is just below critical value of $t_C = 0.0253$ at which the two curves would be identical. Flow rate leads pressure drop and flow amplitude is below maximum.

operate most optimally, in the sense of minimizing the effects of reactance. The mere existence of these conditions is clearly of significant clinical interest because it indicates that if the coronary circulation normally operates at or near a critical value of t_C, then any change in the elasticity or other properties of the system, resulting from disease or clinical intervention, may move the system away from its optimal dynamics.

Furthermore, in the present system if t_L is set equal to 1.0 seconds, which is not an unreasonable estimate following the discussion in Section 4.7, these favourable conditions occur at $t_C = 0.0253$ seconds. While these two values may be inaccurate in *absolute* terms, they indicate that in *relative* terms the capacitive time constant may be two orders of magnitude smaller than the inertial time constant.

4.9 The Concepts of Impedance, Complex Impedance

The concept of "impedance", like that of reactance, arises in the dynamics of the RLC system in *steady state* under a simple oscillatory driving pressure drop and, as emphasized in the previous section, it is only valid, indeed only meaningful, in that context. Again, the concept is borrowed from the flow of alternating current in electric circuits, but it has wide applications in the dynamics of the coronary circulation, specifically in the context of lumped models of the system.

Broadly speaking, impedance is the total impediment to oscillatory flow in the presence of inductance and capacitance effects. We do not use the term "resistance" here, because that term is generally reserved for the effects of viscosity in steady as well as in oscillatory flow. Essentially, impedance embodies both the resistance R and the reactance S discussed in the previous section.

There is a fundamental difference between impedance and the familiar viscous resistance which makes it important not to confuse the two. Viscous resistance dissipates energy which must be replaced constantly from a source of driving energy (pump). Reactance, on the other hand, as seen in the previous section, presents an impediment to flow in the sense of affecting the amplitude and phase of the flow wave, but it does not actually dissipate the flow energy. Under reactive effects flow energy is only *exchanged* between pressure and kinetic energy, as when fluid is accelerated or decelerated, or between pressure inside the capacitive balloon and elastic energy in its walls. Impedance embodies these exchanges as well as the dissipative viscous resistance and therefore it would be inappropriate to describe it simply as "resistance to flow".

The primary reason for using the concept of impedance is that it provides a link beween flow and pressure drop in oscillatory flow, in the same way that resistance provides that link in steady flow. Thus, for steady flow in a tube we have the basic result from Section 2.4 that the flow rate q is simply equal to the pressure drop Δp divided by the resistance R, that is (Eq. 2.4.3)

$$q = \frac{\Delta p}{R} \tag{4.9.1}$$

In oscillatory flow the concept of impedance is introduced in order to be able to write the relation between flow and pressure drop in a similar way, namely

$$q = \frac{\Delta p}{(IMP)} \tag{4.9.2}$$

where IMP is used here as a generic label for impedance until it can be defined more accurately.

In the solution obtained in the previous section the driving oscillatory pressure drop was of the form (Eq. 4.8.2)

$$\Delta p(t) = \Delta p_0 \cos \omega t \tag{4.9.3}$$

while the the flow rate was found to be of the form (Eq. 4.8.5)

$$q(t) = \frac{\Delta p_0}{\sqrt{R^2 + S^2}} \cos (\omega t - \theta) \tag{4.9.4}$$

where Δp_0 is a constant representing the amplitude of the driving oscillatory pressure drop, ω is the (angular) frequency of oscillation, R is resistance, S is reactance, and θ is the phase angle between the pressure and flow waves, as defined in the previous section.

From Eqs.4.9.3,4 it is not possible to write the relation between $q(t)$ and Δp in the form of Eq. 4.9.2. This is because impedance affects both the amplitude and the phase angle of the flow wave, and these effects appear separately in Eq. 4.9.4. The amplitude effect is represented by the term $\sqrt{R^2 + S^2}$, while the phase angle effect is represented by the angle θ. This suggests that impedance itself has an amplitude and phase, which in turn suggests that impedance is a complex quantity with a real and an imaginary part. This indeed turns out to be the case as we shall see below.

To reach this result we begin with a driving pressure drop in complex form, as was done in previous sections, namely

$$\Delta p(t) = \Delta p_0 e^{i\omega t} \tag{4.9.5}$$

A steady state solution with this form of the pressure drop was obtained in details in Section 4.4 (Eqs.4.4.6,9), namely

$$q(t) = K e^{i\omega t} \tag{4.9.6}$$

where

$$K = \frac{\Delta p_0}{R + iS} \tag{4.9.7}$$

$$S = \omega L - \frac{1}{\omega C} \tag{4.9.8}$$

Using these, the equation for the flow rate (Eq. 4.9.6) can then be put in the form

$$q(t) = \frac{\Delta p_0 e^{i\omega t}}{R + iS} \tag{4.9.9}$$

or, using Eq. 4.9.5,

$$q(t) = \frac{\Delta p(t)}{Z} \tag{4.9.10}$$

where

$$Z = R + iS \tag{4.9.11}$$

Eq. 4.9.10 is a relation between flow rate and pressure drop in the basic form of Eq. 4.9.2, therefore we identify Z as the impedance (IMP) in that equation. Furthermore, as anticipated earlier, Z is a complex quantity as defined in Eq. 4.9.11, with the resistance R as its real part and the reactance S as its imaginary part. It is known as "complex impedance".

We note from Eq. 4.9.11 that the amplitude and phase of Z are respectively given by

$$|Z| = \sqrt{R^2 + S^2} \tag{4.9.12}$$

$$\theta = \tan^{-1}\left(\frac{S}{R}\right) \tag{4.9.13}$$

which we recognize as the effects of impedance on the amplitude and phase of the flow wave, as discussed earlier in this section.

To see this more clearly we now reproduce the result in Eq. 4.9.4 which represents the flow rate when the driving pressure drop is a cosine function (Eq. 4.9.9), which is equivalent to the real part of the complex pressure drop in Eq. 4.9.5, that is

$$\Delta p(t) = \Delta p_0 \cos \omega t \tag{4.9.14}$$
$$= \Re\{\Delta p_0 e^{i\omega t}\} \tag{4.9.15}$$

The flow rate corresponding to this pressure drop is therefore the real part of the complex flow rate in Eq. 4.9.9, that is

$$q(t) = \Re\left\{\frac{\Delta p_0 e^{i\omega t}}{R + iS}\right\} \tag{4.9.16}$$
$$= \Delta p_0 \Re\left\{\frac{(\cos \omega t + i \sin \omega t)(R - iS)}{R^2 + S^2}\right\} \tag{4.9.17}$$
$$= \Delta p_0 \left\{\frac{R \cos \omega t + S \sin \omega t}{R^2 + S^2}\right\} \tag{4.9.18}$$

The last expression is identical with the result in Eq. 4.9.4.

Thus, complex impedance offers an elegant way of representing the effects of impedance on both the amplitude and phase of the pressure wave. It also makes it possible to maintain the simple relation between pressure drop and flow, namely that in Eq. 4.9.10.

Indeed, because of the simple relation between q and Δp in Eq. 4.9.10, the analysis of several impedances in series or in parallel remains as simple as the analysis of several resistances in series or in parallel.

For impedances Z_1, Z_2, Z_3 in series, using Eq. 4.9.10 with q as the common flow rate through the system, we have

$$\Delta p = qZ \tag{4.9.19}$$
$$\Delta p_1 = qZ_1 \tag{4.9.20}$$
$$\Delta p_2 = qZ_2 \tag{4.9.21}$$
$$\Delta p_3 = qZ_3 \tag{4.9.22}$$

Then, since the total pressure drop is the sum of the partial pressure drops, we find

$$\Delta p = \Delta p_1 + \Delta p_2 + \Delta p_3 \tag{4.9.23}$$
$$= q(Z_1 + Z_2 + Z_3) \tag{4.9.24}$$

Therefore

$$Z = Z_1 + Z_2 + Z_3 \tag{4.9.25}$$

For impedances Z_1, Z_2, Z_3 *in parallel*, using Eq. 4.9.10 with Δp as the common pressure drop, we have

$$q = \frac{\Delta p}{Z} \tag{4.9.26}$$

$$q_1 = \frac{\Delta p}{Z_1} \tag{4.9.27}$$

$$q_2 = \frac{\Delta p}{Z_2} \tag{4.9.28}$$

$$q_3 = \frac{\Delta p}{Z_3} \tag{4.9.29}$$

Then, since the total flow rate is the sum of the partial flow rates, we find

$$q = q_1 + q_2 + q_3 \tag{4.9.30}$$

$$= \frac{\Delta p}{Z_1} + \frac{\Delta p}{Z_2} + \frac{\Delta p}{Z_3} \tag{4.9.31}$$

$$= \Delta p \left\{ \frac{1}{Z_1} + \frac{1}{Z_2} + \frac{1}{Z_3} \right\} \tag{4.9.32}$$

Therefore

$$\frac{1}{Z} = \frac{1}{Z_1} + \frac{1}{Z_2} + \frac{1}{Z_3} \tag{4.9.33}$$

In particular, for the elements of the RLC system, we have

$$\text{Resistance:} \quad Z_1 = R \tag{4.9.34}$$

$$\text{Inductance:} \quad Z_2 = i\omega L \tag{4.9.35}$$

$$\text{Capacitance:} \quad Z_3 = \frac{1}{i\omega C} \tag{4.9.36}$$

$$\text{Reactance:} \quad Z_4 = Z_2 + Z_3 \tag{4.9.37}$$

$$= i \left(\omega L - \frac{1}{\omega C} \right) \tag{4.9.38}$$

Thus, for the RLC system *in series*

$$Z = Z_1 + Z_2 + Z_3 \tag{4.9.39}$$

$$= Z_1 + Z_4 \tag{4.9.40}$$

$$= R + i \left(\omega L - \frac{1}{\omega C} \right) \tag{4.9.41}$$

And for the RLC system *in parallel*

$$\frac{1}{Z} = \frac{1}{Z_1} + \frac{1}{Z_2} + \frac{1}{Z_3} \tag{4.9.42}$$

$$= \frac{1}{R} + \frac{1}{i\omega L} - \frac{\omega C}{i} \tag{4.9.43}$$

$$= \frac{1}{R} + i \left(\omega C - \frac{1}{\omega L} \right) \tag{4.9.44}$$

Note that the partial impedance of reactance, namely Z_4, is defined as the sum of the partial impedances of inductance and capacitance when these elements are *in series*, therefore Z_4 cannot be used when the two elements are *in parallel*. Instead, we may define

$$\frac{1}{Z_5} = \frac{1}{Z_2} + \frac{1}{Z_3} \tag{4.9.45}$$

$$= \frac{1}{i\omega L} - \frac{\omega C}{i} \tag{4.9.46}$$

$$= i\left(\omega C - \frac{1}{\omega L}\right) \tag{4.9.47}$$

so that for the parallel system we can write

$$\frac{1}{Z} = \frac{1}{Z_1} + \frac{1}{Z_2} + \frac{1}{Z_3} \tag{4.9.48}$$

$$= \frac{1}{Z_1} + \frac{1}{Z_5} \tag{4.9.49}$$

$$= \frac{1}{R} + i\left(\omega C - \frac{1}{\omega L}\right) \tag{4.9.50}$$

which is identical with the result in Eq. 4.9.44.

These are standard results in electric circuit theory, and they have been used extensively in the analysis of lumped models of the coronary circulation. We shall return to them in Chapter 6.

4.10 Summary

The dynamics of the coronary circulation are "forced" in the sense that they are driven by an external force. In free dynamics, the system's behaviour depends on the internal characteristics only. In forced dynamics, by contrast, the behaviour of the system depends on these characteristics as well as on the form of the external driving force.

Free dynamics of the RLC system are governed by the homogeneous part of the solution of the governing equation, which depend on the characteristic properties of the system only. Forced dynamics are governed by the full solution of the equation, including the homogeneous part as well as the particular part of the solution, the so-called "particular solution", which depends on the form of the external driving force.

When the pressure drop Δp driving the forced dynamics of the RLC system is expressed in the form of a complex exponential function, the solution of the governing equation produces two solutions at once: one corresponding to Δp being a sine function and the other to Δp being a cosine function.

In overdamped forced dynamics of the RLC system in series, under an oscillatory driving pressure, the flow begins with a "transient state" in which it moves from a prescribed initial value to a "steady state" in which it oscillates in tandem with the driving oscillatory pressure. Overdamping occurs when $(4t_L/t_C) < 1.0$ which, all else being the same, corresponds to a sufficiently high value of the capacitance C, which in turn corresponds to a more elastic balloon that absorbs the filling without recoil.

In underdamped forced dynamics of the RLC system in series, under an oscillatory driving pressure, flow begins with a "transient state" in which it moves rather erratically to a "steady state" in which it oscillates in tandem with the driving oscillatory pressure. Underdamping occurs when $(4t_L/t_C) > 1.0$ which, all else being the same, corresponds to a sufficiently low value of the capacitance C, which in turn corresponds to a less elastic balloon that recoils in the initial phase.

In critically damped forced dynamics of the RLC system in series, under an oscillatory driving pressure, the flow begins with a "transient state" in which it moves *most expediently* to a "steady state" in which it oscillates in tandem with the driving oscillatory pressure. Critical damping is a singular scenario which occurs when $(4t_L/t_C) = 1.0$. All else being equal, it corresponds to a unique value of the capacitance C that lies precisely between the overdamped and underdamped values.

Lumped model analysis of the coronary circulation is based on only the *steady state dynamics* of the system, transient state dynamics are neglected. The time required for the system to complete the transient state is difficult to estimate, yet it is clearly of clinical importance because it represents the time it would take the system to recover from a dynamic disturbance.

There are singular circumstances under which the reactance of the RLC system vanishes and the system behaves as if the inductance and capacitance do not exist. These circumstances are created by specific combinations of values of the inertial and capacitive time constants. Since the dynamics of the RLC system in series do not accurately represent the dynamics of the coronary circulation, these specific values of the time constants may not be directly relevant to the dynamics of the coronary circulation. However, the mere existence of these unique circumstances in the dynamics of the RLC system suggests strongly that similar circumstances may exist in the dynamics of the coronary circulation and that the system may normally operate at or near these ideal conditions.

Impedance is the total impediment to pulsatile flow, embodying the effects of capacitance and inductance as well as the familiar effect of viscous resistance. Analytically, impedance is a complex quantity of which the real part represents the viscous resistance while the imaginary part represents the reactance which in turn represents the combined effects of capacitance and inductance. In its complex form, impedance conveniently represents that ratio of pressure over flow in pulsatile flow in the same way that resistance represents that ratio in Poiseuille flow.

5

The Analysis of Composite Waveforms

5.1 Introduction

The oscillatory pressure drops used in all previous chapters have been of a particularly simple form, namely that of a trigonometric sine or cosine function. These waveforms have specific properties that make them particularly useful for the study of general dynamics of RLC systems such as those examined in previous chapters. However, the ultimate aim of these studies, in the context of the coronary circulation, is to examine the dynamics of RLC systems under oscillatory pressure drops of more general forms, in particular the forms of pressure waves generated by the heart. In what follows we shall refer to these generically as "composite" wave forms.

An example of a composite waveform is shown in Fig. 5.1.1, compared with a simple sine wave. The first difference to be observed, of course, is the strictly *regular* form of the sine wave compared with the highly irregular form of the composite wave. It is this simple regular form of the sine wave that makes it possible to describe it by a simple sine *function*. How to describe the irregular form of the composite wave is dealt with in the present chapter.

Of course, a composite wave can be described numerically, by tabulating the position of discrete points along the wave, as shown in Fig. 5.1.2 and in Table 5.1.1, but this is a rather awkward method of description. It is certainly not as elegant or efficient as the description of a sine or cosine wave, which can be accomplished by the simple statement $\Delta p = \Delta p_0 \cos \omega t$ which was used repeatedly in previous sections to describe the waveform of the pressure drop Δp (Eq. 4.2.1). More important, the steady state solutions obtained in Chapter 4 were possible only because the pressure drop Δp on the right-hand side of the governing equation (Eq. 3.2.4) was expressed by a simple *analytical* function such as $\sin \omega t$, $\cos \omega t$, or $e^{i\omega t}$. If Δp can only be described in numerical form, the analytical solutions of Chapter 4 would not be possible.

In one of mathematics' most beautiful triumphs this difficulty is completely resolved, using a technique known as Fourier analysis, named for its original author. The theory of Fourier analysis shows that a composite wave such as

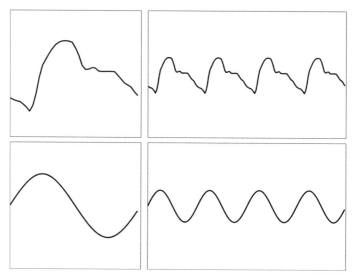

Fig. 5.1.1. Comparison of a composite waveform (top) with the very simple form of the sine wave (bottom). While they are both *periodic*, as seen on the right, the composite wave is highly irregular and is therefore not easy to describe analytically.

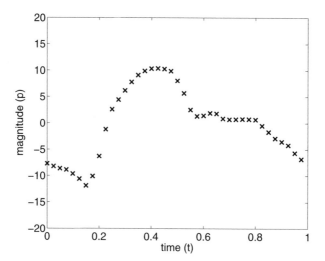

Fig. 5.1.2. A composite wave can be described numerically by tabulating the positions of discrete points along the wave, as shown in Table 5.1.1. The axes are marked generically as t for time and p for pressure. This numerical description is not adequate for obtaining the steady state dynamics associated with this wave, but Fourier analysis shows that the wave can be decomposed into a series of constituent sine and cosine waves for which the dynamics can be obtained, as was done in Chapter 4.

Table 5.1.1. A numerical description of the composite wave shown in Fig. 5.1.2, giving the position (t, p) of each of the discrete points shown along the curve.

t	p	t	p
0.000	-7.7183	0.500	8.0597
0.025	-8.2383	0.525	5.6717
0.050	-8.6444	0.550	2.5232
0.075	-8.8797	0.575	1.3301
0.100	-9.6337	0.600	1.4405
0.125	-10.5957	0.625	1.9094
0.150	-11.8705	0.650	1.8145
0.175	-10.0942	0.675	0.8738
0.200	-6.2839	0.700	0.7055
0.225	-1.1857	0.725	0.7343
0.250	2.6043	0.750	0.7788
0.275	4.4323	0.775	0.7495
0.300	6.1785	0.800	0.6711
0.325	7.8211	0.825	-0.4796
0.350	9.1311	0.850	-1.6541
0.375	9.9138	0.875	-2.8643
0.400	10.3447	0.900	-3.4902
0.425	10.4011	0.925	-4.1714
0.450	10.2807	0.950	-5.6581
0.475	9.8951	0.975	-6.8024

that shown at the top of Fig. 5.1.1 actually consists of a series of sine and cosine waves like the one shown at the bottom of that figure. The composite wave is simply the sum of these so called "harmonics", each of which is a simple sine or cosine wave. This makes it possible to express the composite waveform of the pressure drop Δp in the governing equation (Eq. 3.2.4) as the sum of the sine and cosine functions which constitute that particular composite waveform. The steady state solution of the governing equation can then be obtained for each of these sine and cosine functions *separately*, and then these solutions are collected into a whole. Thus, the steady state solutions obtained in Chapter 4, which were limited to pressure drops of simple sine or cosine waveforms, are not irrelevant to the case of pressure drops of composite waveforms. In fact, they are highly relevant as they actually provide the "building blocks" from which a solution with a pressure drop of a composite waveform is constructed.

The techniques of Fourier analysis are now so well established and so highly developed that it is fair to say that the problem of dealing with composite waves is no longer a problem, it is only a matter of details [28, 197]. While there are now many computer programs that handle these details efficiently, it is not possible to use these reliably without some understanding of the basics of the subject, which is the main purpose of the present chapter.

5.2 Basic Theory

In mathematical language, a wave represents a *periodic function*. A function $p(t)$ is said to be periodic if

$$p(t + T) = p(t) \tag{5.2.1}$$

where T is then called the *period* of that function. An obvious example is the trigonometric function $p(t) = \sin t$ for which

$$p(t + 2\pi) = \sin(t + 2\pi) \tag{5.2.2}$$
$$= \sin t \cos 2\pi + \cos t \sin 2\pi \tag{5.2.3}$$
$$= 1 \times \sin t + 0 \tag{5.2.4}$$
$$= \sin t \tag{5.2.5}$$
$$= p(t) \tag{5.2.6}$$

therefore, $p(t) = \sin t$ is a periodic function with a period $T = 2\pi$. The function is seen graphically in Fig. 5.1.1 (bottom) where the meaning of the period T is quite clear, namely the time interval over which the function assumes a complete cycle of its values. The composite wave seen in Fig. 5.1.1 (top) also represents a periodic function, although the function in this case does not have a simple mathematical form like $\sin \omega t$. Nevertheless, the composite wave in Fig. 5.1.1 represents a periodic function because we can see graphically that the function has a well defined period over which it assumes a complete cycle of its values.

Another example of a periodic function which was used in Chapter 4 and which again has a period $T = 2\pi$ is $p(t) = e^{it}$, because

$$p(t + 2\pi) = e^{i(t+2\pi)} \tag{5.2.7}$$
$$= e^{i2\pi} e^{it} \tag{5.2.8}$$
$$= (\cos 2\pi + i \sin 2\pi) e^{it} \tag{5.2.9}$$
$$= (1 + 0) e^{it} \tag{5.2.10}$$
$$= e^{i\omega t} \tag{5.2.11}$$
$$= p(t) \tag{5.2.12}$$

The theory of Fourier analysis has shown that a periodic function of period T can be expressed as a sum of sine and cosine functions, such that

$$p(t) = \sum_{n=0}^{\infty} A_n \cos\left(\frac{2n\pi t}{T}\right) + \sum_{n=1}^{\infty} B_n \sin\left(\frac{2n\pi t}{T}\right) \tag{5.2.13}$$

$$= A_0 + A_1 \cos\left(\frac{2\pi t}{T}\right) + A_2 \cos\left(\frac{4\pi t}{T}\right) + \dots$$

$$+ B_1 \sin\left(\frac{2\pi t}{T}\right) + B_2 \sin\left(\frac{4\pi t}{T}\right) + \dots \tag{5.2.14}$$

where the A's and B's are constants known as "Fourier coefficients" and are given by

$$A_0 = \frac{1}{T} \int_0^T p(t)\,dt \tag{5.2.15}$$

$$A_n = \frac{2}{T} \int_0^T p(t) \cos\left(\frac{2n\pi t}{T}\right) dt \tag{5.2.16}$$

$$B_n = \frac{2}{T} \int_0^T p(t) \sin\left(\frac{2n\pi t}{T}\right) dt \tag{5.2.17}$$

The infinite series in Eqs. 5.2.13,14 are called Fourier series, and this representation of the function $p(t)$ is then referred to as the Fourier series representation of that function.

We note from the definition of A_0 that it represents the *average value* of the periodic function $p(t)$ over one period. We note further that there are actually *two* series in Eq. 5.2.14 and that, except for A_0, the remaining terms in the two series are *paired*, meaning that the terms in A_1 and B_1 have the same argument, namely $2\pi t/T$, and the next two terms again have the same argument, namely $4\pi t/T$, etc. This makes it possible to combine each pair, using standard trigonometric identities, whereby we can write

$$A_1 \cos\left(\frac{2n\pi t}{T}\right) + B_1 \sin\left(\frac{2n\pi t}{T}\right) = M_1 \cos\left(\frac{2n\pi t}{T} - \phi_1\right) \tag{5.2.18}$$

where M_1, ϕ_1 are two new constants, related to A_1, B_1 by

$$A_1 = M_1 \cos\phi_1 \tag{5.2.19}$$
$$B_1 = M_1 \sin\phi_1 \tag{5.2.20}$$

This pairing process can now be repeated for each pair of terms in Eq. 5.2.14, with the result that the two Fourier series can be combined into one, namely

$$p(t) = A_0 + M_1 \cos\left(\frac{2\pi t}{T} - \phi_1\right) + M_2 \cos\left(\frac{4\pi t}{T} - \phi_2\right)$$
$$+ M_3 \cos\left(\frac{6\pi t}{T} - \phi_3\right) + \dots \tag{5.2.21}$$

or in more compact form

$$p(t) = A_0 + \sum_{n=1}^{\infty} M_n \cos\left(\frac{2n\pi t}{T} - \phi_n\right) \tag{5.2.22}$$

where

$$A_n = M_n \cos \phi_n \qquad (5.2.23)$$

$$B_n = M_n \sin \phi_n \qquad (5.2.24)$$

Eq. 5.2.22 provides a more compact Fourier series representation of the function $p(t)$ because it contains only one series instead of two. In this representation each term except the first is a simple cosine wave with M as its amplitude and ϕ as its phase. Because these waves add up to constitute the function $p(t)$, they are referred to as the "harmonics" of this periodic function.

We note in Eq. 5.2.22 that in the first harmonic the cosine function has the same value at $t = 0$ and at $t = T$, therefore this harmonic has a period T, which is the same as the period of the original function $p(t)$. For this reason it is referred to as the "fundamental harmonic". In the second harmonic, by comparison, the cosine function has the same value at $t = 0$ and at $t = T/2$. Therefore, this harmonic has a period $T/2$ which is half the period of $p(t)$. This pattern continues to higher harmonics.

Because of the reciprocal relation between the period and the frequency of a periodic function, the above pattern can be expressed in terms of the frequencies of the harmonics. Thus, if f is the frequency of the periodic function $p(t)$ in cycles per second (Hz), then

$$f = \frac{1}{T} \qquad (5.2.25)$$

and the *angular frequency* ω is given by

$$\omega = 2\pi f \qquad (5.2.26)$$

$$= \frac{2\pi}{T} \quad \text{radians per second} \qquad (5.2.27)$$

Thus the Fourier series representation in Eq. 5.2.22 can now be put in the form

$$p(t) = A_0 + M_1 \cos(\omega t - \phi_1) + M_2 \cos(2\omega t - \phi_2)$$
$$+ M_3 \cos(3\omega t - \phi_3) + ... \qquad (5.2.28)$$

in which it is seen clearly that the frequency of the first harmonic is ω, the same as the frequency of the original function $p(t)$ and is therefore referred to as the "fundamental frequency". The frequency of the second harmonic is 2ω, and of the third is 3ω, etc. These are important properties of the harmonics of a periodic function which we shall see more clearly later as we consider specific functions. In particular, we shall see that in many cases the first 10 harmonics are sufficient for producing a good representation of a given periodic function such as the composite pressure wave produced by the heart. In that case, the fundamental frequency is the beating frequency of the heart which, under resting conditions, is approximately 1 Hz, thus the frequency of the tenth harmonic would be 10 Hz. It is for this reason that frequencies as high as 10 Hz are considered in the dynamics of the coronary circulation and of the cardiovascular system in general.

5.3 Example: Single-Step Waveform

The theory of Fourier analysis described in the previous section is well established and fairly straightforward, but its application to the analysis of specific composite waveforms involves some tedious calculations and some algebraic intricacies which can only be illustrated by considering a specific example. While mathematical software packages now have specific tools under the heading of Fast Fourier Transforms (FFT) that can handle much of the tedious calculations, these tools cannot be used reliably without a basic understanding of the analytical intricacies involved [28, 197]. For this reason, in this and subsequent sections we consider a series of examples that are intended to illustrate the analytical process involved, starting with a very simple example in this section. In each case, the purpose of the analysis is to find the Fourier series representation of the given waveform, that is to find the series of sine and cosine waves of which the given waveform consists.

Consider the simple waveform consisting of a single step shown in Fig. 5.3.1, which has a period $T = 1$ as seen graphically, and which is defined by the function

$$p(t) = 1, \quad 0 \leq t < \frac{1}{2} \tag{5.3.1}$$

$$= 0, \quad \frac{1}{2} \leq t < 1 \tag{5.3.2}$$

Following the theory presented in the previous section, the Fourier series representation of this periodic function is given by (Eq. 5.2.13)

$$p(t) = \sum_{n=0}^{\infty} A_n \cos\left(\frac{2n\pi t}{T}\right) + \sum_{n=1}^{\infty} B_n \sin\left(\frac{2n\pi t}{T}\right) \tag{5.3.3}$$

$$= A_0 + A_1 \cos\left(\frac{2\pi t}{T}\right) + A_2 \cos\left(\frac{4\pi t}{T}\right) + \dots$$

$$+ B_1 \sin\left(\frac{2\pi t}{T}\right) + B_2 \sin\left(\frac{4\pi t}{T}\right) + \dots \tag{5.3.4}$$

and, using Eqs.5.2.15-17 to find the Fourier coefficients, recalling that $T = 1$ in this case, we have

$$A_0 = \frac{1}{T} \int_0^T p(t)dt \tag{5.3.5}$$

$$= \int_0^1 p(t)dt \tag{5.3.6}$$

$$= \int_0^{1/2} 1 \times dt + \int_{1/2}^1 0 \times dt \tag{5.3.7}$$

$$= \frac{1}{2} \tag{5.3.8}$$

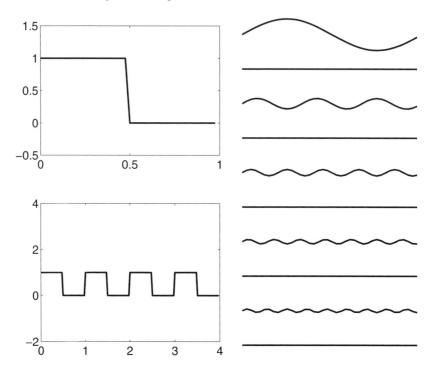

Fig. 5.3.1. A simple waveform consisting of a single step and having a period $T = 1$ as seen in the left two panels. The first ten harmonics of this waveform are shown on the right. The *even* harmonics, namely harmonics $2, 4, 6, 8, 10$ are zero in this case and make no contribution to the Fourier composition of this waveform, as seen on the right. The series is led by the fundamental harmonic which has the same period and hence the same frequency as the original wave, namely the fundamental period and fundamental frequency. The period of the third harmonic is one third of the fundamental period and hence its frequency is three times the fundamental frequency, etc.

Note that by its definition (Eq. 5.3.5), A_0 represents the *average value* of the periodic function over one period, which in Fig. 5.3.1 is seen graphically to be $1/2$, in agreement with the above result.

$$A_n = \frac{2}{T} \int_0^T p(t) \cos\left(\frac{2n\pi t}{T}\right) dt \qquad (5.3.9)$$

$$= 2 \int_0^1 p(t) \cos(2n\pi t) dt \qquad (5.3.10)$$

$$= 2 \int_0^{1/2} 1 \times \cos(2n\pi t) dt + 2 \int_{1/2}^1 0 \times \cos(2n\pi t) dt \qquad (5.3.11)$$

$$= 2 \int_0^{1/2} \cos(2n\pi t) dt \qquad (5.3.12)$$

$$= \frac{\sin(2n\pi t)}{n\pi}\bigg|_0^{1/2} \tag{5.3.13}$$

$$= 0 \quad \text{for all } n \tag{5.3.14}$$

$$B_n = \frac{2}{T}\int_0^T p(t)\sin\left(\frac{2n\pi t}{T}\right)dt \tag{5.3.15}$$

$$= 2\int_0^1 p(t)\sin(2n\pi t)dt \tag{5.3.16}$$

$$= 2\int_0^{1/2} 1 \times \sin(2n\pi t)dt + 2\int_{1/2}^1 0 \times \sin(2n\pi t)dt \tag{5.3.17}$$

$$= 2\int_0^{1/2}\sin(2n\pi t)dt \tag{5.3.18}$$

$$= \frac{-\cos(2n\pi t)}{n\pi}\bigg|_0^{1/2} \tag{5.3.19}$$

$$= \frac{1 - \cos n\pi}{n\pi} \tag{5.3.20}$$

Substituting these values of the Fourier coefficients in Eqs.5.3.4, we obtain the required Fourier series representation of this waveform, namely

$$p(t) = \frac{1}{2} + \sum_{n=1}^{\infty}\left\{\frac{1 - \cos n\pi}{n\pi}\right\}\sin(2n\pi) \tag{5.3.21}$$

$$= \frac{1}{2} + \frac{2}{\pi}\sin(2n\pi) + 0 + \frac{2}{3\pi}\sin(6n\pi) + 0\ldots$$
$$+ \frac{2}{5\pi}\sin(10n\pi) + 0\ldots \tag{5.3.22}$$

$$= \frac{1}{2} + \frac{2}{\pi}\sin(2n\pi) + \frac{2}{3\pi}\sin(6n\pi) + \frac{2}{5\pi}\sin(10n\pi)\ldots \tag{5.3.23}$$

To put the series in the more compact form of Eq. 5.2.22, that is, in terms of the amplitudes M_n and phase angles ϕ_n of the individual harmonics that make up the waveform, we use Eqs.5.2.23,24 to find

$$M_n = \sqrt{A_n^2 + B_n^2} \tag{5.3.24}$$

$$= B_n \quad \text{since } A_n = 0 \text{ for all n} \tag{5.3.25}$$

$$= \left(\frac{1 - \cos n\pi}{n\pi}\right) \tag{5.3.26}$$

and

$$\phi_n = \tan^{-1}\left(\frac{B_n}{A_n}\right) \tag{5.3.27}$$

$$= \pm\frac{\pi}{2} \quad \text{since } A_n = 0 \text{ for all } n \tag{5.3.28}$$

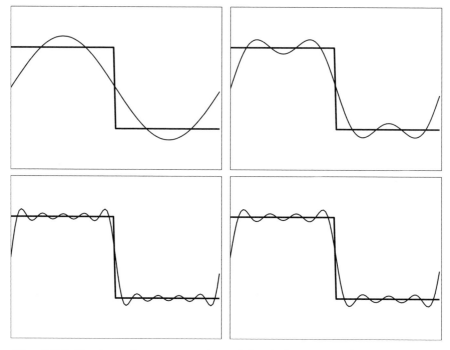

Fig. 5.3.2. Fourier series representations of the single-step waveform, based on the first one, four, seven, and ten harmonics, clockwise from top left corner.

Which of the two values of ϕ is appropriate is determined by satisfying Eq. 5.2.24, namely

$$B_n = M_n \sin \phi_n \qquad (5.3.29)$$

Since $M_n = B_n$ in this case (Eq. 5.3.25), this gives $\sin \phi_n = 1$, and therefore

$$\phi_n = \frac{\pi}{2} \quad \text{for all } n \qquad (5.3.30)$$

Substituting these values of M_n, ϕ_n into Eq. 5.2.22 gives

$$p(t) = A_0 + \sum_{n=0}^{\infty} M_n \cos\left(2n\pi t - \phi_n\right) \qquad (5.3.31)$$

$$= \frac{1}{2} + \sum_{n=0}^{\infty} \left\{ \frac{1 - \cos n\pi}{n\pi} \right\} \cos\left(2n\pi t - \frac{\pi}{2}\right) \qquad (5.3.32)$$

$$= \frac{1}{2} + \sum_{n=0}^{\infty} \left\{ \frac{1 - \cos n\pi}{n\pi} \right\} \sin\left(2n\pi t\right) \qquad (5.3.33)$$

$$= \frac{1}{2} + \frac{2}{\pi} \sin\left(2\pi t\right) + \frac{2}{3\pi} \sin\left(6\pi t\right) + \frac{2}{5\pi} \sin\left(10\pi t\right) \ldots \qquad (5.3.34)$$

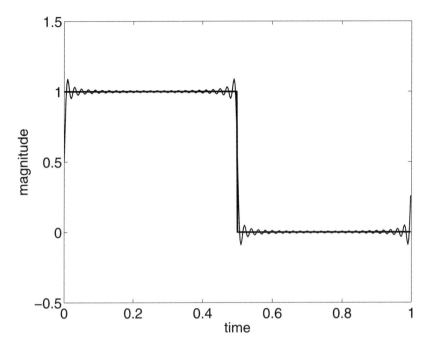

Fig. 5.3.3. Fourier series representation of the single-step waveform based on the first 50 harmonics.

which is identical with the result in Eq. 5.3.23.

Thus, in the present example, because of the very simple form of the wave, the two different forms of Fourier series representation in Eqs. 5.2.13,22 are identical. More precisely, the Fourier series representation of this simple waveform consists of only one series (not two as in Eq. 5.2.13), hence the compact and the non-compact forms of the Fourier represenation are the same. Furthermore, we shall find that the determination of the phase angle ϕ is in general more troublesome than it is in the present simple case. The reason for this is that the range of the inverse tangent function used in Eq. 5.3.27 is limited to the interval $-\pi/2$ to $+\pi/2$ and therefore does not yield all possible angles.

If the harmonics of this waveform are denoted by $p_1(t), p_2(t), p_3(t) \ldots$, then the result in Eq. 5.3.34 can be written as

$$p(t) = \frac{1}{2} + p_1(t) + p_2(t) + p_3(t) \ldots \tag{5.3.35}$$

where the individual harmonics are given by

$$p_1(t) = \frac{2}{\pi} \sin(2\pi t) \tag{5.3.36}$$

$$p_2(t) = 0 \tag{5.3.37}$$

$$p_3(t) = \frac{2}{3\pi} \sin(6\pi t) \qquad (5.3.38)$$

$$p_4(t) = 0 \qquad (5.3.39)$$

$$\vdots$$

It is seen that the first harmonic has the same period and hence the same frequency as the original wave, namely the fundamental period and fundamental frequency. The second and other even-numbered harmonics are zero in this case. The period of the third harmonic is one third of the fundamental period and hence its frequency is three times the fundamental frequency, etc. The first ten harmonics are shown graphically in Fig. 5.3.1.

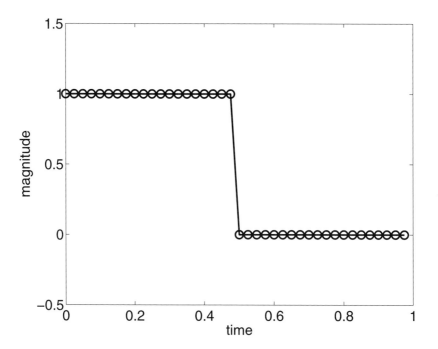

Fig. 5.3.4. Fourier series representation (circles) of the single-step waveform (solid line) based on a Fast Fourier Transform (FFT) program. Such programs, which are available with most mathematical software packages, use an appropriate number of harmonics to produce highly accurate Fourier series representations.

One of the most important pillars of the theory of Fourier analysis is that the *amplitudes* of successive harmonics become successively smaller and hence they make successively smaller contribution to the Fourier representation of the periodic function in hand. This is highly important for practical purposes because the infinite series representing the periodic function can then be truncated at some point without committing large error. This is illustrated graphically in Fig. 5.3.2, where different Fourier series representations are shown, based on the first one, four, seven, and ten harmonics. A Fourier series representation based on the first 50 harmonics is shown in Fig. 5.3.3 where these properties can be observed. We shall see later that a larger number of harmonics does not always produce a more accurate Fourier series representation. Specifically, when the description of a given periodic function is available in only *numerical* form, as in Table 5.1.1 for the cardiac wave, new complications arise which make the optimum number of harmonics dependent on the number of data points available in the numerical description of the waveform.

Computer programs based on the Fast Fourier Transform (FFT) are optimized to use a number of harmonics appropriate for the number of data points available [28, 197], to produce a highly accurate Fourier series representation of a given periodic function, as illustrated in Fig. 5.3.4.

5.4 Example: Piecewise Waveform

Consider next a composite waveform consisting of several steps, as shown in Fig. 5.4.1, and defined by the function

$$p(t) = 4t, \quad 0 \le t < \frac{1}{4} \tag{5.4.1}$$

$$= 1, \quad \frac{1}{4} \le t < \frac{1}{2} \tag{5.4.2}$$

$$-\frac{1}{2}, \quad \frac{1}{2} \le t < \frac{3}{4} \tag{5.4.3}$$

$$= 0, \quad \frac{3}{4} \le t < 1 \tag{5.4.4}$$

Following the theory presented in the Section 5.2, the Fourier series representation of this periodic function is given by (Eq. 5.2.13)

$$p(t) = \sum_{n=0}^{\infty} A_n \cos\left(\frac{2n\pi t}{T}\right) + \sum_{n=1}^{\infty} B_n \sin\left(\frac{2n\pi t}{T}\right) \tag{5.4.5}$$

$$= A_0 + A_1 \cos\left(\frac{2\pi t}{T}\right) + A_2 \cos\left(\frac{4\pi t}{T}\right) + \ldots$$

$$+ B_1 \sin\left(\frac{2\pi t}{T}\right) + B_2 \sin\left(\frac{4\pi t}{T}\right) + \ldots \tag{5.4.6}$$

and, using Eqs. 5.2.15–17 to find the Fourier coefficients, recalling that $T = 1$ in this case, we have

$$A_0 = \frac{1}{T} \int_0^T p(t)dt \tag{5.4.7}$$

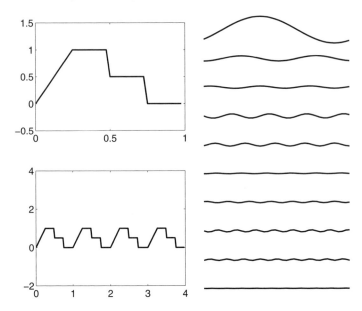

Fig. 5.4.1. A composite "piecewise" waveform consisting of several steps and having a period $T = 1$ as seen in the left two panels. The first ten harmonics of this waveform are shown on the right. The series is led by the fundamental harmonic which has the same period and hence the same frequency as the original wave, namely the fundamental period and fundamental frequency. The periods of the second and third harmonics are one half and one third of the fundamental period, respectively, and hence their frequencies are two and three times the fundamental frequency, etc.

$$= \int_0^1 p(t)dt \tag{5.4.8}$$

$$= \int_0^{1/4} 4t \times dt + \int_{1/4}^{1/2} 1 \times dt + \int_{1/2}^{3/4} \frac{1}{2} \times dt + \int_{3/4}^1 0 \times dt \tag{5.4.9}$$

$$= 2t^2 \Big|_0^{1/4} + t \Big|_{1/4}^{1/2} + \frac{1}{2}t \Big|_{1/2}^{3/4} + 0 \tag{5.4.10}$$

$$= \frac{1}{8} + \frac{1}{4} + \frac{1}{8} \tag{5.4.11}$$

$$= \frac{1}{2} \tag{5.4.12}$$

Again, we note that by its definition (Eq. 5.4.7), the constant A_0 represents the *average value* of the periodic function over one period. The result is seen to be correct from the graphical representation of the waveform in Fig. 5.4.1. For the other Fourier coefficients we have

$$A_n = \frac{2}{T} \int_0^T p(t) \cos\left(\frac{2n\pi t}{T}\right) dt \tag{5.4.13}$$

$$= 2 \int_0^1 p(t) \cos(2n\pi t) dt \tag{5.4.14}$$

$$= 2 \int_0^{1/4} 4t \cos(2n\pi t) dt + 2 \int_{1/4}^{1/2} \cos(2n\pi t) dt$$

$$+ 2 \int_{1/2}^{3/4} \frac{1}{2} \cos(2n\pi t) dt + \int_{3/4}^1 0 \times dt \tag{5.4.15}$$

$$= 4 \left\{ \frac{\cos(2n\pi t)}{2(n\pi)^2} + \frac{t \sin(2n\pi t)}{n\pi} \right\} \Big|_0^{1/4}$$

$$+ \frac{\sin(2n\pi t)}{n\pi} \Big|_{1/4}^{1/2} + \frac{\sin(2n\pi t)}{2n\pi} \Big|_{1/2}^{3/4} + 0 \tag{5.4.16}$$

$$= \frac{2}{(n\pi)^2} \left[\cos\left(\frac{n\pi}{2}\right) - 1 \right] + \frac{1}{2n\pi} \sin\left(\frac{3n\pi}{2}\right) \tag{5.4.17}$$

where the first integral in Eq. 5.4.15 was evaluated using integration by parts, with the standard result [80, 186, 25]

$$\int x \cos kx \, dx = \frac{\cos kx}{k^2} + \frac{x \sin kx}{k} \tag{5.4.18}$$

$$B_n = \frac{2}{T} \int_0^T p(t) \sin\left(\frac{2n\pi t}{T}\right) dt \tag{5.4.19}$$

$$= 2 \int_0^1 p(t) \sin(2n\pi t) dt \tag{5.4.20}$$

$$= 2 \int_0^{1/4} 4t \sin(2n\pi t) dt + 2 \int_{1/4}^{1/2} \sin(2n\pi t) dt$$

$$+ 2 \int_{1/2}^{3/4} \frac{1}{2} \sin(2n\pi t) dt + \int_{3/4}^1 0 \times dt \tag{5.4.21}$$

$$= 4 \left\{ \frac{\sin(2n\pi t)}{2(n\pi)^2} - \frac{t \cos(2n\pi t)}{n\pi} \right\} \Big|_0^{1/4}$$

$$- \frac{\cos(2n\pi t)}{n\pi} \Big|_{1/4}^{1/2} - \frac{\cos(2n\pi t)}{2n\pi} \Big|_{1/2}^{3/4} + 0 \tag{5.4.22}$$

$$= \frac{2}{(n\pi)^2} \sin\left(\frac{n\pi}{2}\right) - \frac{1}{2n\pi} \cos(n\pi) - \frac{1}{2n\pi} \cos\left(\frac{3n\pi}{2}\right) \tag{5.4.23}$$

Here, again, the first integral in Eq. 5.4.21 was evaluated using integration by parts, with the standard result [80, 186, 25]

$$\int x \sin kx \, dx = \frac{\sin kx}{k^2} - \frac{x \cos kx}{k} \tag{5.4.24}$$

Substitution of these expressions for the Fourier coefficients in Eqs.5.4.6 makes the resulting expression for the Fourier series rather cumbersome. Instead, numerical values of A_n, B_n, M_n, ϕ_n can be simply tabulated for the required number of harmonics, as shown in Table 5.4.1.

Table 5.4.1. Numerical values of Fourier coefficients for the first ten harmonics of the piecewise waveform shown in Fig. 5.4.1.

n	A_n	B_n	M_n	ϕ_n (deg)
1	-0.36180	0.36180	0.51166	135
2	-0.10132	0.00000	0.10132	180
3	0.03054	0.03054	0.04318	45
4	0.00000	-0.07958	0.07958	-90
5	-0.03994	0.03994	0.05648	135
6	-0.01126	0.00000	0.01126	180
7	0.01860	0.01860	0.02631	45
8	0.00000	-0.03979	0.03979	-90
9	-0.02019	0.02019	0.02855	135
10	-0.004053	0.00000	0.00405	180

Values of M_n in Table 5.4.1 are determined as before, namely from (Eq. 5.3.24)

$$M_n = \sqrt{A_n^2 + B_n^2} \tag{5.4.25}$$

However, as mentioned in the previous section, the phase angles ϕ_n must satisfy *both* conditions in Eqs.5.2.23,24, namely

$$A_n = M_n \cos \phi_n \tag{5.4.26}$$
$$B_n = M_n \sin \phi_n \tag{5.4.27}$$

These two conditions cannot be replaced by the single condition

$$\phi_n = \tan^{-1}\left(\frac{B_n}{A_n}\right) \tag{5.4.28}$$

because the range of values of the inverse tangent function is limited to the interval $-\pi/2$ to $\pi/2$. For example, using the values of A_1, B_1 from the table, Eq. 5.4.28 gives

$$\phi_1 = \tan^{-1}\left(\frac{B_1}{A_1}\right) \tag{5.4.29}$$

$$= \tan^{-1}\left(\frac{0.3618}{-0.3618}\right) \tag{5.4.30}$$

$$= \tan^{-1}(-1) \tag{5.4.31}$$

$$= -\frac{\pi}{4} \tag{5.4.32}$$

This value of ϕ_1 is incorrect because it does not satisfy Eqs.5.4.26,27. Substituting $\phi_1 = -\pi/4$ in these equations gives

$$A_1 = M_1 \cos(-\pi/4) \tag{5.4.33}$$
$$= 0.51166 \times 0.7071 \tag{5.4.34}$$
$$= 0.3618 \tag{5.4.35}$$

$$B_1 = M_1 \sin(-\pi/4) \tag{5.4.36}$$
$$= 0.51166 \times (-0.7071) \tag{5.4.37}$$
$$= -0.3618 \tag{5.4.38}$$

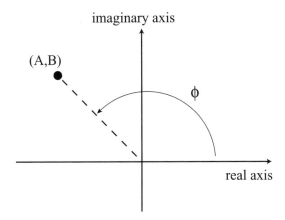

Fig. 5.4.2. The phase angle ϕ of a harmonic with coefficients A, B is correctly obtained as the argument of the complex number $z = A + iB$. This angle, $\phi = \arg(z)$, is measured in an anticlockwise direction from the real axis, as shown, and has the range of values $-\pi$ to π. The example shown here is that of the first harmonic of the piecewise waveform for which the values of the coefficients (from Table 5.4.1) are $A_1 = -0.3618$ and $B_1 = 0.3618$, which are shown in the complex plane above as the coordinates of the complex number z, and which give $\phi = \arg(-0.3618 + i \times 0.3618) = 3\pi/4 = 135°$ as given in Table 5.4.1. The inverse tangent function in this case would give an incorrect value, namely $\phi = \tan^{-1}(B/A) = \tan^{-1}(-1) = \pi/4 = 45°$.

These values of A_1, B_1 are incorrect, the actual values are $A_1 = -0.3618$, $B_1 = 0.3618$ as indicated in the table. The correct value of ϕ_1, that is, a value of ϕ_1 which satisfies both of Eqs.5.4.26,27, is actually $3\pi/4$ or $135°$ as indicated in the table. This value is obtained by writing A_1, B_1 as the real and imaginary parts of a complex number

$$z_1 = A_1 + iB_1 \tag{5.4.39}$$

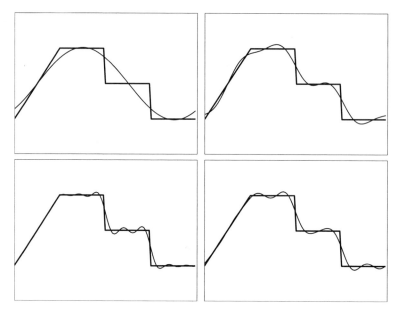

Fig. 5.4.3. Fourier series representations of the piecewise waveform in Fig. 5.4.1, based on the first one, four, seven, and ten harmonics, clockwise from top left corner.

then the correct value of ϕ_1 is obtained as the argument ("arg") of z_1, that is

$$\phi_1 = \arg(z_1) \tag{5.4.40}$$

where the function "arg" is the angle of a complex number in the complex plane or Argand diagram, measured in an anticlockwise direction from the real axis and having the range of values $-\pi$ to π, as illustrated in Fig. 5.4.2.

Numerical values of the coefficients A_1, B_1 can now be extracted from Table 5.4.1 to construct the Fourier series representation of the piecewise waveform in its full form, as in Eq. 5.2.14, giving

$$p(t) = \sum_{n=0}^{\infty} A_n \cos\left(\frac{2n\pi t}{T}\right) + \sum_{n=1}^{\infty} B_n \sin\left(\frac{2n\pi t}{T}\right) \tag{5.4.41}$$

$$= A_0 + A_1 \cos\left(\frac{2\pi t}{T}\right) + A_2 \cos\left(\frac{4\pi t}{T}\right) + ...$$

$$+ B_1 \sin\left(\frac{2\pi t}{T}\right) + B_2 \sin\left(\frac{4\pi t}{T}\right) + ... \tag{5.4.42}$$

$$= 0.5 - 0.3618 \times \cos(2\pi t) - 0.10132 \times \cos(4\pi t)$$

$$+ 0.030536 \times \cos(6\pi t) + 0.3618 \times \sin(2\pi t)$$

$$+ 0.030536 \times \sin(6\pi t) + ... \tag{5.4.43}$$

Or, numerical values of M_n, ϕ_n can be used from Table 5.4.1 to put the series in its compact form, as in Eqs. 5.2.21, 22

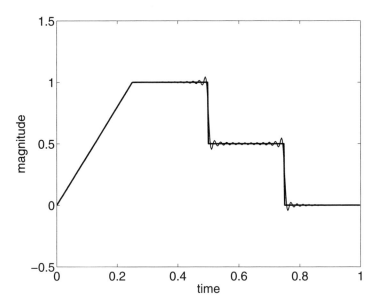

Fig. 5.4.4. Fourier series representation of the piecewise waveform in Fig. 5.4.1, based on the first 50 harmonics.

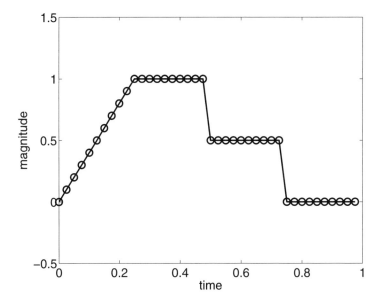

Fig. 5.4.5. Fourier series representation (circles) of the piecewise waveform in Fig. 5.4.1 (solid line) based on a Fast Fourier Transform (FFT) algorithm.

$$p(t) = A_0 + \sum_{n=0}^{\infty} M_n \cos\left(\frac{2n\pi t}{T} - \phi_n\right) \tag{5.4.44}$$

$$= A_0 + M_1 \cos\left(\frac{2\pi t}{T} - \phi_1\right) + M_2 \cos\left(\frac{4\pi t}{T} - \phi_2\right)$$

$$+ M_3 \cos\left(\frac{6\pi t}{T} - \phi_3\right) + \dots \tag{5.4.45}$$

$$= 0.5 + 0.51166 \times \cos\left(2\pi t - 135 \times \pi/180\right)$$
$$+ 0.10132 \times \cos\left(4\pi t - 180 \times \pi/180\right)$$
$$+ 0.043184 \times \cos\left(6\pi t - 45 \times \pi/180\right) + \dots \tag{5.4.46}$$

As in the previous example, if the harmonics of this waveform are denoted by $p_1(t), p_2(t), p_3(t)$ etc., then the individual harmonics are given by

$$p_1(t) = 0.51166 \times \cos\left(2\pi t - 135 \times \pi/180\right) \tag{5.4.47}$$
$$p_2(t) = 0.10132 \times \cos\left(4\pi t - 180 \times \pi/180\right) \tag{5.4.48}$$
$$p_3(t) = 0.043184 \times \cos\left(6\pi t - 45 \times \pi/180\right) \tag{5.4.49}$$

$$\vdots$$

$$p_{10}(t) = 0.0040528 \times \cos\left(20\pi t - 180 \times \pi/180\right) \tag{5.4.50}$$

Again, it is seen that the first harmonic has the same period and hence the same frequency as the original wave, namely the fundamental period and fundamental frequency. The period of the second harmonic is one half of the fundamental period and hence its frequency is twice the fundamental frequency, etc. The first ten harmonics are shown graphically in Fig. 5.4.1.

Fig. 5.4.3 shows the accuracy of this Fourier representation of the piecewise waveform when only the first one, four, seven, and ten harmonics are used. A Fourier representation with the first 50 harmonics is shown in Fig. 5.4.4, and a representation based on a computer based Fast Fourier Transform is shown in Fig. 5.4.5.

5.5 Numerical Formulation

The waveforms considered in the previous two sections were rather artificially constructed in order to illustrate the basic concepts of Fourier analysis and the basic steps involved in its application to a specific waveform. In the context of the coronary circulation, however, the specific waveforms of interest are the pressure and flow waveforms generated by the pumping action of the left ventricle, as in the example shown in Fig. 5.1.2. One important feature of this waveform which is not shared by the examples of the previous two sections is that it cannot be presented in *analytical* form, as in Eqs.5.3.1,2 for the single-step waveform, or in Eqs.5.4.1-4 for the piecewise waveform.

As stated in the introduction to this chapter, a cardiac pressure waveform of the type shown in Fig. 5.1.2 is generally available only in *numerical* form, that is as a set of points, tabulated as in Table 5.1.1 or presented graphically as in Fig. 5.1.2. This is the most natural way in which the waveform would present itself in practice where the set of points would come from pressure or flow measurements at some accessible point within the coronary vasculature, at small time intervals during the oscillatory cycle as shown in Table 5.1.1.

The aim of the present section is to show how such a set of points would be used in the process of Fourier analysis to produce the Fourier series representation of the waveform. Once this representation has been achieved, the waveform becomes like any other waveform, expressed in terms of a series of sine and cosine functions, or in terms of its harmonics as in the examples of the previous sections. Indeed, the data in Table 5.1.1 may be regarded as a *periodic function* like any other we have considered so far, the only difference here is that the function is presented in numerical form rather than analytically. Each pair of values (p, t) in the table represents one point in Fig. 5.1.2, and the entire set of values in the table produce the waveform shown in the figure.

Let the number of points available be denoted by N, which is not to be confused with n which we shall continue to use for the number of harmonics. The Fourier analysis process is considerably easier, of course, when the points are spaced at *regular* intervals of time within the oscillatory cycle, and we shall proceed on that assumption. In fact, if the original set points are not equally spaced in time, it would be best first to place them on a "best-fit" curve and then extract a new set of points from that curve at regular time intervals. It is also easier, particularly when using Fast Fourier Transform (FFT) programs, if N is an *even* number.

If the period of the waveform is denoted by T, and the time interval between successive data points is denoted by Δt, then

$$\Delta t = \frac{T}{N} \tag{5.5.1}$$

In Table 5.1.1 the period has been normalized to $T = 1.0$ and the number of points $N = 40$, therefore $\Delta t = 1/40 = 0.025$ as noted from successive points in the table.

If the time at the beginning of the oscillatory cycle is set at $t = 0$, and if this and subsequent points in time are denoted by t_0, t_1, t_3 etc., then these points are given by

$$t_0 = 0$$
$$t_1 = 1 \times \Delta t$$
$$t_2 = 2 \times \Delta t$$
$$t_3 = 3 \times \Delta t$$
$$\vdots$$

$$t_{N-1} = (N - 1) \times \Delta t \qquad (5.5.2)$$

Note that there are a total of N points in time within one oscillatory cycle. If the corresponding values of $p(t)$ are denoted similarly by p_0, p_1, p_2 etc., then

$$p_0 = p(t_0)$$
$$p_1 = p(t_1)$$
$$p_2 = p(t_2)$$
$$p_3 = p(t_3)$$
$$\vdots$$
$$p_{N-1} = p(t_{N-1}) \qquad (5.5.3)$$

The general form of the Fourier series representation of the cardiac waveform in Fig. 5.1.2 is the same as for other waveforms, namely

$$p(t) = \sum_{n=0}^{\infty} A_n \cos\left(\frac{2n\pi t}{T}\right) + \sum_{n=1}^{\infty} B_n \sin\left(\frac{2n\pi t}{T}\right) \qquad (5.5.4)$$

$$= A_0 + A_1 \cos\left(\frac{2\pi t}{T}\right) + A_2 \cos\left(\frac{4\pi t}{T}\right) + \dots$$

$$+ B_1 \sin\left(\frac{2\pi t}{T}\right) + B_2 \sin\left(\frac{4\pi t}{T}\right) + \dots \qquad (5.5.5)$$

but the Fourier coefficients A_n, B_n in the present case cannot be evaluated by means of integrals as before, because the periodic function $p(t)$ is not available in *analytical* form. But the function is available in *numerical* form, as in Table 5.1.1, therefore the required integrals can be formulated and evaluated numerically in a fairly straightforward manner as shown below.

If each of the N points describing the periodic function $p(t)$ is associated with one time interval Δt, then the N points together cover the entire period T. In the simplest numerical formulation, the value of the function $p(t)$ at t_0, namely p_0, is taken to remain constant over the small time interval Δt associated with t_0, then the value of $p(t)$ at t_1, namely p_1, is taken to remain constant over the next time interval, etc., with the result that the periodic function $p(t)$ is presented graphically as shown in Fig. 5.5.1. This graphical presentation provides the basis for the numerical formulation and evaluation of the Fourier coefficients A_n, B_n.

Briefly, each of the integrals required in the evaluation of the coefficients is reformulated as a *sum*, using standard methods. Thus, for A_0 we have (Eq. 5.2.15)

$$A_0 = \frac{1}{T} \int_0^T p(t)dt \qquad (5.5.6)$$

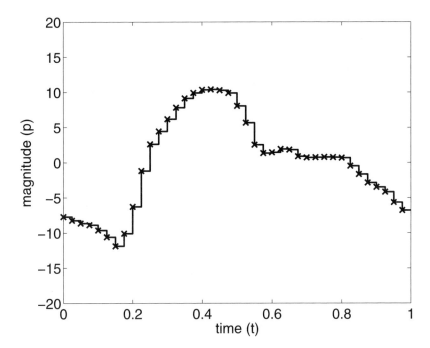

Fig. 5.5.1. Graphical presentation of the periodic function $p(t)$, when the description of the function is available only *numerically*. The data points shown are based on the data in Table 5.1.1 for the cardiac waveform in Fig. 5.1.1. In a numerical formulation of Fourier analysis, each data point is associated with the small time interval Δt between it and the next data point, and over each such Δt the value of p is taken to remain constant as shown in the figure. This allows the numerical formulation and evaluation of the Fourier coefficients A_n, B_n as described in the text.

The integral on the right in fact represents the area under the curve representing the function $p(t)$ over one period. The graphical presentation in Fig. 5.5.1 shows that to a good approximation this area is equal to the sum of the areas of the N long thin rectangles of width Δt rising from the t axis to the curve. This makes it possible to write

$$A_0 = \frac{1}{T}\{p_0\Delta t + p_1\Delta t + p_2\Delta t \ldots p_{N-1}\Delta t\} \tag{5.5.7}$$

$$= \frac{1}{N}\{p_0 + p_1 + p_2 + \ldots + p_{N-1}\}\,\Delta t \tag{5.5.8}$$

$$= \frac{\Delta t}{N}\sum_{k=0}^{N-1} p_k \tag{5.5.9}$$

having used Eq. 5.5.1 in the process.

Numerical expressions for A_n and B_n are obtained in the same way, although the integrals in this case involve the product of $p(t)$ and a sine or cosine function and therefore do not represent simply the area under the $p(t)$ curve. Nevertheless, using the integral expressions for these coefficients from Eqs.5.2.16,17 and converting the integrals involved into sums as for A_0, we find

$$A_n = \frac{2}{T} \int_0^T p(t) \cos\left(\frac{2n\pi t}{T}\right) dt \tag{5.5.10}$$

$$= \frac{2}{T} \left\{ p_0 \cos\left(\frac{2n\pi t_0}{T}\right) \Delta t + p_1 \cos\left(\frac{2n\pi t_1}{T}\right) \Delta t + \ldots \right.$$
$$\left. + p_{N-1} \cos\left(\frac{2n\pi t_{N-1}}{T}\right) \Delta t \right\} \tag{5.5.11}$$

$$= \frac{2}{N} \left\{ p_0 \cos\left(\frac{2n\pi t_0}{T}\right) + p_1 \cos\left(\frac{2n\pi t_1}{T}\right) + \ldots \right.$$
$$\left. + p_{N-1} \cos\left(\frac{2n\pi t_{N-1}}{T}\right) \right\} \Delta t \tag{5.5.12}$$

$$= \frac{2\Delta t}{N} \sum_{k=0}^{N-1} p_k \cos\left(\frac{2n\pi t_k}{T}\right) \tag{5.5.13}$$

$$B_n = \frac{2}{T} \int_0^T p(t) \sin\left(\frac{2n\pi t}{T}\right) dt \tag{5.5.14}$$

$$= \frac{2}{T} \left\{ p_0 \sin\left(\frac{2n\pi t_0}{T}\right) \Delta t + p_1 \sin\left(\frac{2n\pi t_1}{T}\right) \Delta t + \ldots \right.$$
$$\left. + p_{N-1} \sin\left(\frac{2n\pi t_{N-1}}{T}\right) \Delta t \right\} \tag{5.5.15}$$

$$= \frac{2}{N} \left\{ p_0 \sin\left(\frac{2n\pi t_0}{T}\right) + p_1 \sin\left(\frac{2n\pi t_1}{T}\right) + \ldots \right.$$
$$\left. + p_{N-1} \sin\left(\frac{2n\pi t_{N-1}}{T}\right) \right\} \Delta t \tag{5.5.16}$$

$$= \frac{2\Delta t}{N} \sum_{k=0}^{N-1} p_k \sin\left(\frac{2n\pi t_k}{T}\right) \tag{5.5.17}$$

These expressions are valid generally for any periodic function $p(t)$ for which a numerical description is available in terms of N data points as in Table 5.1.1. The expressions are used specifically for that case in the next section.

5.6 Example: Cardiac Waveform

Using the numerical formulation of the previous section and the numerical data in Table 5.1.1 for the cardiac waveform shown in Fig. 5.1.1, we are now in a position to apply Fourier analysis to this wave and to find its harmonics. Essentially, the analysis is the same as for other waves except for the evaluation of the Fourier coefficients A_n, B_n, which in this case must be done numerically. For A_0, using Eq. 5.5.8 and values from Table 5.1.1, we find

$$A_0 = \frac{\Delta t}{N} \{p_0 + p_1 + p_2 + \ldots + p_{N-1}\} \tag{5.6.1}$$

$$= \frac{0.025}{40} \{-7.7183 - 8.2383 - 8.6444 + \ldots - 6.8024\} \tag{5.6.2}$$

$$= 0.025 \times 0.00000475 \tag{5.6.3}$$

$$\approx 0 \tag{5.6.4}$$

We recall from previous examples that A_0 represents the average value of the periodic function $p(t)$ over one complete period. Thus, the fact that this average value is zero in this case indicates that the waveform in Fig. 5.1.2 represents only the *oscillatory* part of a cardiac wave, any constant part has been removed. It is always possible, and in fact desirable, to remove any constant average from a waveform before applying Fourier analysis to it because the analysis is concerned with only the oscillatory part. This principle is illustrated graphically in Fig. 5.6.1.

For the other Fourier coefficients, using Eqs. 5.5.12,16 and the data in Table 5.1.1, and noting that the number of data points $N = 40$ and the period $T = 1.0$, we find

$$A_n = \frac{2}{N} \left\{ p_0 \cos\left(\frac{2n\pi t_0}{T}\right) + p_1 \cos\left(\frac{2n\pi t_1}{T}\right) + \ldots \right.$$
$$\left. + p_{N-1} \cos\left(\frac{2n\pi t_{N-1}}{T}\right) \right\} \Delta t \tag{5.6.5}$$

$$= \frac{1}{20} \{-7.7183 \times \cos(2n\pi \times 0)$$
$$-8.2383 \times \cos(2n\pi \times 0.025) + \ldots$$
$$-6.8024 \times \cos(2n\pi \times 0.975)\} \times 0.025 \tag{5.6.6}$$

$$B_n = \frac{2}{N} \left\{ p_0 \sin\left(\frac{2n\pi t_0}{T}\right) + p_1 \sin\left(\frac{2n\pi t_1}{T}\right) + \ldots \right.$$
$$\left. + p_{N-1} \sin\left(\frac{2n\pi t_{N-1}}{T}\right) \right\} \Delta t \tag{5.6.7}$$

$$= \frac{1}{20} \{-7.7183 \times \sin(2n\pi \times 0)$$
$$-8.2383 \times \sin(2n\pi \times 0.025) + \ldots$$
$$-6.8024 \times \sin(2n\pi \times 0.975)\} \times 0.025 \tag{5.6.8}$$

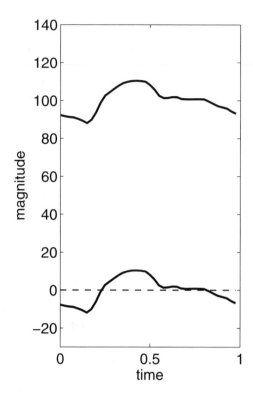

Fig. 5.6.1. A cardiac wave such as the solid curve at the top can always be separated into a purely constant part and a purely oscillatory part. The purely oscillatory part is shown at the bottom and it has the property that its average over one period is zero. In Fourier analysis the constant part of the wave is represented by A_0, thus the result $A_0 = 0$ in Eq. 5.6.3 indicates that the data on which the result is based represents only the oscillatory part of the waveform, any constant part has been removed.

We recall that values of n in these expressions refer to different harmonics. Thus, evaluating these for the first 10 harmonics ($n = 1, 2, \ldots, 10$), the results are shown numerically in Table 5.6.1.

The Fourier representation of this waveform, using the compact form in Eqs.5.2.21,22, is given by

$$p(t) = A_0 + \sum_{n=0}^{\infty} M_n \cos\left(\frac{2n\pi t}{T} - \phi_n\right) \tag{5.6.9}$$

$$= A_0 + M_1 \cos\left(\frac{2\pi t}{T} - \phi_1\right) + M_2 \cos\left(\frac{4\pi t}{T} - \phi_2\right)$$

$$+ M_3 \cos\left(\frac{6\pi t}{T} - \phi_3\right) + \ldots \tag{5.6.10}$$

Table 5.6.1. Values of the Fourier coefficients for the cardiac wave shown in Fig. 5.6.1, using Eqs. 5.6.5,7 with $n = 1, 2 \ldots 10$.

n	A_n	B_n	M_n	ϕ_n (deg)
1	-7.98840	0.15707	7.99000	178.8736
2	-0.42846	-4.41890	4.43960	-95.5381
3	0.88370	0.46246	0.99740	27.6238
4	0.68508	0.28468	0.74187	22.5649
5	-0.35969	0.87460	0.94567	112.3553
6	-0.30961	-0.28316	0.41956	-137.5548
7	-0.53143	-0.20924	0.57114	-158.5089
8	0.26366	-0.15171	0.30419	-29.9153
9	0.02955	0.06432	0.07078	65.3256
10	0.04842	0.16564	0.17258	73.7050

Thus, using values of M_n and ϕ_n from Table 5.6.1, the first 10 harmonics of this waveform are given by, recalling that $T = 1.0$,

$$p_1(t) = M_1 \cos\left(\frac{2\pi t}{T} - \phi_1\right) \tag{5.6.11}$$

$$= 7.99 \times \cos\left(2\pi t - \frac{178.8736 \times \pi}{180}\right) \tag{5.6.12}$$

$$p_2(t) = M_2 \cos\left(\frac{4\pi t}{T} - \phi_2\right) \tag{5.6.13}$$

$$= 4.4396 \times \cos\left(4\pi t - \frac{-95.5381 \times \pi}{180}\right) \tag{5.6.14}$$

$$\vdots$$

$$p_{10}(t) = M_{10} \cos\left(\frac{20\pi t}{T} - \phi_{10}\right) \tag{5.6.15}$$

$$= 0.17258 \times \cos\left(20\pi t - \frac{73.705 \times \pi}{180}\right) \tag{5.6.16}$$

Using these results, the cardiac waveform and its first 10 harmonics are shown in Fig. 5.6.2.

Fig. 5.6.3 shows the accuracy of this Fourier representation of the cardiac waveform when only the first one, four, seven, and ten harmonics are used. It is seen that the representation is fairly accurate with only the first seven harmonics. By contrast, Fourier representations of the single-step and the piecewise waveforms considered in the previous sections were less accurate with as many as fifty harmonics. The reasons for this can be seen clearly in Figs. 5.3.3 and 5.4.4. The presence of *step changes* in those cases, and the behaviour of the Fourier curves in the vicinity of these steps, shows that Fourier

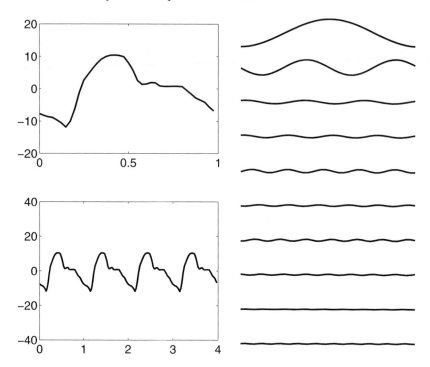

Fig. 5.6.2. The cardiac waveform of Fig. 5.1.1 with its first ten harmonics, using the results in Eqs.5.6.10-15.

series have difficulty replicating step changes. The cardiac waveform does not contain such changes, thus higher accuracy is achieved with a relatively small number of harmonics.

In fact, as mentioned earlier, when the waveform to be represented by a Fourier series is available only in numerical form, the number of harmonics that produces the most accurate representation becomes dependent on the number of data points available in the numerical description of the waveform. Broadly speaking, the theory of Fourier analysis has shown that if the number of data points available is N, then the number of harmonics that produces the most accurate representation is $N/2$ [28, 197]. A smaller *or a larger* number of harmonics produce a less accurate representation, for different reasons. This is an oversimplification of the underlying theory, but it provides a useful guide, indeed a necessary guide, for practical applications of Fourier analysis to specific waveforms.

For the cardiac waveform being considered in this section, the number of data points in the numerical description of the wave (Table 5.1.1) is 40, thus the number of harmonics required to produce the most accurate representation is 20. However, it turns out that this maximum accuracy is reached along a fairly shallow (not sharp) peak, thus the optimum number of harmonics

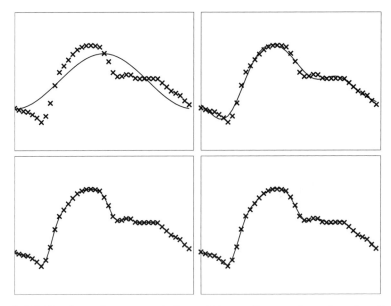

Fig. 5.6.3. Fourier series representations of the cardiac waveform, based on the first one, four, seven, and ten harmonics, clockwise from top left corner.

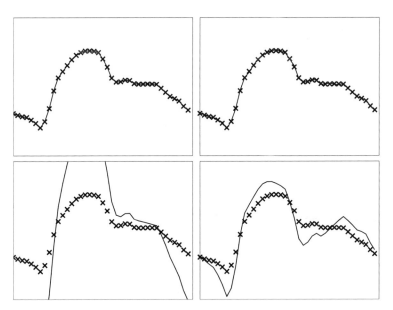

Fig. 5.6.4. Fourier series representation of the cardiac waveform, based on the first 20, 30, 38, and 45 harmonics, clockwise from top left corner.

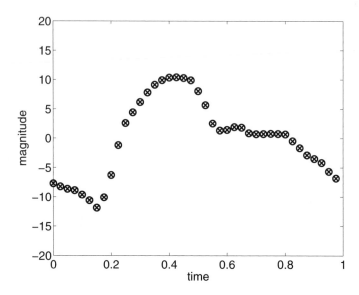

Fig. 5.6.5. Fourier series representation of the cardiac waveform based on a Fast Fourier Transform (FFT) program, where crosses represent the original data points describing the waveform and circles represent the corresponding data points produce by the Fourier series representation. The program precisely replicates the original data points.

need not be treated precisely. In other words, 19 or 21 harmonics will not produce significant differences. In fact, as seen in Fig. 5.6.3, both seven and ten harmonics produce fairly accurate representations, and the difference between them is barely detectable. A Fourier representation with precisely 20 harmonics is shown in Fig. 5.6.4, compared with representations using 30, 38, and 45 harmonics. It is seen that only in the latter two cases the representation breaks down. A representation based on a Fast Fourier Transform (FFT) program is shown in Fig. 5.6.5, where it is seen that the program can produce precisely the data points given in the numerical description of the waveform.

5.7 Summary

The driving pressure in the dynamics of the coronary circulation has a *composite waveform* such as the cardiac pressure wave, while much of the analysis described so far deals with only simple sine or cosine waves. The theory of *Fourier analysis* shows that a composite wave can in fact be expressed as the sum of a series of sine and cosine waves which are referred to as its "harmonics". This makes it possible to extend lumped model analysis, which is based on only simple sine and cosine waves, to now include composite pressure and flow waveforms.

A composite wave is a periodic function of time with period T and angular frequency ω. Its harmonics can be expressed as cosine functions, the first has a period T and angular frequency ω and is referred to as the "fundamental harmonic", the second has a period $T/2$ and angular frequency 2ω, and so on. In many cases the first 10 harmonics are sufficient for producing a good representation of a given periodic function such as the composite pressure wave produced by the heart. In that case, the fundamental frequency is the beating frequency of the heart which, under resting conditions, is approximately 1 Hz, thus the frequency of the tenth harmonic would be 10 Hz. It is for this reason that frequencies as high as 10 Hz are considered in the dynamics of the coronary circulation and of the cardiovascular system in general.

A very simple example of a composite waveform is a single step that is repeated continuously with a period T. The example illustrates how the shape of the step is gradually approximated as more and more harmonics are added.

Another example of a composite waveform is a "piecewise" form consisting of several steps. Fourier analysis of the waveform is essentially the same as for the single-step waveform, but the example illustrates how the Fourier series approximation has difficulties at sharp corners.

Composite waveforms of interest in the dynamics of the coronary circulation, such as the cardiac pressure waveform, can only be expressed in *numerical* form. A numerical formulation of Fourier analysis makes it possible to use the numerical form of the composite wave and proceed to find its harmonics as for other waves.

Typically, 40 data points would normally be sufficient to represent a cardiac waveform numerically. The number of harmonics that produces the most accurate Fourier series representation of the waveform is then 20. A smaller *or larger* number of harmonics produces a less accurate representation. Fourier series replicate the cardiac waveform more easily than they do a piecewise waveform because of the absence of sharp steps in the former.

6

Composite Pressure-Flow Relations

6.1 Introduction

As stated earlier in this book, in the overwhelming majority of heart failures the precipitating factor is a lack of blood supply to the heart itself for its own metabolic needs [83, 206, 14, 128]. In other words, in most cases of what is generally referred to as "heart disease", the heart is not truly diseased in the normal sense of the word, it is simply being deprived of the energy it needs to do its work. And the work of the heart is important, of course, because it is the pump that provides blood supply to all other parts of the body.

As described in Chapter 1, blood supply to the heart, or coronary blood flow, comes via two branches of the aorta as shown in Figs. 1.3.1, 2. Many factors and mechanisms are involved in the short journey of blood from this point at the base of the aorta to points within the myocardium. The one that attracts most attention in medical practice is obstruction of the blood vessels. However, the journey of coronary blood flow involves some of the most complex and highly delicate dynamics and control mechanisms. They can equally disrupt blood supply to the heart. In this chapter and in this book in general, the focus is on these dynamic aspects of coronary blood flow.

The dynamics and control mechanisms of coronary blood flow require as much attention as the obstruction of blood vessels not only because they can and do interfere with orderly blood supply to the heart but because these dynamic aspects of coronary blood flow are not fully known or fully understood and are far less "visible" than an obstruction of a blood vessel. We have seen even in the crude models discussed in previous sections that, because of the pulsatile nature of coronary blood flow, any change that affects capacitance or inductance within the coronary circulation can have significant effects on the dynamics of coronary blood flow. The change may come about as a result of pathology, injury, or the administration of drugs or surgery that alter the properties of the conducting vessels or of the moving fluid. Any of these can disrupt the delicate dynamic balance of the system and thus disrupt blood supply to heart tissue.

Much of the dynamics of coronary blood flow is "hidden" in the sense that many elements of these dynamics cannot be easily measured. Not only are the controlling parameters such as capacitance and resistance inaccessible to direct measurement but direct measurements of pressure and flow which these parameters control are extremely difficult because of the small size of the vessels involved, the violent pulsations of the living heart, and the phase difference between pressure and flow which necessitates simultaneous measurements of both for any meaningful interpretation.

An element of the dynamics of the coronary circulation that is reasonably accessible is the pressure at the base of the aorta or at the entrances to the two main coronary arteries (Fig. 1.3.2). This pressure provides an important key into the coronary circulation because it represents the force that drives flow into the coronary system. It is the external force in the forced dynamics scenarios of the RLC systems examined in Chapter 4. While the driving force is actually the *difference* between this pressure and pressure at exit from the system at the capillary level, it is reasonable to neglect the small pressure at exit and think of the pressure at entry into the system as the driving pressure drop. Thus, if this pressure is denoted by $P(t)$, in this chapter we shall use the notation $P(t)$ for the *pressure drop* driving the flow, instead of Δp used in Chapter 4. This is convenient not only because it simplifies the notation in this chapter but because this pressure represents the pressure wave generated by the heart as it just emerges from the left ventricle, and measurements of this pressure are widely available. We shall refer to it simply as the "cardiac pressure wave".

With the cardiac pressure wave $P(t)$ taken as the force driving coronary blood flow, the ultimate aim of all modelling and experimental studies is to determine the corresponding flow wave generated by this force, which we shall denote by $Q(t)$. In experimental studies, because of lack of access and other difficulties, $Q(t)$ would most likely represent *inflow*, that is flow at entry into the system. It would be measured at the same location as $P(t)$, at the base of the aorta as flow enters the two main coronary ostia, or somewhere further downstream along the two main coronary arteries. In modelling studies, depending on whether the system is being modelled by a parallel or a series circuit, $Q(t)$ may be broken down into inflow and outflow, if they differ, or into partial flows along different branches of the parallel circuit. In physiological or clinical studies, of course, the ultimate question is how much blood flow is reaching the myocardium, therefore one would like $Q(t)$ to represent *outflow* from the system.

It must be remembered that both $P(t)$ and $Q(t)$ are functions of time, *periodic* functions of time, each in general being represented by a composite waveform. Thus, the relation between pressure and flow is itself a function of time, it is different at different points in time within the oscillatory cycle. Only when $P(t)$ is a simple sine or cosine function, that is, when it is a single harmonic, does the relation between pressure and flow remain the same during the oscillatory cycle and $Q(t)$ can be described simply by a fixed amplitude and a fixed phase angle as was done in previous sections. In that case, a fixed relation exists between the amplitudes and phase angles of the pressure and flow waves. When $P(t)$ is a composite waveform, however, $Q(t)$ is also a composite waveform and no such simple relation between the two is possible because the two composite waves cannot be described in terms of single amplitudes and phase angles. But, as we saw in the previous chapter, the two waves can be decomposed into harmonics which *can* be described in this simple way and, as we shall see in this chapter, the simple relation between pressure and flow continues to apply to the *harmonics* of the composite pressure and flow waves.

It must be remembered, also, that the relations between the harmonics of pressure and flow waves, namely the relations based on the concept of impedance introduced in Section 4.9, represent only *steady state dynamics*, not including any transient effects as discussed in Section 4.7. This is not an unreasonable modelling strategy of the coronary circulation because the normal operating mode of the system is that of steady state oscillation, and because a good understanding of the dynamics of the system must be based on this normal steady state mode. Indeed, steady state dynamics form the basis of all lumped models of the coronary circulation, and they form the basis of the pressure-flow relations to be explored in this chapter.

The ultimate aim then is a relation between the form of the composite pressure wave $P(t)$ and the form of the corresponding flow wave $Q(t)$. In medical practice the form of the cardiac pressure wave is highly scrutinized in terms of its graphic details, mainly because these details are interpreted as indicators of the enegertic performance of the cardiac muscle. Yet, this same pressure waveform is responsible for driving coronary blood flow, and its graphic details can equally be interpreted as indicators of the amount of flow going into the coronary circulation. Such interpretation requires an understanding of the relation between the two composite waveforms, however, and this is the subject of the present chapter.

6.2 Composite Pressure-Flow Relations Under Pure Resistance

In this section we consider the relation between a composite pressure waveform and the corresponding flow waveform when the opposition to flow consists of only pure resistance. While, as we shall see, this is a rather trivial case, it provides a good starting point and an important reference for subsequent cases. It also serves to illustrate the analytical steps required to obtain the composite flow waveform $Q(t)$ from a given composite pressure waveform $P(t)$.

We recall that when $P(t)$ is a simple sine or cosine function, as in Eq. 4.8.2,

$$P(t) = P_0 \cos \omega t \tag{6.2.1}$$

and opposition to flow consists of only the resistance R, the corresponding flow wave is given by (Eq. 4.8.7)

$$Q(t) = \frac{P_0}{R} \cos \omega t \tag{6.2.2}$$

$$= \frac{P(t)}{R} \quad \text{when } P(t), Q(t) \text{ are } sinusoidal \text{ waves} \tag{6.2.3}$$

As stated in the introduction to this chapter, $P(t)$ is now being used to denote the pressure drop denoted by Δp in earlier chapters, and P_0 is a constant representing the amplitude of the pressure wave and denoted by Δp_0 in earlier chapters.

It is clear that in this case the relation between the pressure and flow waves is particularly simple. The flow wave has the same form and the same phase angle as the pressure wave, and the amplitude of the flow wave is given by the amplitude of the pressure wave divided by the resistance R, that is

$$\text{phase } \{Q(t)\} = \text{phase } \{P(t)\} \tag{6.2.4}$$

$$\text{amplitude } \{Q(t)\} = \frac{\text{amplitude } \{P(t)\}}{R} \tag{6.2.5}$$

What is important is that this relationship between pressure and flow is fixed, it does not change within the oscillatory cycle. Equally important, however, this relation is only possible when the pressure wave is a simple sine or cosine function. This is clear from the way these solutions were obtained in Chapter 4.

Thus, the relation between pressure and flow cannot be applied to the composite waves of pressure and flow but it can be applied to their individual harmonics because, as we saw in Chapter 5, these harmonics consist of simple sine and cosine waves. Thus, we decompose the pressure wave into its harmonics by writing, as in Chapter 5,

$$P(t) = \bar{p} + p_1(t) + p_2(t) + \ldots + p_n(t) \tag{6.2.6}$$

where $p_1(t) \ldots p_n(t)$ are the n harmonics of the oscillatory part of $p(t)$ and $\bar{p}$ is the (constant) average value of $P(t)$ over one cycle, and if each of these parts of $P(t)$ is now treated separately, we obtain the corresponding series of flow rates, using Eq. 6.2.3,

$$\bar{q} = \frac{\bar{p}}{R} \tag{6.2.7}$$

$$q_1(t) = \frac{p_1(t)}{R} \tag{6.2.8}$$

$$q_2(t) = \frac{p_2(t)}{R} \tag{6.2.9}$$

$$\vdots$$

$$q_n(t) = \frac{p_n(t)}{R} \tag{6.2.10}$$

These components can now be added to give the composite flow wave $Q(t)$ produced by the composite pressure wave $P(t)$, namely

$$Q(t) = \bar{q} + q_1(t) + q_2(t) + \ldots + q_n(t) \tag{6.2.11}$$

There are two important points to emphasize:

1. The sum in Eq. 6.2.11 to obtain the total $Q(t)$ is only possible because the equation governing $Q(t)$, namely Eq. 4.7.1, is *linear*.
2. The simple relation in Eq. 6.2.3 between pressure and flow applies only to the *components* of the pressure and flow waves but not to the composite waves themselves, that is

$$Q(t) \neq \frac{P(t)}{R} \quad \text{when } P(t), Q(t) \text{ are } composite \text{ waves} \tag{6.2.12}$$

6.3 Example: Cardiac Pressure Wave

As a first example of pressure-flow analysis, consider the cardiac waveform given numerically in Table 6.3.1 and shown in Fig. 6.3.1. The waveform represents cardiac pressure in $mm\ Hg$ measured in a 20–kg dog. Note that in this case the pressure data include both the mean and the oscillatory parts of the pressure wave, and we use these below to illustrate the results of the previous section.

To obtain the corresponding flow wave using the scheme outlined in the previous section, the pressure waveform is decomposed into its harmonics as was done in Chapter 5, and the results for the first 10 harmonics are shown numerically in Table 6.3.2. Since the only opposition to flow in this case is the resistance R, then the corresponding harmonics of the flow wave are as in Eq. 6.2.10, namely

$$q_n(t) = \frac{p_n(t)}{R} \quad n = 1, 2, \ldots, 10 \tag{6.3.1}$$

However, the resistance R is not known. Since in this section we are only interested in illustrating the analysis involved, we shall use an estimate of R for this purpose.

In the human cardiovascular system, for a 60–kg man, with a cardiac output of 5 L/min and a mean aortic pressure of 100 $mm\ Hg$, an estimate of total resistance in the systemic circulation would be

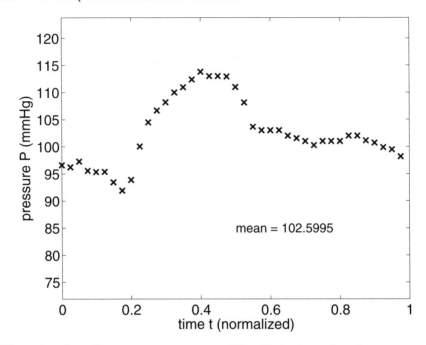

Fig. 6.3.1. A cardiac pressure wave measured in a 20–kg dog and used as an example to illustrate how the corresponding flow wave is obtained from the given pressure waveform.

$$R \text{ (systemic, human)} \approx \frac{100 \, [mm \, Hg]}{5 \, [L/min]} \tag{6.3.2}$$

$$= 20 \, \left[\frac{mm \, Hg}{L/min}\right] \tag{6.3.3}$$

$$\approx \frac{100 \times 1,333}{5 \times 1,000/60} \tag{6.3.4}$$

$$\approx 1,600 \, \left[\frac{dynes \cdot s}{cm^5}\right] \tag{6.3.5}$$

If aortic pressure is taken as the input pressure driving flow into the coronary circulation, and if coronary blood flow is estimated at 5% of cardiac output, then an estimate of total resistance in the coronary circulation is

$$R \text{ (coronary, human)} \approx \frac{100 \, [mm \, Hg]}{0.25 \, [L/min]} \tag{6.3.6}$$

$$= 400 \, \left[\frac{mm \, Hg}{L/min}\right] \tag{6.3.7}$$

$$\approx \frac{100 \times 1,333}{0.25 \times 1,000/60} \tag{6.3.8}$$

$$\approx 32,000 \quad \left[\frac{dynes \cdot s}{cm^5}\right] \qquad (6.3.9)$$

The corresponding estimates for a 20–kg dog, taking the mean cardiac output as 2 L/min, the mean aortic pressure as 100 $mm\ Hg$, and coronary blood flow again as 5% of cardiac output, we find

$$R\ (\text{systemic, dog}) \approx 50 \quad \left[\frac{mm\ Hg}{L/min}\right] \qquad (6.3.10)$$

$$\approx 4,000 \quad \left[\frac{dynes \cdot s}{cm^5}\right] \qquad (6.3.11)$$

$$R\ (\text{coronary, dog}) \approx 1,000 \quad \left[\frac{mm\ Hg}{L/min}\right] \qquad (6.3.12)$$

$$\approx 80,000 \quad \left[\frac{dynes \cdot s}{cm^5}\right] \qquad (6.3.13)$$

Using the estimated value of R for the coronary system of the dog, and the harmonic components of the cardiac pressure wave in Table 6.3.2, where

Table 6.3.1. A numerical description of the cardiac wave shown in Fig. 6.3.1, giving the pressure (P) at different times t within the oscillatory cycle. The oscillatory period has been normalized to 1.0. The pressure data include both the mean and the oscillatory part of the pressure.

t	P	t	P
0.000	96.60	0.500	111.00
0.025	96.21	0.525	108.15
0.050	97.27	0.550	103.65
0.075	95.56	0.575	103.00
0.100	95.34	0.600	103.00
0.125	95.38	0.625	103.00
0.150	93.46	0.650	102.00
0.175	91.92	0.675	101.53
0.200	93.88	0.700	101.00
0.225	100.04	0.725	100.26
0.250	104.50	0.750	101.00
0.275	106.68	0.775	101.00
0.300	108.20	0.800	101.00
0.325	110.00	0.825	102.00
0.350	110.95	0.850	102.00
0.375	112.38	0.875	101.13
0.400	113.80	0.900	100.70
0.425	113.00	0.925	99.86
0.450	113.00	0.950	99.47
0.475	112.93	0.975	98.18

Table 6.3.2. Fourier coefficients of the first 10 harmonics of the pressure wave in Fig. 6.3.1.

n	A_n	B_n	M_n	ϕ_n (deg)
1	-6.5901	0.94298	6.65720	171.8568
2	1.08200	-4.66400	4.78780	-76.9394
3	0.74761	0.62007	0.97129	39.6723
4	0.15931	0.39243	0.42353	67.9050
5	-0.77719	0.93475	1.21560	129.7415
6	-0.28180	-0.56377	0.63028	-116.5580
7	-0.19808	-0.38691	0.43467	-117.1109
8	0.55090	-0.07722	0.55628	-7.9796
9	-0.23510	0.22975	0.32872	135.6601
10	-0.13825	0.18000	0.22696	127.5263

$$p_n(t) = M_n \cos\left(\frac{2\pi nt}{T} - \phi_n\right) \quad n = 1, 2, \ldots, 10 \qquad (6.3.14)$$

we then find

$$\bar{q} = \frac{\bar{p}}{R} \qquad (6.3.15)$$

$$= \frac{102.5995}{1,000} \qquad (6.3.16)$$

$$\approx 0.1026 \quad [L/min] \qquad (6.3.17)$$

where $\bar{p}$ and $\bar{q}$ are the mean values of $p(t)$ and $q(t)$ over one oscillatory cycle. For the oscillatory parts of the flow wave, using Eq. 6.3.14 and Table 6.3.2, and taking the period $T = 1$, we find

$$q_1(t) = \frac{p_1(t)}{R} \qquad (6.3.18)$$

$$= \frac{6.6572}{1,000} \times \cos\left(2\pi t - 171.8568 \times \pi/180\right) \qquad (6.3.19)$$

$$\approx 0.0067 \times \cos\left(2\pi t - 171.8568 \times \pi/180\right) \quad [L/min] \qquad (6.3.20)$$

$$q_2(t) = \frac{p_2(t)}{R} \qquad (6.3.21)$$

$$= \frac{4.7878}{1,000} \times \cos\left(4\pi t + 76.9394 \times \pi/180\right) \qquad (6.3.22)$$

$$\approx 0.0048 \times \cos\left(4\pi t + 76.9394 \times \pi/180\right) \quad [L/min] \qquad (6.3.23)$$

$$\vdots$$

$$q_{10}(t) = \frac{p_{10}(t)}{R} \qquad (6.3.24)$$

$$= \frac{0.22696}{1,000} \times \cos\left(20\pi t - 127.5263 \times \pi/180\right) \qquad (6.3.25)$$

$$\approx 0.0002 \times \cos{(20\pi t - 127.5263 \times \pi/180)} \quad [L/min] \quad (6.3.26)$$

These components of $Q(t)$ can now be added to give the composite flow wave $Q(t)$ produced by the composite pressure wave $P(t)$, that is

$$Q(t) = \bar{q} + q_1(t) + q_2(t) + \ldots + q_{10}(t) \qquad (6.3.27)$$

which is shown graphically together with the pressure wave in Fig. 6.3.2.

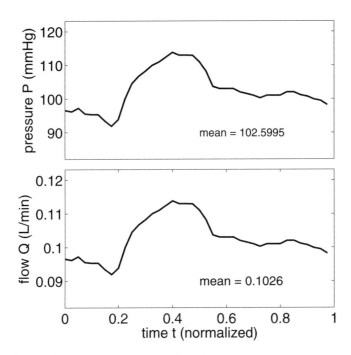

Fig. 6.3.2. Cardiac pressure wave (top) and corresponding flow wave (bottom) when opposition to flow consists of only resistance R which has been estimated at $1,000 \; mm \; Hg \cdot L/min$. We see that in the presence of pure resistance the pressure and flow waves have precisely the same form, the flow wave being only scaled by the value of the resistance R.

It would be convenient to put the pressure and flow waves together, using the same scale, and the results of this section suggest that in order to do so we should plot $P(t)$ and $R \times Q(t)$ (instead of $Q(t)$), as shown in Fig. 6.3.3. This presentation would be useful not only when the opposition to flow consists of pure resistance but also, and particularly, when other elements of the RLC system are present. In such cases, as we shall see, any small change in the form of the flow wave can be detected more easily and can be attributed directly to inertial (L) or capacitive (C) effects only, because the effects of resistance (R) have been scaled out. We shall refer to the product $R \times Q(t)$ as the "R-scaled" flow.

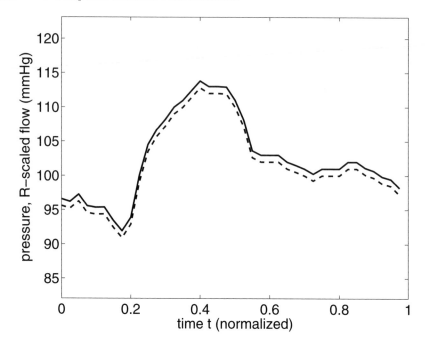

Fig. 6.3.3. When studying pressure-flow relations it is convenient to plot the pressure and flow waves to the same scale so as to compare their waveforms. This can be achieved as seen here by plotting $P(t)$ (solid curve) and the "R-scaled" flow $R \times Q(t)$ (dashed) instead of $Q(t)$. When the opposition to flow consists of only resistance R as it is in this case, the two curves become graphically identical. In this figure they are slightly shifted to make them visibly distinct. The use of R-scaled flow is useful also when other elements of the RLC system are present. In such cases any small change in the form of the flow wave can be detected more easily and can be attributed directly to inertial (L) or capacitive (C) effects only, because the effects of resistance (R) have been scaled out.

6.4 Composite Pressure-Flow Relations Under General Impedance

When opposition to pulsatile flow consists of more than pure resistance, that is, when either inertial or capacitance effects or both are involved, the relation between a composite flow wave and the corresponding flow wave is more complicated than was seen in the previous section. For the purpose of considering this relation here, the term "impedance" shall be used in a general sense here to mean any form of opposition to flow beyond that of pure resistance.

It is convenient in this and in subsequent sections to use the concept of *complex* impedance Z introduced in Section 4.9. It was seen in that section that when the opposition to flow consists of only pure resistance, the complex impedance Z becomes real and equal to R, but when any inertial or capacitive effects are present, that is any "reactive" effects, Z is a complex quantity whose real and imaginary parts depend on the nature and arrangement of the reactive elements. In these terms, our interest in this section is in pressure-flow relations when Z has both a real and an imaginary part.

Let the composite pressure wave under consideration be denoted by $P(t)$ and the corresponding flow wave be denoted by $Q(t)$. In Chapter 5 we saw that $P(t)$ can always be separated into a steady part $\bar{p}$ and a purely oscillatory part, such that

$$P(t) = \bar{p} + p(t) \tag{6.4.1}$$

The mean part of the pressure represents the mean value of $P(t)$ over one oscillatory cycle, namely

$$\bar{p} = \frac{1}{T} \int_0^T P(t)dt \tag{6.4.2}$$

The oscillation part of the pressure, by definition, has a zero mean. This separation of the composite wave is important because the corresponding flow wave will consist similarly of steady and oscillatory parts, which we shall denote by $\bar{q}$ and $q(t)$, respectively, and write

$$Q(t) = \bar{q} + q(t) \tag{6.4.3}$$

The relation between the mean components of the pressure and flow waves, $\bar{p}, \bar{q}$, is fundamentally different from that between the oscillatory components, $p(t), q(t)$, and the two relations must be dealt with separately. For the steady components, the relation between pressure and flow is simply that established in Section 6.2, namely

$$\bar{q} = \frac{\bar{p}}{R} \tag{6.4.4}$$

While in Section 6.2 the opposition to flow consisted of only pure resistance, this relation remains valid in this section even though reactive elements are assumed to be present. The reason for this is that reactive effects come into play only when flow is non-steady. And since here we are able to separate the steady and non-steady parts of the flow, as discussed in Section 6.2, Eq. 6.4.4 can be used for the steady parts of the pressure and flow waves.

One of the most important advantages of using complex impedance is that the relation between the oscillatory parts of the pressure and flow waves can then be put in the general form

$$q(t) = \frac{p(t)}{Z} \tag{6.4.5}$$

However, in this equation the pressure $p(t)$ must be in complex form. And since both $p(t)$ and Z are complex, it follows that $q(t)$ is also complex. Thus, if subscripts r and i are used to denote real and imaginary parts, and we write

$$p(t) = p_r(t) + ip_i(t) \tag{6.4.6}$$

$$Z = z_r + iz_i \tag{6.4.7}$$

$$q(t) = q_r(t) + iq_i(t) \tag{6.4.8}$$

then Eq. 6.4.5 can be put in the form

$$q_r(t) + iq_i(t) = \frac{p_r(t) + ip_i(t)}{z_r + iz_i} \tag{6.4.9}$$

$$= \frac{p_r(t)z_r + p_i(t)z_i + i(p_i(t)z_r - p_r(t)z_i)}{z_r^2 + z_i^2} \tag{6.4.10}$$

and we find

$$q_r(t) = \frac{p_r(t)z_r + p_i(t)z_i}{z_r^2 + z_i^2} \tag{6.4.11}$$

$$q_i(t) = \frac{p_i(t)z_r - p_r(t)z_i}{z_r^2 + z_i^2} \tag{6.4.12}$$

As we saw in earlier sections, the real part of the oscillatory flow rate, namely $q_r(t)$, represents the flow rate when the driving pressure is the real part of $p(t)$, namely $p_r(t)$, and similarly for the imaginary parts of pressure and flow. But as we see clearly from Eqs. 6.4.11, 12

$$q_r(t) \neq \frac{p_r(t)}{z_r} \tag{6.4.13}$$

$$q_i(t) \neq \frac{p_i(t)}{z_i} \tag{6.4.14}$$

The correct pressure-flow relation is

$$q_r(t) = \Re\left\{\frac{p(t)}{Z}\right\} \tag{6.4.15}$$

$$= \Re\left\{\frac{p_r(t) + ip_i(t)}{z_r + iz_i}\right\} \tag{6.4.16}$$

$$q_i(t) = \Im\left\{\frac{p(t)}{Z}\right\} \tag{6.4.17}$$

$$= \Im\left\{\frac{p_r(t) + ip_i(t)}{z_r + iz_iZ}\right\} \tag{6.4.18}$$

which yield the results in Eqs. 6.4.11, 12.

It is important to recall that Eq. 6.4.5 and all the above results that follow from it apply only when the driving pressure, p_r or p_i, is a simple sine or cosine wave. Thus Eq. 6.4.5 cannot be applied directly to a *composite* pressure wave, *but it can be applied to its individual harmonics.* If there are N harmonics and we denote these by $p_n(t)$, then, as we found in Chapter 5, they are given by

$$p_n(t) = M_n \cos\left(\frac{2\pi n t}{T} - \phi_n\right) \quad n = 1, 2, \ldots, N \tag{6.4.20}$$

and the oscillatory part of the pressure wave is given by

$$p(t) = p_1(t) + p_2(t) + \ldots + p_N(t) \tag{6.4.21}$$

where M_n and ϕ_n are (real) constants associated with the Fourier series representation of the composite wave.

Eq. 6.4.5 applies only to each of these harmonics individually, and only if each is considered to be the real or the imaginary part of the complex pressure. Thus, for the first harmonic we can introduce a complex pressure

$$p_1(t) = M_1 \cos\left(\frac{2\pi t}{T} - \phi_1\right) + i M_1 \sin\left(\frac{2\pi t}{T} - \phi_1\right) \tag{6.4.22}$$

and then use Eq. 6.4.5 to get the corresponding harmonic of the flow wave

$$q_1(t) = \frac{p_1(t)}{Z} \tag{6.4.23}$$

This can be done with each harmonic, writing, in general

$$p_n(t) = M_n \cos\left(\frac{2\pi n t}{T} - \phi_n\right) + i M_n \sin\left(\frac{2\pi n t}{T} - \phi_n\right) \tag{6.4.24}$$

which can be put in the more compact exponential notation

$$p_n(t) = M_n e^{i((2\pi n t/T) - \phi_n)} \quad n = 1, 2, \ldots, N \tag{6.4.25}$$

The corresponding harmonics of the oscillatory flow wave are then given by

$$q_n(t) = \frac{p_n(t)}{Z} \quad n = 1, 2, \ldots, N \tag{6.4.26}$$

The oscillatory part of the composite pressure wave is now seen to be the *real* part of the complex pressure $p_n(t)$, which in turn corresponds to the real part of the flow wave, that is, in the notation of Eqs. 6.4.6, 8, where subscripts r and i are used for denoting real and imaginary parts, respectively,

$$p_{nr}(t) = \Re(p_n(t)) \tag{6.4.27}$$

$$= M_n \cos\left(\frac{2\pi n t}{T} - \phi_n\right) \quad n = 1, 2, \ldots, N \tag{6.4.28}$$

$$p_{ni}(t) = \Im(p_n(t)) \tag{6.4.29}$$

$$= M_n \sin\left(\frac{2\pi nt}{T} - \phi_n\right) \quad n = 1, 2, \ldots, N \tag{6.4.30}$$

$$q_{nr}(t) = \Re(q_n(t)) \tag{6.4.31}$$

$$= \Re\left(\frac{p_n(t)}{Z}\right) \quad n = 1, 2, \ldots, N \tag{6.4.32}$$

$$q_{ni}(t) = \Im(q_n(t)) \tag{6.4.33}$$

$$= \Im\left(\frac{p_n(t)}{Z}\right) \quad n = 1, 2, \ldots, N \tag{6.4.34}$$

6.5 Composite Pressure-Flow Relations Under Inertial Effects

Inertial effects are important in coronary blood flow, and in blood flow in general, because of the pulsatile nature of the flow. In pulsatile flow the fluid is repeatedly accelerated and decelerated and hence fluid inertia, or what in previous sections was referred to as the inductance (L), has a significant effect on the relation between pressure and flow. The origin and basic nature of the inertial effect were examined in Chapter 2. In Chapter 3 the effects of inductance on the free dynamics of the RLC system were examined, and the same was done in Chapter 4 for the forced dynamics of the RLC system, using either linear or simple sinusoidal driving pressures. In this section we examine the effects of inertia on pressure-flow relations when the driving pressure has a composite waveform.

As noted in earlier sections, inertial effects in the coronary circulation and in the cardiovascular system in general do not arise in pure form but always in combination with resistance effects and frequently in combination with capacitance effects. Only when a breach occurs within the vascular system is blood able to accelerate and decelerate free from the constraints of the containing vessels and hence free from resistance and capacitance effects. It is therefore not meaningful to study the inertial effect in isolation, that is to study only the inductance element L of the RLC system by itself. In this section we examine the effects of inductance L in combination with resistance R and in later sections within the complete RLC system.

Consider resistance R and inductance L *in series* at first, where the complex impedance, from Eqs. 4.9.34, 35, is given by

$$Z = R + i\omega L \tag{6.5.1}$$

where ω is the angular frequency. Since the individual harmonics of a composite wave have different frequencies, we should strictly write

$$Z_n = R + i\omega_n L \qquad (6.5.2)$$
$$\omega_n = 2\pi n \quad n = 1, 2, \ldots, N \qquad (6.5.3)$$

to highlight the fact that different harmonics will have different impedances. We also note that in general if we write

$$Z_n = z_{nr} + i z_{ni} \quad n = 1, 2, \ldots, N \qquad (6.5.4)$$

where n denotes a particular harmonic, and subscripts r, i denote real and imaginary parts as before, then

$$z_{nr} = R \qquad (6.5.5)$$
$$z_{ni} = \omega_n L \qquad (6.5.6)$$

In other words, as stated earlier, when R, L are *in series*, the real part of the impedance for each harmonic represents pure resistance while the imaginary part represents the inertial effects.

The expression for the real part of the flow rate for individual harmonics can now be put together, using the above notation and the results of the previous section (Eqs. 6.4.11, 28, 30)

$$q_{nr}(t) = \Re\left(\frac{p_n(t)}{Z_n}\right) \qquad (6.5.7)$$

$$= \Re\left(\frac{p_{nr} + i p_{ni}}{z_{nr} + i z_{ni}}\right) \qquad (6.5.8)$$

$$= \frac{p_{nr} z_{nr} + p_{ni} z_{ni}}{z_{nr}^2 + z_{ni}^2} \qquad (6.5.9)$$

$$= \frac{R M_n \cos\left(\frac{2\pi n t}{T} - \phi_n\right) + \omega_n L M_n \sin\left(\frac{2\pi n t}{T} - \phi_n\right)}{R^2 + \omega_n^2 L^2} \qquad (6.5.10)$$

When resistance and inductance are *in parallel*, the above analysis of the flow wave remains intact, with only a change in the form of the impedance. Thus, when R and L are in parallel, we have, from Eqs. 4.9.33–35 and in the notation of the present section,

$$\frac{1}{Z_n} = \frac{1}{R} + \frac{1}{i\omega_n L} \qquad (6.5.11)$$

For easier comparison of the pressure and flow waveforms, it is more convenient to use not the flow rate but the *R-scaled* flow rate introduced earlier, namely

$$R \times q_{nr}(t) = \frac{M_n \cos\left(\frac{2\pi nt}{T} - \phi_n\right) + \omega_n t_L M_n \sin\left(\frac{2\pi nt}{T} - \phi_n\right)}{1 + \omega_n^2 t_L^2} \quad (6.5.12)$$

where

$$t_L = \frac{L}{R} \quad (6.5.13)$$

is the inertial time constant introduced earlier. We saw in the previous section that when opposition to flow consists of pure resistance the forms of the pressure and of the R-scaled flow are identical. Therefore, in the present section where inertial effects are present, any deviation from this identity can be attributed directly and entirely to inertial effects. This highlights the advantage of using the R-scaled flow rate instead of the flow rate for comparison with the pressure waveform. Furthermore, in this way the physical parameters R and L do not need to be specified separately because only their ratio, the inertial time constant t_L is now required.

As discussed at great length in Section 2.5, the inertial time constant t_L is a measure of the time it takes the fluid to respond to a change in the driving pressure. The unit of time in which t_L is expressed depends on the unit of time used for the angular frequency ω_n in Eq. 6.5.11. When the frequency is in radians per second, t_L must be expressed in seconds so that the product $\omega_n t_L$ in Eq. 6.5.11 becomes nondimensional as it should be. Minutes can be used in the same way. In what follows and in much of this book we express t_L in seconds. As discussed in earlier sections, while the actual value of t_L in the coronary circulation is not known, its order of magnitude is clearly seconds rather than minutes.

Eq. 6.5.11 produces the individual harmonics of the oscillatory part of the R-scaled flow wave. The complete oscillatory flow wave is finally obtained by adding these harmonics, that is

$$R \times q_r(t) = R \times q_{1r}(t) + R \times q_{2r}(t) \ldots + R \times q_{Nr}(t) \quad (6.5.14)$$

where N is the number of harmonics. The complete R-scaled flow wave (corresponding to real part of driving pressure) is then given by

$$R \times Q(t) = R \times \bar{q} + R \times q_r(t) \quad (6.5.15)$$

Results, comparing the R-scaled flow wave with the corresponding pressure wave at different values of t_L, and using the one-step, piecewise, and cardiac pressure waves, are shown in Fig. 6.5.1–3. It is seen that when t_L is very small, inertial effects are insignificant and the forms of the pressure and the R-scaled flow rate become identical. At the other extreme, at higher values of t_L, inertial effects become increasingly more significant as evidenced by the considerable difference they produce between the forms of the pressure and the R-scaled flow wave. Increasingly, the oscillatory part of the flow wave diminishes, leaving only the steady part, as the value of t_L increases.

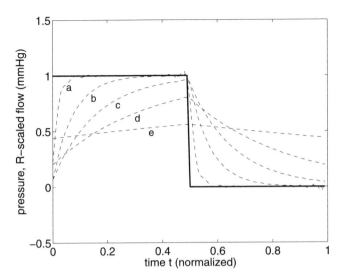

Fig. 6.5.1. Pressure (solid) and R-scaled flow waves (dashed) through a resistance R and inductance L in series, and for different values of the inertial time constant t_L in seconds: (a) 0.02, (b) 0.075, (c) 0.16, (d) 0.35, (e) 2.0. At the lowest value of t_L, R-scaled flow wave is close to that of pressure, indicating that oscillatory flow is little affected by the inertia of the fluid, while at the highest value of t_L the opposite is true and oscillatory flow is reduced almost to zero, leaving mainly the steady part of the flow.

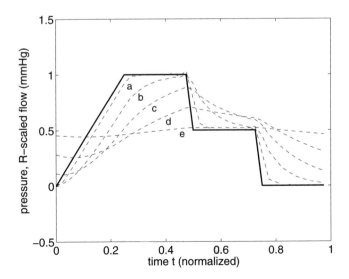

Fig. 6.5.2. See caption for Fig. 6.5.1.

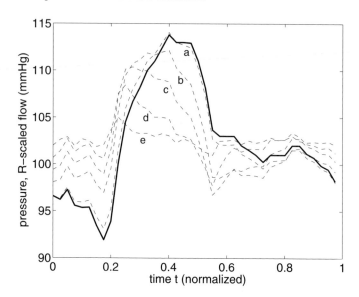

Fig. 6.5.3. See caption for Fig. 6.5.1.

so that

$$Z_n = \frac{i\omega_n LR}{R + i\omega_n L} \tag{6.5.16}$$

$$= \frac{\omega_n^2 L^2 R + i(\omega_n LR^2)}{R^2 + \omega_n^2 L^2} \tag{6.5.17}$$

and the real and imaginary parts of the impedance are thus given by

$$z_{nr} = \frac{\omega_n^2 L^2 R}{R^2 + \omega_n^2 L^2} \tag{6.5.18}$$

$$z_{ni} = \frac{\omega_n LR^2}{R^2 + \omega_n^2 L^2} \tag{6.5.19}$$

These expressions for the complex impedance are now used instead of those in Eqs. 6.5.5, 6 and all subsequent steps are repeated to find that the R-scaled flow wave in this case is given by

$$R \times q_{nr}(t) = \frac{\omega_n t_L M_n \cos\left(\frac{2\pi nt}{T} - \phi_n\right) + M_n \sin\left(\frac{2\pi nt}{T} - \phi_n\right)}{\omega_n t_L} \tag{6.5.20}$$

Results for different values of the inertial time constant t_L, and using the one-step, piecewise, and cardiac pressure wave, are shown in Figs. 6.5.4–6. In contrast with the case of RL in series, it is seen that in this case the forms of the pressure and R-scaled flow waves become identical at *higher* values of

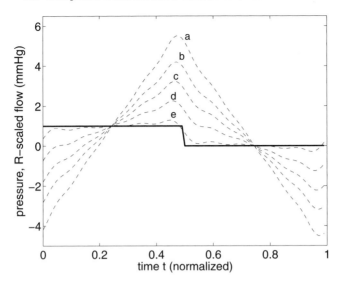

Fig. 6.5.4. Pressure (solid) and R-scaled flow waves (dashed) through a resistance R and inductance L in parallel, and for different values of the inertial time constant t_L in seconds: (a) 0.025, (b) 0.035, (c) 0.05, (d) 0.09, (e) 0.5. Much larger swings in flow rate are observed in this case, compared with the case of R, L in series, because flow has the option of accelerating or decelerating through the inductor without being constrained by the resistor.

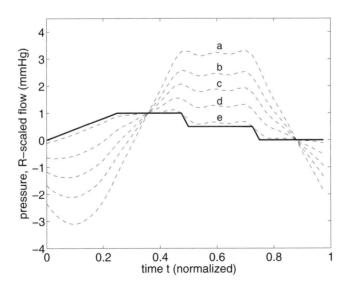

Fig. 6.5.5. See caption for Fig. 6.5.4.

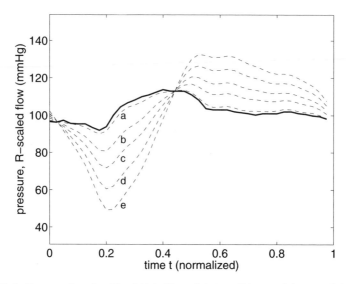

Fig. 6.5.6. See caption for Fig. 6.5.4. Here (a) 0.5, (b) 0.1, (c) 0.06, (d) 0.04, (e) 0.03.

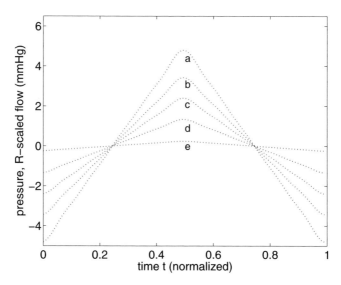

Fig. 6.5.7. R-scaled inductive flow only, produced by the one-step pressure wave of Fig. 6.5.4, and for different values of the inertial time constant t_L in seconds: (a) 0.025, (b) 0.035, (c) 0.05, (d) 0.09, (e) 0.5. At the highest value of t_L inductive flow is near zero, leaving flow mainly through the resistor. At the lowest value of t_L inductive flow is high and is in addition to flow through the resistor. Total flow rate through the system is shown in Fig. 6.5.4.

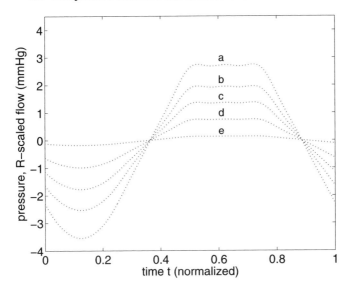

Fig. 6.5.8. R-scaled inductive flow only, produced by the piecewise pressure wave of Fig. 6.5.5, and for different values of the inertial time constant t_L in seconds: (a) 0.025, (b) 0.035, (c) 0.05, (d) 0.09, (e) 0.5. At the highest value of t_L inductive flow is near zero, leaving flow mainly through the resistor. At the lowest value of t_L inductive flow is high and is in addition to flow through the resistor. Total flow rate through the system is shown in Fig. 6.5.5.

t_L, which represent higher inertial effects. The reason for this is that when R and L are in parallel, flow has the option of going through the resistor rather than the inductor, so that when inertial effects become very high the entire flow goes through the resistor and, as seen Section 6.3, the pressure and R-scaled flow waveforms become identical. At the other extreme, when inertial effects are very low, more of the flow goes through the inductor. Under these conditions much larger swings in flow rate are produced within the oscillatory cycle than are produced when R and L are in series. In that case the flow is constrained because it has to go through the resistor even when values of t_L are low.

These observations can be seen more clearly in Figs. 6.5.7–9 where only flow through the inductor is shown. It is seen that flow through the inductor is highest when inertial effects are low, while it is near zero when inertial effects are high. The steady part of the flow, of course, does not contribute to flow through the inductor.

It is interesting to note in Figs. 6.5.4–9 that when R, L are in parallel, flow through the inductor can accelerate or decelerate freely, unimpeded by the presence of the resistor. Thus, within an oscillatory cycle, total flow through the system first increases in response to a positive pressure gradient and then decreases as the gradient changes sign, or vice versa, depending on the form of

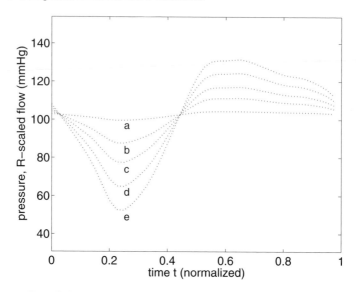

Fig. 6.5.9. R-scaled inductive flow only, produced by the cardiac pressure wave of Fig. 6.5.6, and for different values of the inertial time constant t_L in seconds: (a) 0.5, (b) 0.1, (c) 0.06, (d) 0.04, (e) 0.03. At the highest value of t_L inductive flow is near zero, leaving flow mainly through the resistor. At the lowest value of t_L inductive flow is high and is in addition to flow through the resistor. Total flow rate through the system is shown in Fig. 6.5.6.

the pressure wave. The result is that inductive flow will typically go through zero twice within the oscillatory cycle, as seen in Figs. 6.5.7–9.

6.6 Composite Pressure-Flow Relations Under Capacitance Effects

Capacitance effects are important in coronary blood flow because of the pulsatile nature of the flow and because of the elasticity and hence compliance of the coronary vessels. While this statement is equally true in the cardiovascular system as a whole, capacitance effects play a more critical role in the dynamics of the coronary circulation because of the violent compression effects which the cardiac muscle exerts on coronary vasculature imbedded within the myocardium. The way this so called "tissue pressure" or "intramyocardial pressure" is intermingled with normal capacitance effects due to vessel compliance is far from fully understood, which makes the role of capacitance in the coronary circulation all the more critical. Indeed, this is one of the major problems in the modelling of the coronary circulation which we shall consider later. In this section we consider the capacitance effect without this added complication, with the aim of understanding the effect pure capacitance has on pressure-flow relations.

The origin and basic nature of the capacitance effect were examined in Chapter 2, where the effect was likened to that of flow going into an elastic balloon. In Chapter 3, the effects of capacitance on the free dynamics of the RLC system were examined, and the same was done in Chapter 4 for the forced dynamics of the RLC system, using either linear or simple sinusoidal driving pressures. In this section we examine the effects of capacitance on pressure-flow relations when the driving pressure has a composite waveform.

Using the analogy of flow into an elastic balloon, it was discussed in Section 3.2 that when the balloon is *in series* with other elements of the RLC system, flow through the system is clearly limited by the capacity of the balloon. It is clear from that section, therefore, that the effect of capacitance in the coronary circulation would only arise *in parallel* with other elements of the RLC system. Indeed, the nature of this effect in the physiological system enforces this view: flow through the coronary vasculature has the option of inflating the vessels or simply flowing through. The two options are clearly in parallel, in the sense that they can be independent of each other. Nevertheless, in this section we examine the capacitance effect in combination with resistance, both in series and in parallel, in order to compare the two cases.

Consider resistance R and capacitance C *in series* at first, where the complex impedance, from Eqs. 4.9.34, 36, is given by

$$Z = R + \frac{1}{i\omega C} \tag{6.6.1}$$

$$= R - \frac{i}{\omega C} \tag{6.6.2}$$

where ω is the angular frequency. As in the previous section, for the individual harmonics of a composite wave, since they have different frequencies, we write

$$Z_n = R - \frac{i}{\omega_n C} \tag{6.6.3}$$

$$\omega_n = 2\pi n \quad n = 1, 2, \ldots, N \tag{6.6.4}$$

where N is the number of harmonics. The real and imaginary parts of the complex impedance are given by, using subscripts r, i to denote real and imaginary as before,

$$z_{nr} = R \tag{6.6.5}$$

$$z_{ni} = \frac{-1}{\omega_n C} \tag{6.6.6}$$

We see again that the real part of the impedance represents pure resistance while the imaginary part in the present case represents the capacitance effects.

Following the same steps as in the previous section, we find the real parts of individual harmonics of the flow wave

$$q_{nr}(t) = \Re\left(\frac{p_n(t)}{Z_n}\right) \tag{6.6.7}$$

$$= \Re\left(\frac{p_{nr} + ip_{ni}}{z_{nr} + iz_{ni}}\right) \tag{6.6.8}$$

$$= \frac{p_{nr}z_{nr} + p_{ni}z_{ni}}{z_{nr}^2 + z_{ni}^2} \tag{6.6.9}$$

$$= \frac{R(\omega_n C)^2 M_n \cos\left(\frac{2\pi n t}{T} - \phi_n\right) - \omega_n C M_n \sin\left(\frac{2\pi n t}{T} - \phi_n\right)}{1 + (R\omega_n C)^2} \tag{6.6.10}$$

and for the corresponding harmonics of the *R-scaled* flow wave

$$R \times q_{nr}(t) = \frac{(\omega_n t_C)^2 M_n \cos\left(\frac{2\pi n t}{T} - \phi_n\right) - \omega_n t_C M_n \sin\left(\frac{2\pi n t}{T} - \phi_n\right)}{1 + (\omega_n t_C)^2} \tag{6.6.11}$$

where t_C is the inertial time constant, given by

$$t_C = RC \tag{6.6.12}$$

We recall from Section 2.6 that C and t_C are measures of the *compliance* of the vascular system or the balloon used as a model. The higher the value of C or t_C, the more compliant the system is, allowing a greater change in its volume and hence more flow into it. This factor is extremely important in the coronary circulation because it intervenes between flow entering the system at the root of the main coronary arteries and flow leaving the system at the capillary end. In the absence of compliance the two flows would be equal at all times, thus a measurement of flow at entry gives a measure of the flow being delivered at the all-important receiving end. In the presence of compliance, the connection between inflow and outflow is lost, and while over one or more oscillatory cycles the two flows will normally be equal, at any one moment within an oscillatory cycle they are unequal.

Thus, a good model of the role of capacitance in the dynamics of the coronary circulation is essential, and much of the research work in this area has been directed at this problem. In particular, efforts have been directed at obtaining an estimate of the value of the capacitance C. From the definition of C in Section 2.6 (Eq. 2.6.6), the dimensions of C are seen to be the dimensions of volume over pressure which indeed represents the change in volume obtained from a given change in pressure. The higher the value of C the higher the change in volume obtained for a given change in pressure, and hence the more "compliant" the system is. In a series of experiments by Judd et al [97, 98] it was estimated that for the dog heart the value is 0.002 *ml/mm Hg* per 100 *g* of heart tissue. If, for the purpose of discussion we consider the dog's heart to actually be 100 g ($= 0.5\%$ of body weight), then the value of C for the entire coronary system of that dog is 0.002 $ml/mm\ Hg$.

Consistent with previous chapters, in what follows we shall continue to use the capacitive time constant t_C as a measure of capacitance or compliance, rather than C. The dimensions of t_C are, of course, the dimensions of time, and we shall use seconds for the units of t_C as we did for the inertial time constant t_L. Using the above estimate for C, and the estimate for the resistance R obtained in Section 6.3 for the coronary system of a 20 kg dog, (Eq. 6.3.12), we find $t_C \approx 0.12\ s$ for that system. We shall use this value as a guide to the range of values of t_C used in what follows.

With the individual harmonics of the oscillatory part of the R-scaled flow wave obtained from Eq. 6.6.11, the complete oscillatory flow wave is finally obtained by adding these harmonics, that is

$$R \times q_r(t) = R \times q_{1r}(t) + R \times q_{2r}(t) \ldots + R \times q_{Nr}(t) \qquad (6.6.13)$$

where N is the number of harmonics. The complete R-scaled flow wave (corresponding to real part of driving pressure) is then given by

$$R \times Q(t) = R \times \bar{q} + R \times q_r(t) \qquad (6.6.14)$$

Results, comparing the R-scaled flow wave with the corresponding pressure wave at different values of t_C, and using the one-step, piecewise, and cardiac pressure waves, are shown in Fig. 6.6.1–3. It is seen that when t_C is high, indicating high compliance, the R-scaled flow curve comes close to the pressure curve. Recalling that here the resistance and capacitance are in series, the result indicates that the capacitor is not having much effect on the flow rate, the latter being close to what it would be in the absence of the capacitor. By contrast, when t_C is low, which corresponds to low compliance, the R-scaled flow rate is reduced to almost steady flow at the average value of the driving pressure. The capacitor in this case has a considerable effect on the flow. It is, effectively, damping the oscillatory part of the flow.

When resistance and capacitance are *in parallel*, the complex impedance, from Eqs. 4.9.33, 34, 36 and in the notation of the present section, is given by

$$\frac{1}{Z_n} = \frac{1}{R} + i\omega_n C \qquad (6.6.15)$$

so that

$$Z_n = \frac{R}{1 + iR\omega_n C} \qquad (6.6.16)$$

$$= \frac{R - i\omega_n CR^2}{1 + (R\omega_n C)^2} \qquad (6.6.17)$$

and the real and imaginary parts of the complex impedance are thus given by

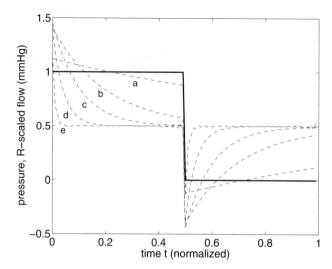

Fig. 6.6.1. Pressure (solid) and R-scaled flow waves (dashed) through a resistance R and capacitance C in series, and for different values of the capacitive time constant t_C in seconds: (a) 1.0, (b) 0.2, (c) 0.1, (d) 0.04, (e) 0.01. At the highest value of t_C the R-scaled flow curve is closest to the pressure curve, indicating that the capacitor is not having much effect on the flow rate, the latter being close to what it would be in the absence of the capacitor. When t_C is low, by contrast, the R-scaled flow rate is reduced to almost steady flow at the average value of the driving pressure. The capacitor in this case is effectively damping the oscillatory part of the flow.

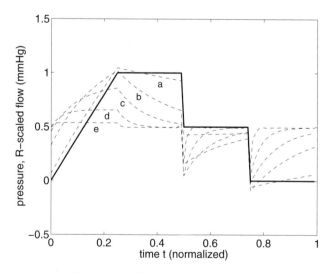

Fig. 6.6.2. See caption for Fig. 6.6.1.

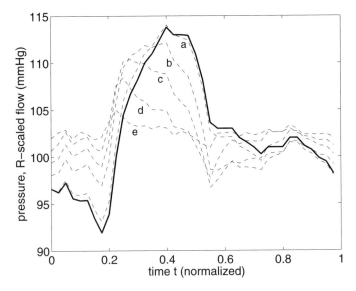

Fig. 6.6.3. See caption for Fig. 6.6.1.

$$z_{nr} = \frac{R}{1 + (R\omega_n C)^2} \tag{6.6.18}$$

$$z_{ni} = \frac{-\omega_n C R^2}{1 + (R\omega_n C)^2} \tag{6.6.19}$$

Following the same steps as in the previous section, the R-scaled flow wave is given by

$$R \times q_{nr}(t) = M_n \cos\left(\frac{2\pi n t}{T} - \phi_n\right) - \omega_n t_C M_n \sin\left(\frac{2\pi n t}{T} - \phi_n\right) \tag{6.6.20}$$

Results for a low value of the capacitance time constant ($t_C = 0.01\ s$) are presented in Figs. 6.6.4–6 where R-scaled total flow and R-scaled capacitive flow are shown separately. In this case, because of the low value of t_C, which indicates low compliance, and because the capacitance is in *parallel* with the resistance, most flow is through the resistor with very little flow going through the capacitor. The situation is the reverse of that in Figs. 6.6.1–3 where the resistance and capacitance are in series.

Results for a high value of the capacitive time constant ($t_C = 0.3$) are shown in Figs. 6.6.7–9. In this case, because of the high compliance within the system, and because this compliance is in *parallel* with the resistance, very high flow rates are drawn into the capacitor and very little (by comparison) into the resistor. R-scaled total flow waveform is very far from the pressure waveform, consistent with large capacitive effects.

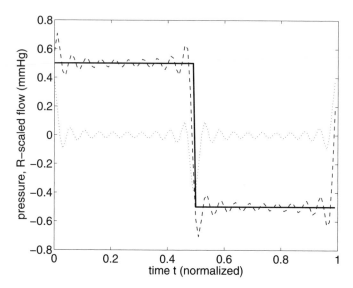

Fig. 6.6.4. R-scaled total flow (dashed) and R-scaled capacitive flow (dotted) through a resistance and capacitance in parallel and under the driving composite pressure wave shown by the solid curve and with a low value of the capacitive time constant ($t_C = 0.01$). Because of low compliance, capacitive flow is near zero, hence total flow is mostly through the resistor as indicated by the closeness of the total flow and pressure curves.

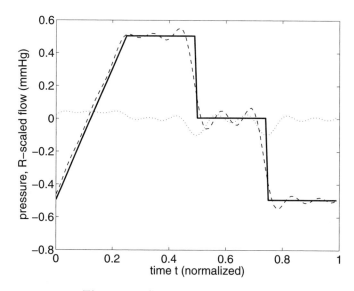

Fig. 6.6.5. See caption for Fig. 6.6.4.

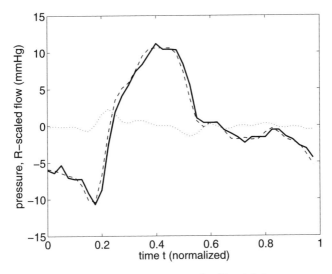

Fig. 6.6.6. See caption for Fig. 6.6.4.

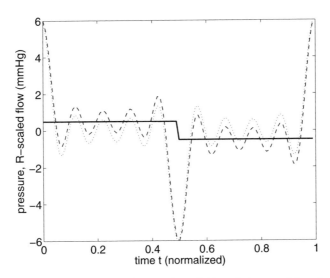

Fig. 6.6.7. R-scaled total flow (dashed) and R-scaled capacitive flow (dotted) through a resistance and capacitance in parallel and under the driving composite pressure wave shown by the solid curve and with a high value of the capacitive time constant ($t_C = 0.3$). Because of the high compliance, very high flow rates are drawn into the capacitor, with very little (by comparison) drawn into the resistor. Pressure and total flow waveforms are far apart, consistent with large capacitive effects.

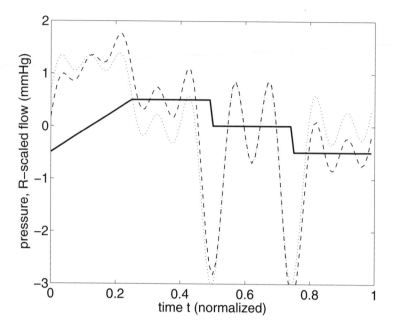

Fig. 6.6.8. See caption for Fig. 6.6.7.

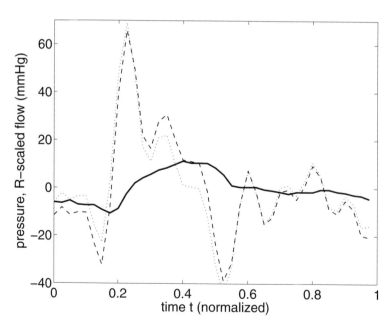

Fig. 6.6.9. See caption for Fig. 6.6.7.

6.7 Composite Pressure-Flow Relations Under RLC in Series

The RLC system in series provides a basic model in which the elements of resistance, inductance, and capacitance are all present but their effects are constrained by each other because of their series arrangement. Free and forced dynamics of this system have been examined in previous chapters using zero, linear, or sinusoidal driving pressures. In this section we examine pressure flow relations under this system, using composite driving pressure waves.

The complex impedance for R, L, C in series was found in Section 4.9, Eq. 4.9.41, namely

$$Z = R + i \left(\omega L - \frac{1}{\omega C} \right) \tag{6.7.1}$$

where ω is the angular frequency of oscillation of the driving pressure. For a composite pressure wave consisting of N harmonics the impedance will be different for each of these harmonics because of their different frequencies, and we shall use the notation

$$Z = R + i \left(\omega_n L - \frac{1}{\omega_n C} \right) \tag{6.7.2}$$

$$\omega_n = 2\pi n \qquad n = 1, 2, \ldots, N \tag{6.7.3}$$

where n denotes a particular harmonic and ω_n is the angular frequency of that harmonic. For the real and imaginary parts of the impedance, using subscripts r, i as before, we have in this case

$$z_{nr} = R \tag{6.7.4}$$

$$z_{ni} = \omega_n L - \frac{1}{\omega_n C} \tag{6.7.5}$$

In the notation of previous sections, consider the composite pressure wave

$$P(t) = \bar{p} + p(t) \tag{6.7.6}$$

where $\bar{p}$ is the mean value of $P(t)$ over one oscillatory cycle, which is also referred to as the "steady part" of $P(t)$, and $p(t)$ is the purely oscillatory part of $P(t)$, meaning, as we saw earlier, that the mean value of $p(t)$ over one oscillatory cycle is zero. Being generally a *composite* wave, this part of the driving pressure wave is represented in terms of its harmonics which we shall denote by $p_n(t)$, where $n = 1, 2, \ldots, N$ and N is the total number of harmonics used in that representation. It was seen in Chapter 5 that for any composite waveform these harmonics can be put in the form

$$p_n(t) = M_n \cos \left(\omega_n t - \phi_n \right) \qquad n = 1, 2, \ldots, N \tag{6.7.7}$$

where M_n are the Fourier coefficients and ϕ_n the phase angles discussed in Chapter 5, and where time t has been normalized such that the oscillatory cycle is in the interval $t = 0$ to $t = 1$, which is equivalent to taking the period of oscillation $T = 1$. We shall use this normalization throughout this chapter. Furthermore, the pressure-flow analysis is greatly simplified if the harmonics in Eq. 6.7.7 are seen as the real parts of the corresponding complex set of harmonics, namely

$$p_n(t) = p_{nr}(t) \tag{6.7.8}$$

$$p_{nr}(t) = \Re\left\{ M_n e^{i(\omega_n t - \phi_n)} \right\} \tag{6.7.9}$$

$$= M_n \cos(\omega_n t - \phi_n) \qquad n = 1, 2, \ldots, N \tag{6.7.10}$$

The corresponding imaginary parts of these harmonics, which are required in the analysis below, are then given by

$$p_{ni}(t) = \Im\left\{ M_n e^{i(\omega_n t - \phi_n)} \right\} \tag{6.7.11}$$

$$= M_n \sin(\omega_n t - \phi_n) \qquad n = 1, 2, \ldots, N \tag{6.7.12}$$

The advantage of using the complex form of the harmonics of the pressure wave is that it makes the relation between these and the corresponding harmonics of the flow wave particularly simple, namely

$$\text{complex flow harmonic} = \frac{\text{complex pressure harmonic}}{Z_n} \tag{6.7.13}$$

and since the driving pressure wave in this presentation is composed of the *real* parts of the complex pressure harmonics (Eq. 6.7.9), the resulting flow wave is composed of the real parts of the complex flow harmonics, that is

$$q_{nr}(t) = \Re\left(\frac{M_n e^{i(\omega_n t - \phi_n)}}{Z_n} \right) \tag{6.7.14}$$

$$= \frac{p_{nr} z_{nr} + p_{ni} z_{ni}}{z_{nr}^2 + z_{ni}^2} \tag{6.7.15}$$

$$= \frac{R M_n \cos(\omega_n t - \phi_n) + \left(\omega_n L - \frac{1}{\omega_n C}\right) M_n \sin(\omega_n t - \phi_n)}{R^2 + \left(\omega_n L - \frac{1}{\omega_n C}\right)^2} \tag{6.7.16}$$

and the corresponding R-scaled flow harmonics are

$$R \times q_{nr}(t) = \frac{M_n \cos(\omega_n t - \phi_n) + \left(\omega_n t_L - \frac{1}{\omega_n t_C}\right) M_n \sin(\omega_n t - \phi_n)}{1 + \left(\omega_n t_L - \frac{1}{\omega_n t_C}\right)^2} \tag{6.7.17}$$

where $t_L \, (= L/R)$ and $t_C \, (= CR)$ are the inertial and capacitive time constants. Both constants are involved in this case because both inertial and capacitive effects are present. The complete R-scaled oscillatory flow wave is finally obtained by adding its individual harmonics, that is

$$R \times q_r(t) = R \times q_{1r}(t) + R \times q_{2r}(t) \ldots + R \times q_{Nr}(t) \qquad (6.7.18)$$

and the complete R-scaled flow wave (corresponding to real part of driving pressure) is finally given by

$$R \times Q(t) = R \times \bar{q} + R \times q_r(t) \qquad (6.7.19)$$

where $\bar{q}$ is the steady part of the flow wave corresponding to the steady part of the pressure wave $\bar{p}$, if any, the two being related by

$$\bar{q} = \frac{\bar{p}}{R} \qquad (6.7.20)$$

Results, comparing the R-scaled flow wave with the corresponding pressure wave, with $t_L = 0.1\ s$ and a range of values of t_C, and using the one-step, piecewise, and cardiac pressure waves, are shown in Fig. 6.7.1–3. They demonstrate that at the highest value of t_C the flow wave exhibits a behaviour similar to that of the overdamped dynamics of the LRC system observed in Sections 3.3 and 4.4–6. It must be remembered, however, that overdamped or underdamped conditions relate to the dynamics of the RLC system in the *transient*

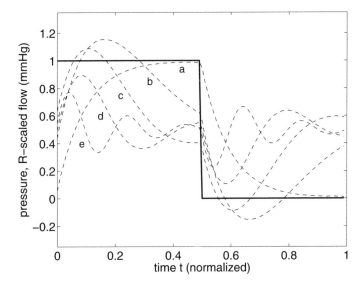

Fig. 6.7.1. Pressure (solid) and R-scaled flow waves (dashed) through the RLC system in series, with the inertial time constant $t_L = 0.1\ s$ and different values of the capacitive time constant t_C in seconds: (a) 5.0, (b) 0.2, (c) 0.1, (d) 0.04, (e) 0.01. The highest value of t_C corresponds to low compliance (more rigid balloon) and hence flow is dominated by inertial effects, and at low values of t_C the reverse is true.

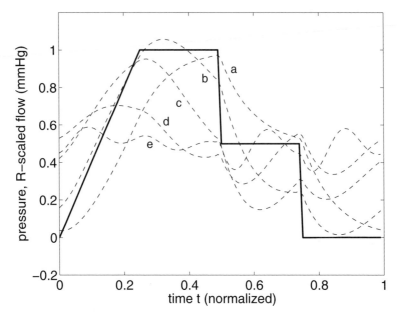

Fig. 6.7.2. See caption for Fig. 6.7.1.

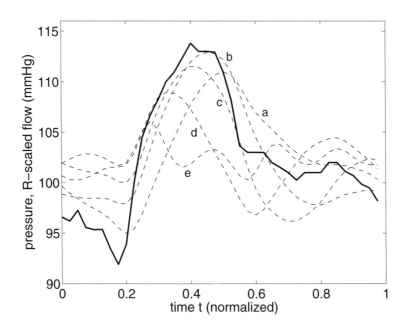

Fig. 6.7.3. See caption for Fig. 6.7.1.

state only, while the results in this section and in the context of lumped models in general relate to the system in *steady state* only. Nevertheless, the mechanisms underlying the different behaviour observed in Figs. 6.7.1, 2 are similar to those underlying overdamped and underdamped behaviour. Here, at the highest value of t_C, which indicates low compliance (or a more rigid balloon), the balloon filling occurs without a recoil or oscillations, while at lower values of t_C, because of higher compliance of the balloon, the reverse is true. This phenomenon is seen more clearly in Figs. 6.7.1, 2 because the composite pressure waves in these two figures contain step changes in pressure. It is as if these step changes create "micro transient states" within the oscillatory cycle. In Fig. 6.7.3 this does not occur as distinctly because there are no step changes in pressure, although here too, the curve with the highest value of t_C is the one with the least oscillations.

Another way of looking at the results in Figs. 6.7.1–3, is that at higher values of t_C capacitance effects are small and the dynamics of the system are dominated by inertial effects, so that the results resemble those of resistance and inertial effects in series observed in Section 6.5. At lower values of t_C capacitive effects become more significant to the point of dominating the dynamics, and the results then resemble those of resistance and capacitance effects in series observed in Section 6.6. In that section, however, inertial effects were entirely absent while Figs. 6.7.1–3 still contain inertial effects, with

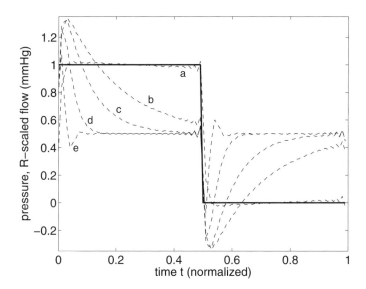

Fig. 6.7.4. Pressure (solid) and R-scaled flow waves (dashed) through the *RLC* system in series, as in Figs. 6.7.1–3 but with the inertial time constant $t_L = 0.01\,s$. This change reduces the inertial effects to an insignificant level, making the system resemble that of resistance and capacitance in series considered in Section 6.6 and seen in Figs. 6.6.1–3.

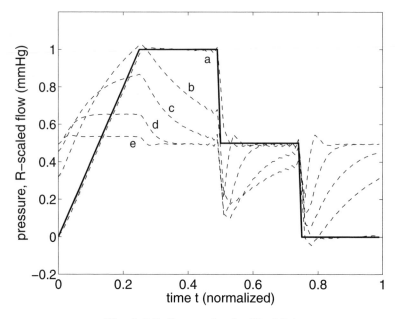

Fig. 6.7.5. See caption for Fig. 6.7.4.

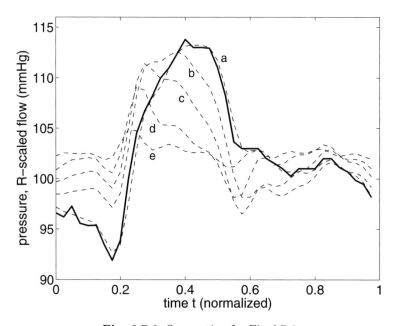

Fig. 6.7.6. See caption for Fig. 6.7.4.

$t_L = 0.1\,s$. To make the two situations more comparable we must reduce the inertial effect yet further. Indeed, Figs. 6.7.4–6 show results with the same values of t_C as in Figs. 6.7.1–3 but with $t_L = 0.01$. The flow waves now closely resemble those in Section 6.6, Figs. 6.6.1–3.

6.8 Composite Pressure-Flow Relations Under RLC in Parallel

The RLC system in parallel again provides a model in which the elements of resistance, inductance, and capacitance, are all present, but here their effects are independent of each other because of their parallel arrangement. However, while flow through each element is not affected by the other two elements, total flow through the system is of course affected by all three, thus each element continues to affect total pressure-flow relation. Again, while free and forced dynamics of this system have been examined in previous chapters using zero, linear, or sinusoidal driving pressures, in this section we examine pressure flow relations using composite waves.

The complex impedance for R, L, C in parallel was found in Section 4.9, Eq. 4.9.44, namely

$$\frac{1}{Z} = \frac{1}{R} + i\left(\omega C - \frac{1}{\omega L}\right) \tag{6.8.1}$$

so that

$$Z = \frac{R - iR^2\left(\omega C - \frac{1}{\omega L}\right)}{1 + R^2\left(\omega C - \frac{1}{\omega L}\right)^2} \tag{6.8.2}$$

where ω is the angular frequency of oscillation of the driving pressure. For a composite pressure wave consisting of N harmonics the impedance will be different for each of these harmonics because of their different frequencies, and we shall use the notation

$$Z_n = \frac{R - iR^2\left(\omega_n C - \frac{1}{\omega_n L}\right)}{1 + R^2\left(\omega_n C - \frac{1}{\omega_n L}\right)^2} \tag{6.8.3}$$

$$\omega_n = 2\pi n \qquad n = 1, 2, \ldots, N \tag{6.8.4}$$

where n denotes a particular harmonic and ω_n is the angular frequency of that harmonic. For the real and imaginary parts of the impedance, using subscripts r, i as before, we have in this case

$$z_{nr} = \frac{R}{1 + R^2 \left(\omega_n C - \frac{1}{\omega_n L}\right)^2} \tag{6.8.5}$$

$$z_{ni} = \frac{-R^2 \left(\omega_n C - \frac{1}{\omega_n L}\right)}{1 + R^2 \left(\omega_n C - \frac{1}{\omega_n L}\right)^2} \tag{6.8.6}$$

Following the same steps as in the previous section, and omitting some of the details, we find the harmonics of the oscillatory flow wave

$$q_{nr}(t) = \Re \left(\frac{M_n e^{i(\omega_n t - \phi_n)}}{Z_n}\right) \tag{6.8.7}$$

$$= \frac{p_{nr} z_{nr} + p_{ni} z_{ni}}{z_{nr}^2 + z_{ni}^2} \tag{6.8.8}$$

$$= \frac{M_n \cos\left(\omega_n t - \phi_n\right) - R\left(\omega_n C - \frac{1}{\omega_n L}\right) M_n \sin\left(\omega_n t - \phi_n\right)}{R} \tag{6.8.9}$$

the corresponding R-scaled flow harmonics

$$R \times q_{nr}(t) = M_n \cos\left(\omega_n t - \phi_n\right)$$
$$- \left(\omega_n t_C - \frac{1}{\omega_n t_L}\right) M_n \sin\left(\omega_n t - \phi_n\right) \tag{6.8.10}$$

and the complete R-scaled oscillatory flow wave

$$R \times q_r(t) = R \times q_{1r}(t) + R \times q_{2r}(t) \ldots + R \times q_{Nr}(t) \tag{6.8.11}$$

using the same notation as in the previous section.

Results, comparing the R-scaled flow wave with the corresponding pressure wave, taking $t_L = 0.1\,s$ and $t_C = 0.01\,s$, and using the one-step, piecewise, and cardiac pressure waves, are shown in Fig. 6.8.1–3. Also shown separately in these figures are the R-scaled resistive, inductive, and capacitive flows, which we shall denote by $q_{res}(t), q_{ind}(t), q_{cap}(t)$, respectively, and which together make up the total flow wave, that is,

$$q(t) = q_{res}(t) + q_{ind}(t) + q_{cap}(t) \tag{6.8.12}$$

Since these three elements of the flow are in parallel in this case, their harmonics, to be denoted by $q_{n,res}(t), q_{n,ind}(t), q_{n,cap}(t)$, are subject to different impedances, which we shall denote by $Z_{n,res}, Z_{n,ind}, Z_{n,cap}$ and which are given by (Eqs. 4.9.34–36)

$$Z_{n,res} = R \tag{6.8.13}$$
$$Z_{n,ind} = i\omega_n L \tag{6.8.14}$$
$$Z_{n,cap} = \frac{1}{i\omega_n C} \tag{6.8.15}$$

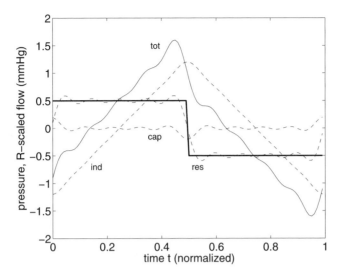

Fig. 6.8.1. Pressure (heavy solid) and R-scaled total flow wave (tot) through the RLC system in parallel, with the inertial time constant $t_L = 0.1\,s$ and the capacitive time constant $t_C = 0.01\,s$. The dashed curves show the resistive (res), inductive (ind), and capacitive (cap) flows. It is seen that with these values of the time constants, flow through the system is dominated by the inductive and resistive components, capacitive flow is small by comparison. To facilitate the comparison, only the oscillatory parts of the pressure and flow waves are shown.

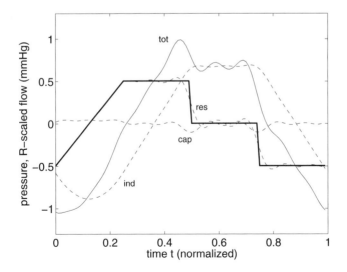

Fig. 6.8.2. See caption for Fig. 6.8.1.

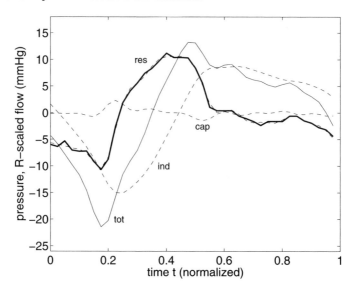

Fig. 6.8.3. See caption for Fig. 6.8.1.

The relation between the pressure and flow harmonics for the individual parallel flows is the same as that for the harmonics of total flow, that is, as in Eq. 6.8.7, here

$$q_{nr,res}(t) = \Re \left(\frac{M_n e^{i(\omega_n t - \phi_n)}}{Z_{n,res}} \right) \qquad (6.8.16)$$

$$q_{nr,ind}(t) = \Re \left(\frac{M_n e^{i(\omega_n t - \phi_n)}}{Z_{n,ind}} \right) \qquad (6.8.17)$$

$$q_{nr,cap}(t) = \Re \left(\frac{M_n e^{i(\omega_n t - \phi_n)}}{Z_{n,cap}} \right) \qquad (6.8.18)$$

Evaluating these by following the step as before, and omitting the details, we find

$$R \times q_{nr,res}(t) = M_n \cos(\omega_n t - \phi_n) \qquad (6.8.19)$$

$$R \times q_{nr,ind}(t) = \left(\frac{M_n \sin(\omega_n t - \phi_n)}{\omega_n t_L} \right) \qquad (6.8.20)$$

$$R \times q_{nr,cap}(t) = -\omega_n t_C M_n \sin(\omega_n t - \phi_n) \qquad (6.8.21)$$

The individual parallel flow waves shown in Figs. 7.3.1–3 are obtained by adding the harmonics of each flow, that is,

$$R \times q_{r,res}(t) = R \times q_{1r,res} + R \times q_{2r,res} + \ldots + R \times q_{Nr,res} \qquad (6.8.22)$$

$$R \times q_{r,ind}(t) = R \times q_{1r,ind} + R \times q_{2r,ind} + \ldots + R \times q_{Nr,ind} \quad (6.8.23)$$
$$R \times q_{r,cap}(t) = R \times q_{1r,cap} + R \times q_{2r,cap} + \ldots + R \times q_{Nr,cap} \quad (6.8.24)$$

Only the oscillatory parts of pressure and flow waves are shown, the steady parts are omitted to make the graphical comparison easier by using the same scale.

The results in Figs. 6.8.1–3 indicate that with $t_L = 0.1\,s$ and $t_C = 0.01\,s$ capacitance effects are small and total flow is dominated by inertial effects. By contrast, when inductance and capacitance are in *series*, the same values of the time constants lead to a flow dominated by capacitive effects as seen earlier in Figs. 6.7.1–3. If the capacitive time constant is increased to $t_C = 0.3\,s$, which means higher compliance (a more elastic balloon), keeping the inertial time constant unchanged, total flow becomes dominated by capacitive effects as seen in Figs. 6.8.4–6. Because the two elements are in parallel, however, this does not change the resistive and the inductive flows.

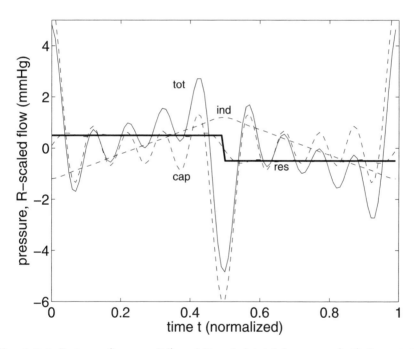

Fig. 6.8.4. Pressure (heavy solid) and R-scaled total flow wave (tot) through the *RLC* system in parallel, as in Figs. 6.8.1–3, with the value of the inertial time constant unchanged at $t_L = 0.1\,s$ but the value of the capacitive time constant is increased to $t_C = 0.3\,s$. The increase in the value of t_C corresponds to higher compliance (a more elastic balloon), hence total flow is seen to be dominated by capacitive effects. The resistive (res) and inductive (ind) flows are unaffected by this change because the elements are in parallel.

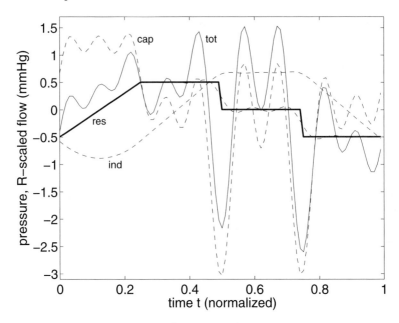

Fig. 6.8.5. See caption for Fig. 6.8.4.

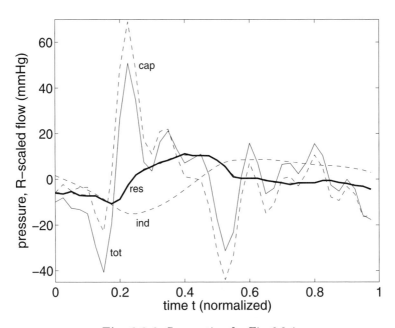

Fig. 6.8.6. See caption for Fig. 6.8.4.

6.9 Summary

Coronary blood flow may be disrupted by an obstruction in the *vessels* conveying the flow or by a disruption in the delicate *dynamics* of the flow, the dynamics of the coronary circulation. The focus in this book is on the latter. At the core of the dynamics of the coronary circulation is the relation between the form of the composite pressure wave driving the flow and the form of the resulting flow wave.

The simple relation between the amplitudes and phase angles of pressure and flow sine and cosine waves cannot be applied when pressure and flow are *composite* waves, but they can be applied to their individual harmonics. Furthermore, the flow contributions from individual harmonics can then be *added* to obtain the composite flow wave because the equation governing the flow is *linear*.

When the opposition to flow consists of pure resistance, a composite pressure wave produces a composite flow wave of identical form but shifted by only the units of pressure and flow. For the study of pressure-flow relations it is convenient to remove this shift so that the two waves actually coincide. This can be done by scaling the flow wave by the resistance. The resulting "R-scaled" flow wave is useful not only when the resistance to flow consists of pure resistance but also, and particularly, when inertial and capacitive effects are involved too.

The flow wave associated with a composite pressure wave cannot be obtained directly and in full when opposition to flow consists of general impedance rather than pure resistance. In this case, direct pressure-flow relations exist only between the *individual harmonics* of pressure and flow. Thus, harmonics of the flow wave must be obtained individually first, then these are used to construct the flow wave as a whole.

Inertial effects can lead to large swings in the R-scaled flow waveform, away from the corresponding pressure waveform. When the inductor is in series with a resistor, as is the case in the physiological system under normal circumstances, these swings are controlled by the presence of the resistor. When the two elements are in parallel, however, flow swings may be very large as they are free from the effects of the resistor. While in the coronary circulation a parallel inductive flow can only occur in the presence of a breach within the coronary vasculature, the results indicate clearly that any change in effective inductance L, locally or of the system as a whole, may upset the delicate dynamics of the system and hence the relation between pressure and flow. Drugs affecting the consistency of blood or the caliber of blood vessels, for example, which are generally considered as targeting only the resistance to flow, will actually also alter inertial effects within the system.

Like inertial effects, capacitive effects can lead to large swings in the R-scaled flow waveform, away from the corresponding pressure waveform. When the capacitance is in series with the resistance, these swings are constrained by the presence of the resistance, but when the two elements are in parallel, the flow swings may be very large because they are free from the effects of the resistor. In the coronary circulation, and in the cardiovascular system in general, capacitance is always in parallel with resistance because it is provided by the elasticity of the conducting vessels while resistance is provided by viscosity of the fluid, thus the flow always has the option of flowing through *or* inflating the vessels. The two options are in parallel. Any change in this property of the coronary system, which may be brought about, for example, by vasodilator drugs which cause the conducting vessels to fully inflate and thereby lose their ability to provide any further compliance, may disrupt the normal dynamics of the system because of the absence of normal capacitive effects. Similarly, vascular spasm, whether it is induced by drugs or by regulatory mechanisms, will also alter the normal compliance of the system and thereby disrupt its normal dynamics.

The dynamics of the RLC system in series depend not only on the *relative* values of t_C and t_L but on their individual values too. The reason for this is that even when compliance is high, capacitive effects cannot dominate the flow because they are still constrained by any remaining inertial effects in the system. Only when the latter are reduced to an insignificant level do capacitance effects dominate. While in the coronary circulation, as stated previously, capacitance effects are always in parallel, the RLC system in series provides an important "ground state" reference for parallel and hybrid lumped models.

Capacitive effects are much more pronounced when capacitance is in *parallel* with other elements of the RLC system. This is important because capacitive effects in the coronary circulation are in fact in parallel, as they are caused by elasticity of the conducting vessels. Therefore, coronary flow has the (parallel) option of flowing through or inflating the vessels. Thus, again, a change in the capacitive property of the coronary arteries, by drugs, spasm, or disease can drastically change the character of the flow wave.

7

Lumped Models

7.1 Introduction

A succession of lumped models, guided by a series of observations and sometimes heroic experiments, have been the principal means by which an understanding of the dynamics of the coronary circulation has evolved to this point. The body of work associated with this effort has become so large and its thread so intricate that it has become almost impossible to give an accurate account of it without committing some historic or material errors. The following sampling, in chronological order, provides some sign posts which will lead the keen reader to many more references: [85, 86, 83, 169, 19, 22, 111, 49, 59, 65, 128, 40, 121, 32, 110, 102, 157, 24, 33, 130, 195, 29, 107, 90, 115, 97, 98, 183, 131, 162, 163, 47].

The early notion of the coronary circulation as a "windkessel", a combination of resistance and capacitance, was a natural off-shoot from the dynamics of the *systemic* circulation as it was understood at the time [135, 141, 153], but the special characteristics of the coronary circulation soon became apparent [83]. The mechanically hostile milieu in which coronary vasculature is embedded, the peculiar and mostly diastolic coronary flow wave, the high coronary flow reserve, and the severe and multifaceted regulatory environment in which the coronary circulation operates were special issues that had to be dealt with. While some progress has been made in each case, essentially the same issues remain outstanding today. Whether because a certain characteristic cannot be accurately modelled or because it can be modelled in more than one way, an all-encompassing model, lumped or otherwise, able to deal with these issues as they combine in the dynamics of the coronary circulation is yet to emerge. The subject remains very much a "work-in-progress".

Most lumped models of the coronary circulation to date have been based on essentially three types of elements, namely the elements of the RLC system. However, the total number of elements used in a given model, the number of each type, and the number of different ways in which these can be arranged have provided the scope for a wide range of different models. The purpose of this chapter is not to enumerate these models but to return to the foundations on which they stand. We return to the basic elements of the RLC system and proceed from there in a systematic manner to examine the way in which they may give rise to some of the characteristic features observed in the dynamics of the coronary circulation. The intention here is to provide not complete models but the conceptual ingredients from which such models would be constructed.

As we have seen in earlier chapters, in the RLC system the resistance R is taken to represent the viscous resistance between the moving fluid and the vessel wall, the inductance L to represent the inertia of the moving fluid, and the capacitance C the elasticity of the vessel wall. These three effects provide an appropriate starting point because it is known on purely physical grounds that these effects do exist in the coronary circulation and must therefore play a role in its dynamics. Of course, other effects exist too: viscoelasticity within the vessel wall, intramyocardial pressure surrounding coronary vessels, wave reflections within the coronary vascular tree, and myogenic vasomotor activity and other control mechanisms, but these are all seen to exist at a higher level of complexity. Models of the coronary circulation presented in the past have been mostly at this higher level. In this chapter we focus on the foundations on which these models are based.

Specifically, we examine four different arrangements of the RLC elements which deal with different dynamical issues in somewhat increasing degree of complexity. In a sense they are *basic lumped models* which may be referred to as $LM0, LM1, LM2, LM3$. As a shorthand notation we present the elements of the model inside curly brackets, with a *comma* representing a *parallel* connection between two elements and a *plus sign* representing a *series* connection. In this notation the four models are given by:

$$LM0 : \{R, C\} \qquad \text{"windkessel"} \qquad (7.1.1)$$

$$LM1 : \{R_1, \{R_2 + C\}\} \qquad \text{viscoelastic, viscoelasticity} \quad (7.1.2)$$

$$LM2 : \{\{R_1 + L\}, \{R_2 + C\}\} \qquad \text{inertia (inductance)} \qquad (7.1.3)$$

$$LM3 : \{\{R_1 + (P_b)\}, \{R_2 + C\}\} \qquad \text{back pressure} \qquad (7.1.4)$$

The models are examined in more detail below. In particular, pressure-flow relations under these models are examined, using the cardiac pressure wave as the driving pressure.

7.2 LM0: {R,C}

The simplest lumped model of the coronary circulation, which we shall refer to as $LM0$, is the so-called "windkessel" model, which was first devised for the cardiovascular system as a whole [135, 141, 153]. In that context, it was recognized early that the cardiac pressure pulse generated by the left ventricle is not transmitted directly to the periphery but is first absorbed by the compliance of the aorta and its major branches. It is as if the energy of the pulse is first expended on inflating a balloon, and then as the pulse abates, the balloon deflates and returns this energy to drive the flow downstream somewhat more gently than the original pulse. The same scenario is believed to occur in the coronary circulation.

Since the two main coronary arteries that bring blood supply into the coronary circulation have their origin at the base of the ascending aorta just as it leaves the left ventricle (Figs. 1.3.1, 2), flow entering these arteries is subject to the full force of the cardiac pressure pulse. And it is well established that the main coronary arteries have a considerable degree of compliance, which can in fact be easily observed in the course of coronary cine-angiography. Thus, the ingredients for a windkessel scenario, namely a pulsating pressure and compliant vessels, are present in the coronary circulation as they are in the cardiovascular system as a whole. Indeed, compliance, or capacitance effects, within the coronary circulation are believed to be the result of not only the elasticity of the coronary vessels but also the enormous contraction and relaxation of the cardiac muscle tissue in which many of the coronary vessels are embedded. Thus, capacitance effects rank high in the dynamics of the coronary circulation.

As discussed at great length in Chapter 2, capacitance effects in the coronary circulation and in the cardiovascular system as a whole do not occur in isolation but in combination with the ever-present resistance to flow due to viscous effects between moving blood and the vessel wall. If flow in the coronary circulation were *steady*, the most elementary model of the circulation would consist of only a driving pressure and the effect of that resistance, as discussed in Section 2.4. But because of the pulsatile nature of the driving pressure, the most elementary model of the coronary circulation must take into account the effects of capacitance, and thus include both capacitance and resistance. Furthermore, the nature of capacitance effects in the coronary circulation is such that flow has the option of moving against the resistance *or* capacitance, that is, the option of moving forward *or* inflating the vessels. In other words, the effects of capacitance and resistance are *in parallel*. In the context of previous sections, therefore, the most elementary model of the coronary circulation is RC in parallel, as shown in Fig. 7.2.1.

The complex impedance for RC in parallel was found in Section 4.9, Eqs.4.9.34,35, namely

$$\frac{1}{Z} = \frac{1}{R} + i\omega C \qquad (7.2.1)$$

so that

$$Z = \frac{R}{1 + iR\omega C} \qquad (7.2.2)$$

where ω is the angular frequency of oscillation of the driving pressure. For a composite pressure wave consisting of N harmonics the impedance will be

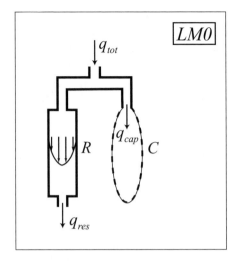

Fig. 7.2.1. The most elementary model of the coronary circulation is a resistance R and capacitance C in parallel, originally known as the "windkessel" model, and which we shall refer to as $LM0 : \{R,C\}$.

different for each harmonic because of their different frequencies and, as before, we shall use the notation

$$Z_n = \frac{R}{1 + iR\omega_n C} \tag{7.2.3}$$

$$\omega_n = 2\pi n \qquad n = 1, 2, \ldots, N \tag{7.2.4}$$

where n denotes a particular harmonic and ω_n is the angular frequency of that harmonic.

Following the same steps as in Chapter 6, and omitting some of the details, we find the harmonics of the oscillatory flow wave as

$$q_{nr}(t) = \Re\left(\frac{p_n(t)}{Z_n}\right) \qquad n = 1, 2, \ldots, N \tag{7.2.5}$$

where $p_n(t)$ are the harmonics of the driving pressure in their complex exponential form, namely

$$p_n(t) = M_n e^{i(\omega_n t - \phi_n)} \qquad n = 1, 2, \ldots, N \tag{7.2.6}$$

Thus, Eq. 7.2.5 gives

$$q_{nr}(t) = \Re\left\{\frac{M_n e^{i(\omega_n t - \phi_n)}}{R/(1 + iR\omega_n C)}\right\} \tag{7.2.7}$$

$$= \frac{1}{R} M_n \cos(\omega_n t - \phi_n) - \omega_n C M_n \sin(\omega_n t - \phi_n) \tag{7.2.8}$$

$$n = 1, 2, \ldots, N$$

and the corresponding R-scaled flow harmonics are

$$R \times q_{nr}(t) = M_n \cos{(\omega_n t - \phi_n)} - \omega_n t_C M_n \sin{(\omega_n t - \phi_n)} \qquad (7.2.9)$$
$$n = 1, 2, \ldots, N$$

where t_C is the capacitive time constant ($= RC$). The complete R-scaled oscillatory flow wave (corresponding to real part of driving pressure) is finally given by

$$R \times q_r(t) = R \times q_1(t) + R \times q_2(t) \ldots + R \times q_N(t) \qquad (7.2.10)$$

Eq. 7.2.10 gives total flow into the parallel system (corresponding to real part of driving pressure), denoted by q_{tot} in Fig. 7.2.1. This total flow consists of resistive and capacitive components, denoted by q_{res}, q_{cap}, respectively, in that figure, that is

$$q(t) = q_{tot} = q_{res}(t) + q_{cap}(t) \qquad (7.2.11)$$

Since here these elements of the flow are in parallel, their harmonics, to be denoted by $q_{n,res}(t), q_{n,cap}(t)$, are subject to different impedances, which we shall denote by $Z_{n,res}, Z_{n,cap}$ and which are given by (Eqs.4.9.34,36)

$$Z_{n,res} = R \qquad (7.2.12)$$

$$Z_{n,cap} = \frac{1}{i\omega_n C} \qquad (7.2.13)$$

The relation between the pressure and flow harmonics for the individual parallel flows is the same as that for the harmonics of total flow, that is, as in Eq. 7.2.5, here we have

$$q_{nr,res}(t) = \Re\left\{ \frac{p_n(t)}{Z_{n,res}} \right\} \qquad (7.2.14)$$

$$= \Re\left\{ \frac{M_n e^{i(\omega_n t - \phi_n)}}{R} \right\} \qquad (7.2.15)$$

$$q_{nr,cap}(t) = \Re\left\{ \frac{p_n(t)}{Z_{n,cap}} \right\} \qquad (7.2.16)$$

$$= \Re\left\{ \frac{M_n e^{i(\omega_n t - \phi_n)}}{1/(i\omega_n C)} \right\} \qquad (7.2.17)$$

Evaluating these by following the same step as before, and omitting the details, we find

$$R \times q_{nr,res}(t) = M_n \cos{(\omega_n t - \phi_n)} \qquad (7.2.18)$$

$$R \times q_{nr,cap}(t) = -\omega_n t_C M_n \sin{(\omega_n t - \phi_n)} \qquad (7.2.19)$$

The individual parallel flow waves are obtained by adding the harmonics of each flow, that is

$$R \times q_{r,res}(t) = R \times q_{1r,res} + R \times q_{2r,res} + \ldots + R \times q_{Nr,res} \quad (7.2.20)$$
$$R \times q_{r,cap}(t) = R \times q_{1r,cap} + R \times q_{2r,cap} + \ldots + R \times q_{Nr,cap} \quad (7.2.21)$$

As before, only the oscillatory parts of the pressure and flow waves are shown, the steady parts being omitted to make the graphical comparison of these waves easier by using the same scale.

Results, comparing total and individual R-scaled flow waves, using the cardiac pressure wave and a relatively low value of the capacitive time constant, namely $t_C = 0.01\,s$, are shown in Fig. 7.2.2. At this value of t_C, which corresponds to low compliance (the balloon is stiff), capacitive flow is small. Resistive flow, which is independent of capacitive effects because of the parallel arrangement, is identical in form with the pressure waveform on an R-scaled basis and is thus represented by the same curve as the pressure wave in Fig. 7.2.2. Total flow, again on an R-scaled basis, is only slightly different in form from the pressure waveform, the difference being due to the small capacitive flow.

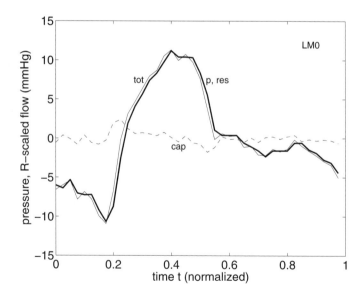

Fig. 7.2.2. Pressure-flow relations under $LM0$, with a relatively low value of the capacitive time constant, namely $t_C = 0.01\,s$. Heavy solid curve (p, res) represents both the pressure wave and the R-scaled resistive flow, which are identical because the resistance is in parallel. Thin solid curve (tot) represents total flow into the parallel system, and the dashed curve (cap) represents capacitive flow. Total flow is dominated by resistive flow and is only slightly affected by capacitance.

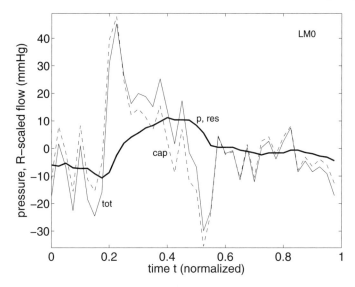

Fig. 7.2.3. Pressure-flow relations under $LM0$ as in Fig. 7.2.2, but here with a considerably higher value of the capacitive time constant, namely $t_C = 0.2\,s$. Total flow (tot) is now dominated by capacitive effects and follows the curve of capacitive flow (cap).

At a higher value of the capacitive time constant, namely $t_C = 0.2\,s$, the situation is drastically changed as seen in Fig. 7.2.3. Here compliance is considerably higher (the balloon is more elastic), allowing correspondingly higher capacitive flow. The form of the R-scaled total flow wave is considerably different from that of the pressure wave because total flow is dominated by capacitive flow. Resistive flow is the same as in Fig. 7.2.2, but here, because of much higher capacitive flow, resistive flow is a relatively less significant part of total flow.

It is important to recall that capacitive flow is driven not by a pressure difference but by the rate of change of a pressure difference, as discussed at great length in Section 2.6. In present notation, where $p(t)$ actually represents a pressure difference, this means that capacitive flow depends not on $p(t)$ but on the derivative of $p(t)$. Indeed, following Eq. 2.6.8, here we have

$$q_{cap}(t) = C\frac{dp(t)}{dt} \qquad (7.2.22)$$

which in R-scaled form gives

$$R \times q_{cap}(t) = t_C\frac{dp(t)}{dt} \qquad (7.2.23)$$

This equation indicates clearly that R-scaled capacitive flow is in fact proportional to the *slope* of the pressure curves in Figs. 7.2.2, 3, the constant of

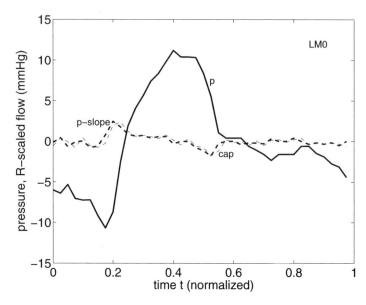

Fig. 7.2.4. Flow under $LM0$, with $t_C = 0.01\,s$. Capacitive flow (cap) is driven not by a pressure difference but by the rate of change of a pressure difference. Graphically, the form of the capacitive flow curve is dictated not by the form of the pressure curve (p) but by the slope of that curve (p-slope).

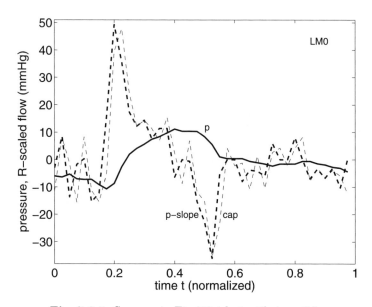

Fig. 7.2.5. Same as in Fig. 7.2.4 but with $t_C = 0.2\,s$.

proportionality being the capacitive time constant t_C. Thus, if we compare the R-scaled capacitive flow, $R \times q_{cap}(t)$ not with the pressure curve but with its slope, we expect agreement between the two. This comparison is shown in Figs. 7.2.4, 5.

7.3 LM1: {R₁,{R₂+C}}

It is clear from results of the previous section that capacitive effects in the coronary circulation are not likely to be present in isolation, or in pure form, because this would lead to large swings in the flow waveform which are not observed in the physiological system. This has led to the view that the compliance of blood vessels, which is responsible for capacitance effects, is produced by viscoelasticity rather than pure elasticity of the vessel wall [79, 204, 78, 52, 55, 4]. Thus, capacitive flow, the filling of the balloon, is resisted not only by the elasticity of the vessel wall but also by some viscoelastic forces within the wall.

We recall from Section 2.6 that in the case of purely elastic capacitance the relation between the pressure $p(t)$ inside a balloon and the rate of flow $q(t)$ into it is such that the flow rate depends on the *rate of change of pressure*, dp/dt, rather than on the pressure itself. We refrain from calling this a "pressure gradient" because it is a rate of change of pressure with *time*, not to be confused with a rate of change of pressure along a tube. The relation is such that

$$\frac{dp}{dt} = \frac{1}{C}q \qquad \text{(purely elastic capacitance)} \qquad (7.3.1)$$

where C is the capacitance constant. The flow rate (into the balloon) and the rate of change of pressure have the same sign, so that when the rate of change of pressure is positive (pressure is increasing), flow rate into the balloon is positive, and vice versa. When the rate of change of pressure is zero, capacitive flow, flow into or out of the balloon, is zero.

When the balloon wall is not purely elastic but has some viscoelastic component, the relation between pressure and flow is of the form

$$\frac{dp}{dt} = \frac{1}{C}q + B\frac{dq}{dt} \qquad \text{(viscoelastic capacitance)} \qquad (7.3.2)$$

where B is a constant relating to the viscoelastic property of the wall. Here the rate of change of pressure required to maintain flow into or out of the balloon depends not only on the flow rate but on the rate of change of flow rate. The consequence of this in modelling the coronary circulation is that it changes the complex impedance for a capacitive element.

We recall from Section 4.9 that the complex impedance for a purely elastic capacitor is given by (Eq. 4.9.36)

$$Z = \frac{1}{i\omega C} \quad \text{(purely elastic capacitance)} \quad (7.3.3)$$

Following the same step as in Section 4.9 to find the complex impedance for a viscoelastic capacitor, we consider an oscillatory pressure in complex exponential form

$$p(t) = p_0 e^{i\omega t} \quad (7.3.4)$$

where p_0 is a constant, and substitute in Eq. 7.3.2 to get

$$B\frac{dq}{dt} + \frac{1}{C}q = ip_0\omega e^{i\omega t} \quad (7.3.5)$$

To find the complex impedance we solve this equation for $q(t)$, recalling from Section 4.9 that only the particular part of the solution is required. Following the same steps as in Section 4.9, we readily find

$$q(t) = \frac{ip_0\omega e^{i\omega t}}{iB\omega + \frac{1}{C}} \quad (7.3.6)$$

Using Eq. 7.3.4 for $p(t)$ and rearranging, this becomes

$$q(t) = \frac{p(t)}{B + \frac{1}{i\omega C}} \quad (7.3.7)$$

or

$$q(t) = \frac{p(t)}{Z} \quad (7.3.8)$$

where

$$Z = B + \frac{1}{i\omega C} \quad (7.3.9)$$

Comparing this with the results in Section 4.9, we see that the complex impedance for viscoelastic capacitance is the same as that for a purely elastic capacitance *plus a resistance in series*, as in Eq. 6.6.1, namely

$$Z = R + \frac{1}{i\omega C} \quad \text{(viscoelastic capacitance)} \quad (7.3.10)$$

Thus, to incorporate viscoelasticity into the windkessel model we simply add a resistance in series with the purely elastic capacitance in that model, which leads to the model shown in Fig. 7.3.1. In current notation the resulting model is $\{R_1 + \{R_2 + C\}\}$ and we shall refer to it as $LM1$. It is sometimes referred to as the viscoelastic windkessel model.

The two resistances R_1, R_2 in this model are clearly different and for modelling purposes must be allowed to assume different values. We shall find that if we continue to use the concept of R-scaled flow as being the product of

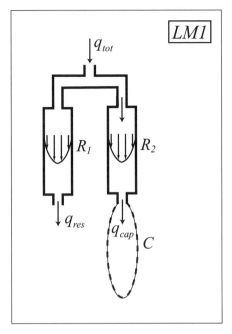

Fig. 7.3.1. A modified (viscoelastic) "windkessel" model in which capacitance is assumed to be not purely elastic as in $LM0$ but viscoelastic, which is equivalent to having a purely elastic capacitance C in series with a resistance R_2. Total flow through the system then consists of q_{res} along one of the two parallel branches and q_{cap} along the other.

resistance and flow, namely $R \times q$, and if we now use R_1 in this product, that is define R-scaled flow as $R_1 \times q$, then only the ratio of the two resistances is required subsequently. In other words, if we introduce

$$\lambda = \frac{R_2}{R_1} \qquad (7.3.11)$$

then only the value of λ is required in subsequent analysis.

The complex impedances along the two branches of $LM1$, using results in Section 4.9, are given by

$$Z_{res} = R_1 \qquad (7.3.12)$$

$$Z_{cap} = R_2 + \frac{1}{i\omega C} \qquad (7.3.13)$$

For convenience, we continue to refer to the two branches of the model as the "resistive" branch and the "capacitive" branch even though there is resistance in both branches. The complex impedance for the entire system is then given by

$$\frac{1}{Z} = \frac{1}{Z_{res}} + \frac{1}{Z_{cap}} \qquad (7.3.14)$$

$$Z = \frac{R_1 + i\omega R_1 R_2 C}{1 + i\omega C(R_1 + R_2)} \qquad (7.3.15)$$

where ω is the angular frequency of oscillation of the driving pressure. For the individual harmonics of a composite pressure wave consisting of N harmonics, in the notation of the previous section, we then have

$$Z_{n,res} = R_1 \qquad (7.3.16)$$

$$Z_{n,cap} = R_2 + \frac{1}{i\omega_n C} \qquad (7.3.17)$$

$$Z_n = \frac{R_1 + i\omega_n R_1 R_2 C}{1 + i\omega_n C(R_1 + R_2)} \qquad (7.3.18)$$

where n denotes a particular harmonic and ω_n is the angular frequency of that harmonic.

For a composite pressure wave consisting of the following harmonics (in their complex exponential form)

$$p_n(t) = M_n e^{i(\omega_n t - \phi_n)} \qquad n = 1, 2, \dots, N \qquad (7.3.19)$$

we obtain the corresponding harmonics of the flow wave, as in Chapter 6,

$$q_{nr}(t) = \Re\left\{ \frac{p_n(t)}{Z_n} \right\} \qquad (7.3.20)$$

$$= \Re\left\{ \frac{M_n e^{i(\omega_n t - \phi_n)}}{(R_1 + i\omega_n R_1 R_2 C)/(i\omega_n R_1 C)} \right\} \qquad (7.3.21)$$

$$= [(1 + \omega_n^2 C^2 R_2(R_1 + R_2))M_n \cos(\omega_n t - \phi_n) \\ -\omega_n R_1 C M_n \sin(\omega_n t - \phi_n)]/[R_1(1 + (\omega_n R_2 C)^2)] \qquad (7.3.22)$$

or in R-scaled form

$$R_1 \times q_{nr}(t) = [(1 + \omega_n^2 t_C^2 \lambda(1 + \lambda))M_n \cos(\omega_n t - \phi_n) \\ -\omega_n t_C M_n \sin(\omega_n t - \phi_n)]/[1 + (\omega_n \lambda t_C)^2] \qquad (7.3.23)$$

where t_C is the capacitive time constant ($= R_1 C$). Similarly, and omitting the details, we find the R-scaled resistive and capacitive flows

$$q_{nr,res}(t) = \Re\left\{ \frac{p_n(t)}{Z_{n,res}} \right\} \qquad (7.3.24)$$

$$R_1 \times q_{nr,res}(t) = M_n \cos(\omega_n t - \phi_n) \qquad (7.3.25)$$

$$q_{nr,cap}(t) = \Re\left\{\frac{p_n(t)}{Z_{n,cap}}\right\} \tag{7.3.26}$$

$$R_1 \times q_{nr,cap}(t) = \frac{\omega_n^2 t_C^2 \lambda M_n \cos(\omega_n t - \phi_n) - \omega_n t_C M_n \sin(\omega_n t - \phi_n)}{1 + (\omega_n \lambda t_C)^2}$$
$$\tag{7.3.27}$$

Finally, the total and the two parallel flow waves are obtained by adding the harmonics of each flow, that is, in R-scaled format

$$R_1 \times q_r(t) = R_1 \times q_{1r}(t) + R_1 \times q_{2r}(t)\ldots + R_1 \times q_{Nr}(t) \tag{7.3.28}$$

$$R_1 \times q_{r,res}(t) = R_1 \times q_{1r,res} + R_1 \times q_{2r,res} + \ldots + R_1 \times q_{Nr,res} \tag{7.3.29}$$

$$R_1 \times q_{r,cap}(t) = R_1 \times q_{1r,cap} + R_1 \times q_{2r,cap} + \ldots + R_1 \times q_{Nr,cap} \tag{7.3.30}$$

As before, only the oscillatory parts of the pressure and flow waves are used, the steady parts being omitted to make the graphical comparison of these waves easier by using the same scale.

Results, demonstrating the effects of viscoelasticity on the relation between pressure and flow waves are shown in Figs.7.3.2-4. In all three figures the value

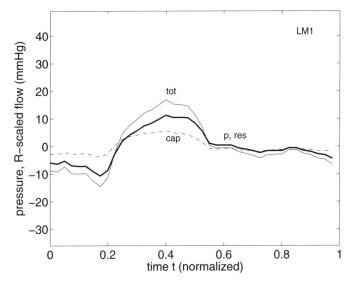

Fig. 7.3.2. Pressure-flow relations under $LM1$ with the same value of the capacitive time constant as in Fig. 7.2.3 of the purely elastic windkessel model, namely $t_C = 0.2\,s$, but here with the addition of viscoelastic effects in the vessel wall. A measure of these effects is the value of the parameter $\lambda = R_2/R_1$ where R_2 is a resistance in series with a purely elastic capacitance and R_1 is a resistance in parallel with it. Results in this figure are based on $\lambda = 2.0$, which seems sufficient to produce dramatic change in the forms of the R-scaled capacitive (cap) and total (tot) flow waves compared with those in Fig. 7.2.3. The heavy solid curve represents the pressure wave (p) as well as the R-scaled resistive flow wave (res).

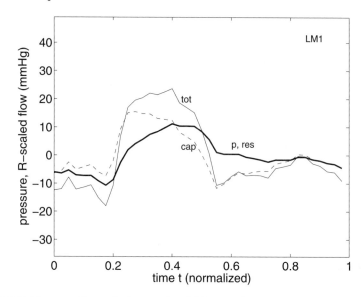

Fig. 7.3.3. Pressure-flow relations under $LM1$ as in Fig. 7.3.2 but here with reduced viscoelastic effect, namely $\lambda = R_2/R_1 = 0.5$.

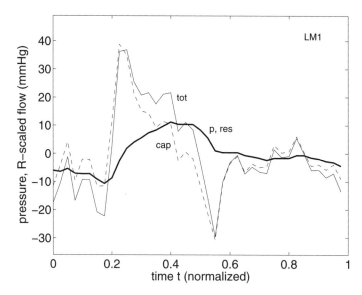

Fig. 7.3.4. Flow under $LM1$ but with considerably reduced viscoelastic effect, namely $\lambda = R_2/R_1 = 0.1$. Viscoelastic effects at this value of λ are practically insignificant, and the total flow wave is again dominated by an erratic form of the capacitive flow as in Fig. 7.2.3 of the purely elastic windkessel model.

of the capacitive time constant is taken to be the same as that in Fig. 7.2.3 of the purely elastic windkessel model in the previous section, namely $t_C = 0.2\,s$. In addition to this, the value of $\lambda\,(= R_2/R_1)$ is required in the present model as a measure of the degree of viscoelasticity. In Fig. 7.3.2, results are shown with $\lambda = 2.0$. Comparison of these results with those in Fig. 7.2.3 show dramatically the effects of viscoelasticity on the form of the flow wave. Chaotic wave swings are entirely eliminated, and the flow wave follows the form of the pressure wave rather than the form of its slope as in Fig. 7.2.5. This dramatic effect diminishes as the value of λ is reduced, as shown in Figs. 7.3.3, 4. A reduction in the value of λ corresponds to a reduction in the value of R_2 relative to that of R_1, which in turn corresponds to a reduction in the viscoelastic property of the vessel wall.

7.4 LM2: {{R₁+L},{R₂+C}}

The modified windkessel model of the previous section lacks an important element of the RLC system, namely the inductor L which represents inertial effects as discussed in Section 2.5. With a few exceptions [203, 188, 75], inductance has not usually been included in lumped models of the coronary circulation, possibly because inertial effects are usually associated with the transient dynamics of the system rather than the steady state dynamics which lumped models deal with. It must be remembered, however, that even within steady state dynamics, within the oscillatory cycle, fluid is being accelerated and decelerated and therefore inertial effects will play a role. It is a different role from that which inertial effects play in the transient state, where the effects are represented by exponential functions that die out as time goes on. In steady state, inertial effects have a permanent cyclic presence which does not die out and which, as we shall see, affects the dynamics of the system and the relation between pressure and flow waveforms.

The position of an inductor in a lumped model of the coronary circulation is fairly clear because acceleration and deceleration of the flow within the oscillatory cycle occur along the conducting vessels rather than within the capacitor. Acceleration and deceleration associated with capacitance is usually minimal because it involves mainly local inflation and deflation of the vessels. Flow along the vessels, on the other hand, involves considerable acceleration and deceleration and hence considerable inertial effects. And these inertial effects are coupled with the viscous resistance within the conducting vessels such that the flow is subject to both, without an "either/or" option. By contrast, the capacitance effects and the viscous resistance are "either/or" options because the flow may inflate the vessels *or* move along them. Thus, the capacitance C in a lumped model of the coronary circulation is appropriately placed *in parallel* with the viscous resistance R_1, as in $LM0$ and $LM1$. Inductance L, on the other hand, is appropriately placed *in series* with that resistance as shown in Fig. 7.4.1, which we shall refer to as $LM2$.

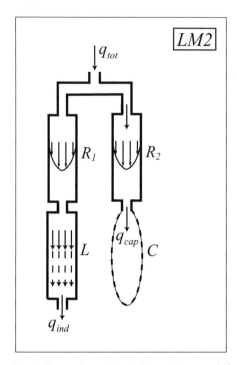

Fig. 7.4.1. A viscoelastic lumped model with an inductor (L) representing inertial effects of fluid acceleration and deceleration within the oscillatory cycle. Total flow into the system (q_{tot}) consists of two parallel flows: inductive flow (q_{ind}) and capacitive flow (q_{cap}). The inductor L is in series with a resistor R_1 representing the viscous resistance along the conducting vessels, while the capacitor C is in series with a resistor R_2 representing viscoelasticity within the vessel wall.

The complex impedances along the two parallel branches of $LM2$, which we shall refer to as the capacitive (cap) and inductive (ind) branches, are obtained from the results of Section 4.9, namely

$$Z_{ind} = R_1 + i\omega L \tag{7.4.1}$$

$$Z_{cap} = R_2 + \frac{1}{i\omega C} \tag{7.4.2}$$

and the complex impedance for the entire system is then given by

$$\frac{1}{Z} = \frac{1}{Z_{ind}} + \frac{1}{Z_{cap}} \tag{7.4.3}$$

$$Z = \frac{(R_1 + i\omega L)(1 + i\omega R_2 C)}{1 + i\omega C(R_2 + R_1 + i\omega L)} \tag{7.4.4}$$

where ω is angular frequency of the oscillatory driving pressure. For the individual harmonics of a composite pressure wave consisting of N harmonics, in the notation of the previous section, we then have

$$Z_{n,ind} = R_1 + i\omega_n L \tag{7.4.5}$$

$$Z_{n,cap} = R_2 + \frac{1}{i\omega_n C} \tag{7.4.6}$$

$$Z_n = \frac{(R_1 + i\omega_n L)(1 + i\omega_n R_2 C)}{1 + i\omega_n C(R_2 + R_1 + i\omega_n L))} \tag{7.4.7}$$

where n denotes a particular harmonic and ω_n is the angular frequency of that harmonic.

For a composite pressure wave consisting of the following harmonics (in their complex exponential form)

$$p_n(t) = M_n e^{i(\omega_n t - \phi_n)} \qquad n = 1, 2, \ldots, N \tag{7.4.8}$$

we obtain the corresponding harmonics of the R-scaled flow waves, as in Chapter 6, but omitting a considerable amount of algebra

$$q_{nr,ind}(t) = \Re\left\{ \frac{p_n(t)}{Z_{n,ind}} \right\} \tag{7.4.9}$$

$$R_1 \times q_{nr,ind}(t) = \frac{M_n \cos(\omega_n t - \phi_n) + \omega_n t_L M_n \sin(\omega_n t - \phi_n)}{1 + \omega_n^2 t_L^2} \tag{7.4.10}$$

$$q_{nr,cap}(t) = \Re\left\{ \frac{p_n(t)}{Z_{n,cap}} \right\} \tag{7.4.11}$$

$$R_1 \times q_{nr,cap}(t) = \frac{\omega_n^2 t_C^2 \lambda M_n \cos(\omega_n t - \phi_n) - \omega_n t_C M_n \sin(\omega_n t - \phi_n)}{1 + \omega_n^2 t_C^2 \lambda^2} \tag{7.4.12}$$

$$q_{nr}(t) = \Re\left\{ \frac{p_n(t)}{Z_n} \right\} \tag{7.4.13}$$

$$R_1 \times q_{nr}(t) = \frac{[1 + \omega_n^2 t_C^2 \lambda]\{M_n \cos(\omega_n t - \phi_n) + \omega_n t_L M_n \sin(\omega_n t - \phi_n)\}}{(1 + \omega_n^2 t_L^2)(1 + \omega_n^2 t_C^2 \lambda^2)}$$

$$+ \frac{[1 + \omega_n^2 t_L^2]\{\omega_n^2 t_C^2 \lambda M_n \cos(\omega_n t - \phi_n) - \omega_n t_C M_n \sin(\omega_n t - \phi_n)\}}{(1 + \omega_n^2 t_L^2)(1 + \omega_n^2 t_C^2 \lambda^2)} \tag{7.4.14}$$

where, as before (Eq. 7.3.11), $\lambda = R_2/R_1$. Finally, the total and the two parallel flow waves are obtained by adding the harmonics of each flow, that is, in R-scaled format

$$R_1 \times q_r(t) = R_1 \times q_{1r}(t) + R_1 \times q_{2r}(t) \ldots + R_1 \times q_{Nr}(t) \qquad (7.4.15)$$

$$R_1 \times q_{r,ind}(t) = R_1 \times q_{1r,ind} + R_1 \times q_{2r,ind} + \ldots + R_1 \times q_{Nr,ind} \quad (7.4.16)$$

$$R_1 \times q_{r,cap}(t) = R_1 \times q_{1r,cap} + R_1 \times q_{2r,cap} + \ldots + R_1 \times q_{Nr,cap} \quad (7.4.17)$$

As before, only the oscillatory parts of the pressure and flow waves are used, the steady parts being omitted to make the graphical comparison of these waves easier by using the same scale.

Results based on this model require that three parameters be specified, namely the capacitive and inertial time constants t_C, t_L, and the ratio λ $(= R_2/R_1)$. To examine particularly the role of inertial effects, results for a relatively high value of t_L are compared with results for a relatively low value in Figs. 7.4.2, 3. It is seen that at a high value of t_L flow along the inductive branch of the system is considerably reduced, with the result that total flow into the system is dominated by flow along its capacitive branch. At the other extreme, where t_L is low, flow along the inductive branch is fairly high to the extent that total flow into the system is now closer in form to that of the pressure wave rather than that of the capacitive wave.

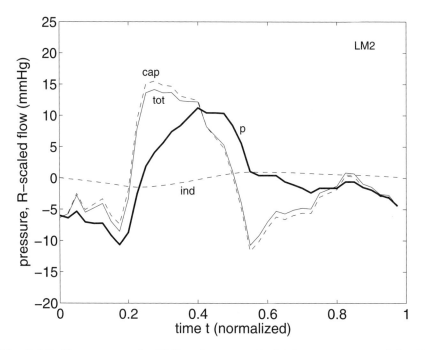

Fig. 7.4.2. Flow waves under $LM2$ and a composite driving pressure wave (p), with $t_C = 0.2\,s$, $\lambda = 0.5$, and a relatively high value of the inertial time constant, namely $t_L = 1.0\,s$. Flow along the inductive branch (ind) of the system is small and total flow (tot) is dominated by flow along the capacitive branch (cap) of the system.

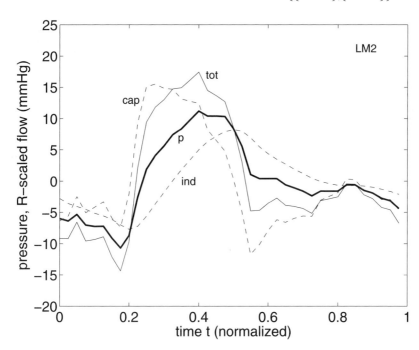

Fig. 7.4.3. Flow waves under $LM2$ and a composite driving pressure wave (p), as in Fig. 7.4.2 but with a relatively low value of the inertial time constant, namely $t_L = 0.1\,s$. Inductive (ind) and capacitive (cap) flows are both high, but their forms are complimentary, their highs and lows nearly balancing each other, leaving total flow (tot) closer in form to that of the composite pressure wave (p).

The results in Fig. 7.4.3 in fact suggest the possibility that the most important role which inertial effects may play in the dynamics of the coronary circulation is that of acting as a balance for capacitive effects. Thus, with the combination of parameter values in Fig. 7.4.3 it is seen that the highs and lows of the capacitive and inductive flows almost balance each other, suggesting that a combination of parameters may exist at which the two flows *exactly* balance each other. Indeed, a close scrutiny of Eq. 7.4.14 shows that when

$$\lambda = 1.0 \quad \text{and} \quad t_L = t_C \tag{7.4.18}$$

the equation reduces to

$$R_1 \times q_{nr}(t) = M_n \cos\left(\omega_n t - \phi_n\right) \tag{7.4.19}$$

$$= \Re\left\{M_n e^{i(\omega_n t - \phi_n)}\right\} \tag{7.4.20}$$

which means that the R-scaled total flow wave has become identical in form to that of the driving pressure. Under these conditions it is as if the inductor

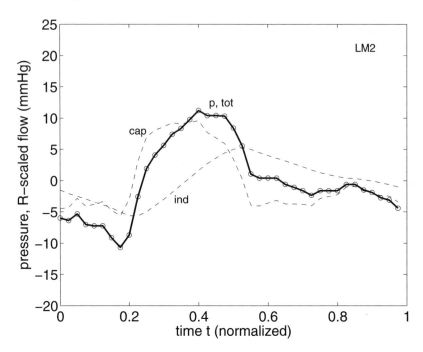

Fig. 7.4.4. Flow waves under $LM2$ and a composite driving pressure wave (p), with $\lambda = 1.0$ and $t_L = t_C = 0.2\,s$. This unique combination of parameter values reduces the total impedance of the system to a simple resistance R_1. Inductive (ind) and capacitive (cap) flows precisely balance each other, thereby reducing the reactance of the system to zero. The R-scaled total flow wave (tot) becomes identical in form to the pressure waveform (p). The two waves are here represented by the heavy solid curve and the open circles to make them visually distinguishable.

and capacitor do not exist. Also, the expression for the harmonic impedances Z_n for the system as a whole in Eq. 7.4.7 can be put in the form

$$\frac{Z_n}{R_1} = \frac{(1 + i\omega_n t_L)(1 + i\omega_n \lambda t_C)}{1 + i\omega_n t_C(\lambda + 1 + i\omega_n t_L))} \tag{7.4.21}$$

from which we see that when $\lambda = 1.0$ and $t_L = t_C$ (as in Eq. 7.4.18), this expression reduces to

$$Z_n = R_1 \tag{7.4.22}$$

which indicates that total impedance has been reduced to simple resistance R_1. Therefore, the R-scaled flow wave becomes identical in form to that of the driving pressure, as in Eq. 7.4.19. This unique situation is illustrated in Fig. 7.4.4 where the results are based on $\lambda = 1.0$ and $t_C = t_L = 0.2\,s$. With these parameter values inertial and capacitive effects exactly balance each other so that R-scaled total flow into the system follows precisely the form of the driving pressure wave.

When in an RLC system the effects of capacitance and inductance precisely "cancel" each other, which means that the system reactance is zero, as discussed in Section 4.8, the system behaves as if the reactive elements L, C do not exist. As we saw in Section 4.8, when the driving pressure is a single harmonic (not a composite wave), the conditions under which this unique situation occurs depend not only on the values of L and C but also on the angular frequency ω (Eq. 4.8.11). The phenomenon is well known in the study of electric circuits, being referred to as "series resonance" or "parallel resonance" depending on the configuration of the RLC system [43]. In that context, interest is particularly in the value of the *frequency* at which the unique conditions occur, hence the reference to resonance. In the context of the coronary circulation, by contrast, the phenomenon is of interest in relation to the optimum operation of the system rather than to resonance. This is so particularly because the driving force in electric circuits is of a single harmonic form, while in the coronary circulation the driving force is of a composite waveform. Since a composite wave consists of many single harmonics which operate at different frequencies, the combination of values of R, L, ω at which the unique conditions occur will generally be different for each harmonic. What the results of this section demonstrate is that, depending on the configuration of the RLC system, these conditions may become independent of the frequency and thus apply equally to all harmonics as in the present case.

7.5 LM3: {{R₁+(pb)},{R₂+C}}

A feature of coronary blood flow that distinguishes it from flow in other parts of the cardiovascular system is an apparent discrepancy in the relation between pressure and flow within the oscillatory cycle. More specifically, during the systolic phase of the oscillatory cycle, when driving pressure is rising rapidly to a peak, coronary blood flow is highly diminished or even reversed, while during the diastolic phase of the oscillatory cycle, when driving pressure is coming down from its peak, coronary blood flow is at its highest [101]. In other words, coronary blood flow occurs mostly in diastole when driving pressure is diminishing rather than in systole when driving pressure is rising. The reasons for this discrepancy are not fully understood, although some possible mechanisms have been suggested.

It is important to note in this discussion that by "coronary blood flow" is meant *input flow*, that is flow entering the system. The reason for this is that from a practical standpoint this is the only flow to which there is reasonable access for measurements. From a clinical standpoint, of course, of more interest is flow at exit from the system, that is flow at the capillary end of the coronary circulation. The relation between this output flow and flow at entry into the system is not known because it depends on what goes on inside the system, which is what lumped models attempt to uncover. Of course, conservation of

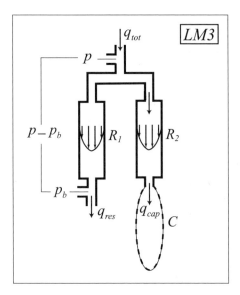

Fig. 7.5.1. A modified windkessel lumped model with a provision for back pressure p_b to simulate the effects of surrounding tissue pressure on coronary vessels imbedded within the cardiac muscle tissue. Back pressure p_b is assumed to affect flow q_{res} in the resistive branch of the system only. Both the primary driving pressure p and the back pressure p_b are oscillatory, thus resistive flow is driven by the oscillatory difference between them, namely $p - p_b$. Remaining features of the model are the same as those of $LM1$ in Fig. 7.3.1.

mass requires that *on average*, inflow and outflow must be the same. That is, under normal circumstances, the amount of fluid entering the system must equal that leaving the system in the course of one oscillatory cycle. It is at any particular moment within the oscillatory cycle that inflow and outflow are usually different. We return to this question later in this chapter. The point of raising this issue here is only to emphasize that the term "coronary blood flow" in the present context refers to flow at entry into the system.

In this section we consider one possible mechanism that may be responsible for the apparent discrepancy in the relation between (input) pressure and (input) coronary blood flow within the oscillatory cycle, namely that of so called "tissue pressure", or "intramyocardial pressure". Under this mechanism, it is postulated that during the systolic phase of the oscillatory cycle, although input driving pressure is relatively high, cardiac muscle tissue is contracting and exerting high pressure on coronary blood vessels imbedded within this tissue [84, 20, 10, 189, 68, 187, 151, 6, 41, 87, 113, 165, 202, 101]. The effect of this is an increase in pressure within the lumen of these vessels, which leads to the notion of a "back pressure" created within the system during systole [97, 98]. The input driving pressure must overcome this back pressure *in addition* to pre-existing resistance within the system, hence the reduction

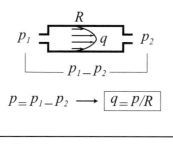

Fig. 7.5.2. Back pressure p_b affects the flow by simply changing the pressure difference driving the flow. In the notation of the present chapter, p actually represents a *pressure difference*, which in the absence of back pressure is simply the difference $(p_1 - p_2)$ between pressures at the input and output ends of the system as shown at the top. The addition of a back pressure at the output end then simply means that the pressure difference driving the flow is now $p - p_b$ as shown at the bottom, where p continues to represent the *pre-existing* pressure difference, namely $p = p_1 - p_2$.

in flow. The situation is similar to that of obstructing the output end of a trumpet, or of the exhaust system of a car, thus creating higher pressure at that end with familiar consequences. In the coronary circulation this may occur momentarily within the oscillatory cycle as the cardiac muscle contracts in the course of its pumping action, leading to reduced flow, possibly even reverse flow. The challenge for lumped model analysis is to use a configuration of RCL components in such a way as to reproduce this scenario, in particular to reproduce the possibility of drastically reduced or reverse flow during the systolic phase of the oscillatory cycle.

In a model that addresses this challenge, which is shown in Fig. 7.5.1 and which we shall refer to as *LM3*, it is postulated that tissue pressure will affect the pressure drop driving the flow along the resistive branch of the parallel system as illustrated schematically in Fig. 7.5.2. The source of the back pressure p_b is taken to be a combination of pressure within the left ventricle and a constant residual pressure within the cardiac tissue which exists independently of ventricular pressure [97, 98]. The resulting combination is the pressure waveform shown in Fig. 7.5.3 along with the primary driving pressure waveform.

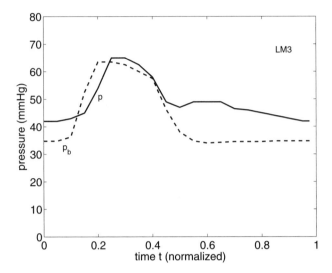

Fig. 7.5.3. Primary pressure wave p and back pressure wave p_b used in $LM3$. The source of the back pressure p_b is assumed to be a combination of pressure within the left ventricle and a constant residual pressure within the cardiac tissue which exists independently of ventricular pressure.

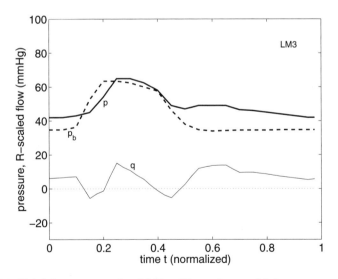

Fig. 7.5.4. Total flow wave under $LM3$, with a primary driving pressure wave $p(t)$ and a back pressure wave $p_b(t)$. Parameter values of the system are $\lambda = 1.0$ and $t_C = 0.1\,s$. The results demonstrate that back pressure can produce considerably reduced and some reverse flow within the oscillatory cycle.

The complex impedances along the two branches of this model are the same as those of $LM1$ discussed in Section 7.3, namely

$$Z_{res} = R_1 \tag{7.5.1}$$

$$Z_{cap} = R_2 + \frac{1}{i\omega C} \tag{7.5.2}$$

where the subscripts res and cap refer to the resistive and capacitive branches, respectively. The complex impedance for the entire system is given by

$$\frac{1}{Z} = \frac{1}{Z_{res}} + \frac{1}{Z_{cap}} \tag{7.5.3}$$

$$Z = \frac{R_1 + i\omega R_1 R_2 C}{1 + i\omega C(R_1 + R_2)} \tag{7.5.4}$$

where ω is the angular frequency of oscillation of the driving pressure. For the individual harmonics of a composite pressure wave consisting of N harmonics we have, as before,

$$Z_{n,res} = R_1 \tag{7.5.5}$$

$$Z_{n,cap} = R_2 + \frac{1}{i\omega_n C} \tag{7.5.6}$$

$$Z_n = \frac{R_1 + i\omega_n R_1 R_2 C}{1 + i\omega_n C(R_1 + R_2)} \tag{7.5.7}$$

where n denotes a particular harmonic and ω_n is the angular frequency of that harmonic.

The essence of the present model is that along the capacitive branch of the parallel system the oscillatory pressure driving flow is $p(t)$, but along the resistive branch the driving pressure is $p(t) - p_b(t)$. Both pressures are oscillatory and, therefore, to follow the same analysis as before, they must be presented in their complex exponential form

$$p_n(t) = M_n e^{i(\omega_n t - \phi_n)} \qquad n = 1, 2, \ldots, N \tag{7.5.8}$$

$$p_{bn}(t) = M_{bn} e^{i(\omega_{bn} t - \phi_{bn})} \qquad n = 1, 2, \ldots, N \tag{7.5.9}$$

where subscript b refers to properties of the back pressure wave p_b.

The harmonics of the R-scaled flow rates along the two branches of the parallel system are then given by, following the same steps as for previous models and omitting some of the details,

$$q_{nr,res}(t) = \Re\left\{\frac{p_n(t) - p_{bn}(t)}{Z_{n,res}}\right\} \tag{7.5.10}$$

$$R_1 \times q_{nr,res}(t) = M_n \cos(\omega_n t - \phi_n) - M_{bn} \cos(\omega_{bn} t - \phi_{bn}) \tag{7.5.11}$$

$$q_{nr,cap}(t) = \Re\left\{\frac{p_n(t)}{Z_{n,cap}}\right\} \tag{7.5.12}$$

$$R_1 \times q_{nr,cap}(t) = \frac{\omega_n^2 t_C^2 \lambda M_n \cos(\omega_n t - \phi_n) - \omega_n t_C M_n \sin(\omega_n t - \phi_n)}{1 + (\omega_n \lambda t_C)^2} \tag{7.5.13}$$

The harmonics of total flow into the system are then simply given by

$$R_1 \times q_{nr}(t) = R_1 \times q_{nr,res}(t) + R_1 \times q_{nr,cap}(t) \tag{7.5.14}$$

Finally, the actual flow waves are obtained by adding their harmonics to obtain the oscillatory parts of the waves, plus the steady part of each flow if any,

$$R_1 \times q_{r,res}(t) = R_1 \times \bar{q} + R_1 \times q_{1r,res} + R_1 \times q_{2r,res} + \ldots + R_1 \times q_{Nr,res} \tag{7.5.15}$$

$$R_1 \times q_{r,cap}(t) = R_1 \times q_{1r,cap} + R_1 \times q_{2r,cap} + \ldots + R_1 \times q_{Nr,cap} \tag{7.5.16}$$

$$R_1 \times q_r(t) = R_1 \times q_{r,res}(t) + R_1 \times q_{r,cap}(t) \tag{7.5.17}$$

Steady flow through the system is included this time because it is relevant to present discussion. The steady flow rate $\bar{q}(t)$ is given by

$$\bar{q}(t) = \frac{\bar{p}(t) - \bar{p}_b(t)}{R_1} \tag{7.5.18}$$

where the overline bar indicates average over one oscillatory cycle. The steady part of the flow, of course, is part of the flow in the *resistive branch* of the parallel system because the capacitive branch of the system supports oscillatory flow only.

Total flow under $LM3$, with primary driving pressure $p(t)$ and back pressure $p_b(t)$ is shown in Fig. 7.5.4. It is seen that the model meets the objective of demonstrating considerably reduced and some reverse flow through the system. Results using the same model parameter values but in the absence of any back pressure are shown in Fig. 7.5.5, which indicate clearly that the reduced and reverse flow observed in Fig. 7.5.4 are due entirely to the back pressure in this model. Furthermore, the individual flow waves along the two parallel branches with and without back pressure are shown in Figs. 7.5.6, 7, which confirm that the effect of back pressure is confined to the resistive part of the flow only.

The effects of cardiac muscle contraction on the dynamics of the coronary circulation have also given rise to a number of other concepts regarding the possible mechanisms involved. In the "waterfall" concept, the vessels imbedded within the cardiac muscle tissue are believed to collapse under muscle contraction, leading to an interval (in systole) when pressure is rising inside the vessels but there is no flow, as in a waterfall before the water level in the reservoir has reached the mouth of the fall [161, 51, 111, 205, 60]. Flow

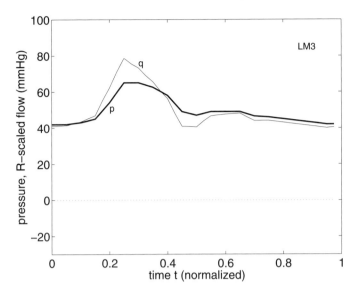

Fig. 7.5.5. Total flow wave under $LM3$ with a primary driving pressure wave $p(t)$ and the same parameter values as in Fig. 7.5.4 but in the absence of any back pressure, demonstrating that the reduced and reverse flow observed in that figure are due entirely to the effects of back pressure.

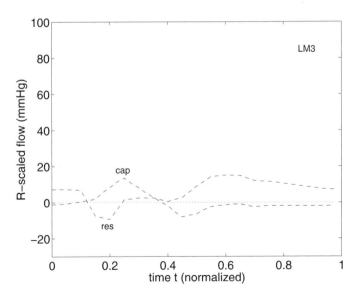

Fig. 7.5.6. The resistive (res) and capacitive (cap) flow waves under $LM3$ with back pressure, as in Fig. 7.5.4.

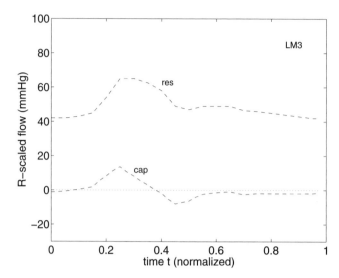

Fig. 7.5.7. The resistive (*res*) and capacitive (*cap*) flow waves under *LM*3 but in the absence of back pressure, as in Fig. 7.5.5.

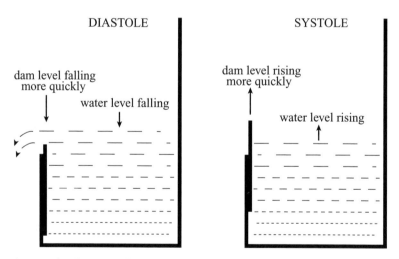

Fig. 7.5.8. A schematic illustration of the waterfall concept in which the water level in the reservoir is analogous to the pressure inside the coronary vessels while the level of the dam is analogous to pressure outside the vessels caused by cardiac muscle contraction. In systole, water level in the reservoir is rising but the level of the dam is rising more quickly and no flow is possible. In diastole, water level in the reservoir is falling but the level of the dam is falling more quickly, and a point is reached where flow becomes possible.

resumes only when pressure inside the vessels exceeds the pressure outside, which strangely occurs in diastole when inside pressure is actually decreasing but at the same time the cardiac muscles are relaxing and the external pressure is decreasing more steeply. Thus, to state the waterfall analogy correctly it has to be said that flow resumes when the water level in the reservoir is actually falling but the height of the dam is falling more quickly (Fig. 7.5.8).

In other concepts the effects of muscle contraction on the dynamics of the coronary circulation have been described as an "intramyocardial pump" effect [29] or "variable elastance" effect [189, 113, 114, 202].

7.6 Inflow-Outflow

One of the most important aspects of the dynamics of the coronary circulation is the fact that at almost any point in time within the oscillatory cycle, total inflow into the system is not equal to total outflow. This difference between inflow and outflow is particularly important because, as mentioned earlier, flow measurements in the coronary circulation are extremely difficult and at best would provide only *inflow* data. We recall that inflow into the coronary circulation occurs via the two main coronary ostia at the root of the ascending aorta (Figs. 1.3.1, 2). Outflow from the system of course occurs at the capillary bed, hence the difficulty of obtaining an actual measure of it. Yet, this outflow is what is most relevant clinically.

It is therefore important to understand the nature and source of the difference between inflow and outflow in the dynamics of the coronary circulation, at different times within the oscillatory cycle and under different circumstances, which we examine in this section. It must be emphasized, of course, that this difference is an *oscillatory function of time* that has a *zero mean*. In other words, under normal circumstances inflow and outflow are equal *on average*, the average being taken over one or at most a few cycles. Differences occur largely *within the oscillatory cycle*. These differences are important, nevertheless, because they are indicators of the interplay between the dynamics of different elements of the coronary circulation. More specifically, the difference between inflow and outflow, as we shall see, is a measure of the interplay between and the relative effects of capacitance, resistance, and inductance, thus, a change in any of these properties of the coronary circulation will lead to a change in the delicate balance between inflow and outflow within the oscillatory cycle.

In the windkessel system ($LM0$) considered in Section 7.2, for example, when compliance is low, with a value of the capacitive time constant $t_C = 0.01$ s, capacitive flow is very low and total flow through the system is dominated by resistive flow as seen in Fig. 7.2.2. Under these circumstances inflow into the system is very nearly the same as outflow, as shown in Fig. 7.6.1 where inflow is plotted against outflow at different points in time within the oscillatory cycle. As compliance increases, however, with $t_C = 0.2$ s, capacitive flow

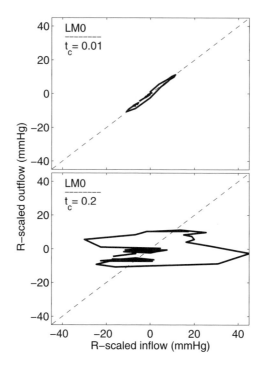

Fig. 7.6.1. Inflow-outflow during the oscillatory cycle (as the closed curve is traced around once) through the $LM0$ lumped model of Section 7.2 with two different values of the capacitive time constant t_C. At the lower value of t_C (top), capacitance and capacitive flow are low and flow through the system is mostly resistive flow, thus inflow and outflow are nearly equal at different points in time within the oscillatory cycle. In purely resistive flow the curve would remain on the dashed line as inflow and outflow would be exactly equal at all times. At the higher value of t_C (bottom), capacitance and capacitive flow are higher and are more involved in the dynamics of the system. Inflow and outflow are rarely equal at any time during the oscillatory cycle.

increases accordingly as seen in Fig. 7.6.3, and under these circumstances the time course of inflow into the system is widely different from that of outflow, as shown in Fig. 7.6.1. The figure indicates that in this case when inflow is high, outflow is low, and vice versa.

In the viscoelastic windkessel system ($LM1$) considered in Section 7.3, when viscoelasticity is moderately high, with $\lambda = R_2/R_1 = 2.0$, capacitive flow is again constrained to the extent that flow through the system is mostly resistive flow as seen in Fig. 7.6.3. Inflow and outflow are nearly equal throughout the oscillatory cycle as shown in Fig. 7.6.2. As the effect of viscoelasticity is reduced, with $\lambda = 0.1$, however, capacitive flow plays an increased role in the dynamics of the system as seen in Fig. 7.6.3, and changes the interplay between inflow and outflow as shown in Fig. 7.6.2.

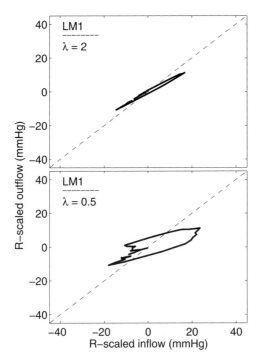

Fig. 7.6.2. Inflow-outflow during the oscillatory cycle as in Fig. 7.6.2 but here through the $LM1$ lumped model of Section 7.3 with two different values of the parameter λ ($= R_2/R_1$) which is a measure of the effect of viscoelasticity within the vessel wall. At the higher value of λ (top), viscoelastic effects are high, reducing capacitive flow and the effect of capacitance within the system, with the result that inflow and outflow are nearly equal throughout the oscillatory cycle. At the lower value of λ (bottom), capacitance and capacitive flow are higher and are more involved in the dynamics of the system. Inflow and outflow are rarely equal at any time during the oscillatory cycle.

Finally, in the inductive system ($LM2$) considered in Section 7.4, when inertial effects are high, with a value of the inertial time constant $t_L = 1.0$ s, flow through the system is mostly capacitive flow as seen in Fig. 7.4.2, and the corresponding interplay between inflow and outflow is shown in Fig, 7.6.3. In this extreme case the dynamics of the system consist of mostly inflow which goes towards inflating and deflating the balloon, with barely any outflow from the system. As inertial effects are reduced, with $t_L = 0.1$ s, more flow can go through the inductive branch of the system as seen in Fig. 7.4.4, and the corresponding interplay between inflow and outflow is shown in Fig. 7.6.3. Thus, the inertial effects in $LM2$ act as a balance against the capacitive effects as discussed in Section 7.4.

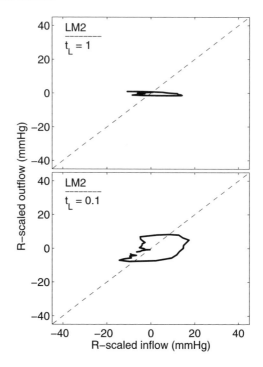

Fig. 7.6.3. Inflow-outflow during the oscillatory cycle as in Fig. 7.6.1 but here through the $LM2$ lumped model of Section 7.4 with two different values of the inertial time constant t_L. At the higher value of t_L (top), inertial effects are high, thus reducing resistive flow and causing flow through the system to be mostly capacitive flow. Under these conditions the dynamics of the system consist of no more than filling and emptying the balloon, thus only inflow is involved with very little outflow (as the curve is traced back and forth nearly horizontally). At the lower value of t_L (bottom), inertial effects are reduced, thus allowing more inductive flow and making total flow less dominated by capacitive flow. Inflow and outflow are rarely equal at any time during the oscillatory cycle.

7.7 Summary

Many lumped models have been proposed in the past, but issues characterizing the dynamics of the coronary circulation remain largely unresolved, whether because a certain characteristic cannot be accurately modelled or because it can be modelled in more than one way. Four basic lumped models based on elements of the RLC system illustrate some of the main issues.

$LM0$: {R,C} represents the most simple lumped model of the coronary circulation, generally referred to as the "windkessel", incorporating the two fundamental effects of resistance and capacitance. It is clear from the results that capacitive effects can play a significant role in the dynamics of the coronary circulation. It is equally clear, however, that the way in which capacitance is present in this elementary model of the system is rather crude because it produces erratic swings in the flow waveform which are not usually observed in the physiological system. These erratic swings, as we have seen, result from the direct relation between capacitive flow and the slope of the pressure curve, and this relation, in turn, results from the isolated presence of the capacitor in this elementary model.

$LM1 : \{R1,\{R2+C\}\}$ represents a modified windkessel model, sometimes referred to as a "viscoelastic windkessel". It takes into account viscoelasticity within the vessel wall by adding a resistance R_2 in series with the capacitance C and in parallel with a resistance R_1 representing viscous effects as before. The results indicate that this has a dramatic effect on the form of the flow wave. Chaotic wave swings observed in $LM0$ are entirely eliminated as the flow wave now follows the form of the pressure wave rather than the form of its slope. The resulting more regular flow curve is closer to what is observed in the physiological system, which suggests that viscoelastic effects do play a significant role in the dynamics of the coronary circulation.

In LM2: $\{\{R1+L\},\{R2+C\}\}$ the effects of fluid inertia are added in the form of an inductor in series with the viscous resistance and in parallel with the capacitive branch of the system. The results indicate that the role of inertial effects in the dynamics of the coronary circulation may extend far beyond the simple effects of acceleration and deceleration of the fluid in the course of the oscillatory cycle. In the presence of capacitance effects, the two effects may combine to a dynamical advantage of the system. It is not unreasonable to speculate that the coronary circulation may be designed to take advantage of this. Indeed, this possibility may be used, as in Section 4.8, to estimate the lumped parameter values of the system, on the assumption that the system may be designed to operate at or near these values. On that assumption, we are led again to conclude that any intervention that moves the system away from this optimum state may upset the normal dynamics of the coronary circulation. Thus, a change in the values of dynamical parameters t_C, t_L, or λ, which may come about by disease or by clinical or surgical intervention, may produce changes in the form of the flow wave which can be as detrimental as the occlusion of an artery. Furthermore, while the occlusion of an artery is fairly conspicuous and can be easily detected, a change in the dynamical parameters of the system would be highly inconspicuous.

In LM3: $\{\{R1+(pb)\},\{R2+C\}\}$ an attempt is made to account for the effect of cardiac muscle contraction on coronary vasculature imbedded within the cardiac muscle tissue by introducing an element of "back pressure" into the model. The results demonstrate that this "intramyocardial pressure" as it is called, is a plausible mechanism for the highly reduced and some reverse coronary blood flow during the systolic phase of the cardiac cycle. Other mechanisms that have been proposed include the so-called "waterfall", "intramyocardial pump", and "variable elastance" mechanisms.

In the coronary circulation as modelled by lumped RLC systems, inflow into and outflow from the system are rarely equal at any time during the oscillatory cycle. The main player in this is capacitive flow which goes towards inflating and deflating the balloon. In the extreme case where capacitive flow is dominant, there is no outflow at all from the system although inflow is present at all times during the oscillatory cycle. More commonly, viscoelasticity and inductance counteract the capacitive effect to produce a more balanced interplay between inflow and outflow within the oscillatory cycle. The significance of these results in the dynamics of the coronary circulation is that a change in any of these properties of the system will change the nature of this interplay.

8

Elements of Unlumped-Model Analysis

8.1 Introduction

In the lumped-model analysis of previous chapters the coronary circulation was seen, in essence, as a closed "black box". In unlumped-model analysis, to use the same language, the box is opened. In lumped-model analysis an attempt is made to infer the dynamic properties of the coronary circulation in terms of only input and output. In unlumped-model analysis an attempt is made to infer the dynamic properties in terms of internal properties and structure of the system.

The most important internal property of the coronary circulation is the vast geometrical structure of its conducting vessels. This aspect of the coronary circulation is "invisible" to lumped-model analysis. In unlumped-model analysis, by contrast, vascular structure provides the main grounds for analysis as well as the most daunting challenge. Furthermore, the intricate branching architecture of coronary vasculature gives the coronary circulation a *space dimension* which again is invisible to lumped-model analysis. Indeed, pressure and flow waves discussed in previous chapters were considered as functions of time only, because a space dimension does not exist in lumped-model analysis.

Another important property of the coronary circulation that is invisible to lumped-model analysis is that of local flow phenomena within the conducting vessels. Flow in a single tube, for example, is used in lumped-model analysis simply as the basis for electrical and mechanical analogies. The analysis cannot include flow phenomena within the tube if these are outside the confines of the analogies. In particular, flow *propagation* within the coronary arteries, which occurs because of the pulsatile nature of the flow and the elasticity of the arterial wall, is well outside the scope of lumped-model analysis. This is because wave propagation has a strong space dimension and involves flow events such as wave reflections that are strongly related to the branching architecture of the coronary network. In general there are strictly no theoretical grounds for extending the single tube analogy to flow in a vast network of tubes, yet the validity of lumped-model analysis rests heavily on the validity of that extension. In unlumped model analysis the basic approach is to start from the level of a single tube and move up to larger numbers of tubes.

These differences between the lumped and unlumped models of the coronary circulation indicate clearly that both approaches have their strengths and weaknesses, and each has its own challenges in attempting to unravel the dynamics of the coronary circulation. Having said that, however, there is no doubt that future work in this endeavor lies with the unlumped model.

Interestingly, in medical practice today the coronary circulation is treated as an unlumped system in the sense that the main focus is on whether flow within individual coronary arteries is obstructed in any way. But, the system is also treated as a black box in the sense that flow is assumed to be guaranteed once it has been established that vessels are not obstructed. This latter view is essentially that of a "plumbing system" devoid of any internal dynamics, which is highly inconsistent with the intricate dynamics of the coronary circulation.

8.2 The Streamwise Space Dimension

Flow in a tube forms the basis of lumped- and unlumped-model analysis of the coronary circulation. In lumped-model analysis, flow in a tube is used as an analogue for flow in the coronary system as a whole. In unlumped-model analysis, flow in a tube is used as the micro building block from which a model of the system as a whole would be built.

The most important difference between the two approaches, however, has to do with the way the tube is viewed in each case. In lumped-model analysis, the tube is viewed as having no space dimension, being represented by only its resistance and/or capacitance. While the length and diameter of the tube may be used in determining its resistance, these space properties are not part of the lumped-model analysis. In unlumped-model analysis, by contrast, the tube is given its streamwise space dimension and this dimension is not only part of unlumped-model analysis but is the central part of that analysis. Indeed, this new dimension, which we shall denote by x, is the most important distinguishing difference between the lumped- and the unlumped-model approaches to the coronary circulation, as illustrated in Fig. 8.2.1.

In previous chapters we saw that in lumped-model analysis changes in pressure and flow in a tube are considered to be only changes in time t. In unlumped-model analysis changes in pressure and flow are considered both in space and time. The addition of space dimension is particularly important because pulsatile flow in an elastic tube, and hence pulsatile flow in the coronary circulation, is a phenomenon in space and time. This is because pulsatile flow in an elastic tube actually *propagates* as a wave along the tube, much like a wave on the surface of a lake, and wave propagation is a phenomenon in space and time as illustrated in Fig. 8.2.2. At any *fixed position* within the tube, pressure and flow vary in time in an oscillatory manner in the form of a wave, usually a composite wave such as those considered in previous chapters.

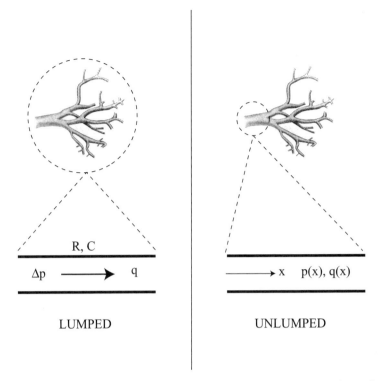

Fig. 8.2.1. In lumped-model analysis, flow in a tube is used as an analogue for flow in the coronary system as a whole, with properties of the tube such as resistance R and capacitance C being ascribed to the system as a whole. Pressure drop Δp and flow rate q may vary in time but not along the tube. In unlumped-model analysis, flow in a tube is used as a model for flow in each tube segment within the system. Furthermore, in addition to being functions of time t as in the lumped model, pressure and flow are here considered as functions of the streamwise space coordinate x. The space dimension x does not exist in lumped-model analysis.

But now, in addition, at any *fixed point in time*, pressure and flow vary also in space, again in the form of a wave, but now a wave in space, along the streamwise space dimension of the tube, a dimension which does not exist in lumped-model analysis.

The addition of a space dimension in the analysis of the coronary circulation is important not only because wave propagation has a space dimension but also because the coronary circulation itself has a space dimension. Coronary vasculature is highly intricate in its space distribution and branching pattern, and this aspect of the coronary circulation is as much a part of the system as is the dynamics of the coronary circulation. Indeed, an ultimate aim in the study of the coronary circulation is to determine pressure and flow properties at different levels of the vasculature or at different regions of the heart tissue, which requires that coronary vasculature be considered in its

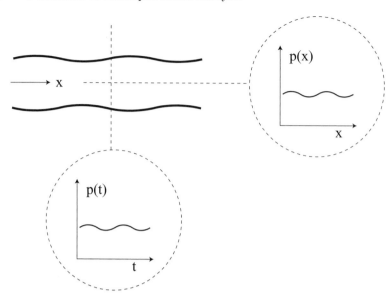

Fig. 8.2.2. Wave propagation in an elastic tube consists of two oscillations, one in space and the other in time. At any *fixed position* within the tube, pressure varies in time in an oscillatory manner in the form of a wave ($p(t)$). At any *fixed point in time*, pressure varies again in an oscillatory manner but now in the form of a wave in space ($p(x)$), along the streamwise space dimension of the tube.

space distribution, unlumped. But to do so requires knowlege of the branching architecture of coronary vasculature. Indeed, the accuracy and utility of unlumped-model analysis depends as much on an accurate mapping of coronary vasculature within the myocardium as it does on an accurate description of the associated dynamics. Thus, the acquisition of data on the branching form and distribution of coronary vasculature constitutes a considerable part of unlumped-model analysis. Lumped-model analysis, by contrast, does not require any information on coronary vasculature. We see again, in conclusion, that differences between the lumped- and the unlumped-model approaches to the coronary circulation make the two methods compliment rather than compete with each other.

8.3 Steady Flow along Tube Segments

Consider flow in a single tube at first, in which we set a coordinate x along the axis of the tube, being zero at the entrance and positive in the direction of the flow. In contrast with the analysis in Section 2.3 where flow q is expressed in terms of a pressure difference Δp between the two ends of the tube, the aim here is to consider both the flow and pressure as being functions of position x along the tube, as shown in Fig. 8.3.1. To do this we use Eqs. 2.3.2, 4

$$\frac{dp}{dx} = -\frac{8\mu}{\pi a^4} q \tag{8.3.1}$$

which, upon integration, gives

$$p(x) = p(0) - \frac{8\mu}{\pi a^4} qx \tag{8.3.2}$$

where $p(0)$ is the pressure at the tube entrance ($x = 0$), a is the tube radius and μ is the viscosity of the fluid. If the tube length is l, then the pressure at the other end of the tube is

$$p(l) = p(0) - \frac{8\mu}{\pi a^4} ql \tag{8.3.3}$$

which is the same as the result obtained in Section 2.3, Eq. 2.3.4, noting that in that section a constant pressure gradient was used, defined by

$$k = \frac{\Delta p}{l} = \frac{p(l) - p(0)}{l} \tag{8.3.4}$$

The main focus in this section is on Eq. 8.3.2 in which the pressure is seen as a function of position x, and on using this equation to track the pressure distribution in a system of tubes. For this purpose, consider two tube segments in series now, to be identified by subscripts 0 and 1. Allowing the lengths and radii of the two segments to be different but using the same flow rate, then Eq. 8.3.2 gives

$$p_0(x_0) = p_0(0) - \frac{8\mu}{\pi a_0^4} qx_0 \tag{8.3.5}$$

$$p_1(x_1) = p_1(0) - \frac{8\mu}{\pi a_1^4} qx_1 \tag{8.3.6}$$

where x_0 and $p_0(x_0)$ are the streamwise coordinate and corresponding pressure in the first tube only, and similarly for the second tube identified by subscript 1, as illustrated in Fig. 8.3.1. In these expressions we are clearly assuming that the idealized conditions of Poiseuille flow prevail along the full length of each tube segment, neglecting deviations from these conditions at entry and exit regions and at the junction between two tube segments. Thus, the results to follow will be inaccurate $locally$ in these regions, but our interest here and in subsequent sections is primarily in the global pressure distribution along a large number of tube segments connected in series or in a branching pattern. The junction between the two tube segments occurs at

$$x_0 = l_0 \qquad \text{or} \qquad x_1 = 0 \tag{8.3.7}$$

and on the assumption of pressure continuity at that point, we have

$$p_1(0) = p_0(l_0) \tag{8.3.8}$$

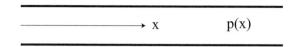

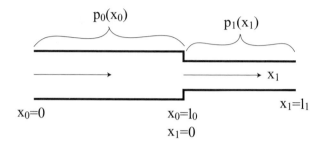

Fig. 8.3.1. In unlumped-model analysis the pressure p in a tube is considered as a function of streamwise position coordinate x measured from the tube entrance. In a sequence of tube segments in series, the pressure and the position coordinate are re-defined in each tube segment and are confined to that tube only.

It is important to note that the domain of p_0 is restricted to the first tube segment only, and the domain of p_1 is restricted to the second tube segment only. Thus, $p_1(0)$ is a function of x_1 only, and $p_1(0)$ is the value of p_1 at $x_1 = 0$ with no ambiguity. Similarly, p_0 is a function of x_0 only, and $p_0(0)$ and $p_0(l_0)$ are values of p_0 at $x_0 = 0$ and at $x_0 = l_0$, respectively, again without ambiguity.

Eqs. 8.3.5, 8 then give

$$p_1(0) = p_0(0) - \frac{8\mu}{\pi a_0^4} q l_0 \tag{8.3.9}$$

which indicates how the pressure at the beginning of the second tube segment is determined by the pressure at the end of the first tube segment. Using this result in Eq. 8.3.6, we find that the pressure *at any point* within the second tube segment is then determined by

$$p_1(x_1) = p_0(l_0) - \frac{8\mu}{\pi a_1^4} q x_1 \tag{8.3.10}$$

These expressions can be generalized in a straightforward way, so that in the presence of a third tube segment we would have

$$p_2(0) = p_1(0) - \frac{8\mu}{\pi a_1^4} q l_1 \tag{8.3.11}$$

$$p_2(x_2) = p_1(l_1) - \frac{8\mu}{\pi a_2^4} q x_2 \tag{8.3.12}$$

and in general, if there are $n + 1$ segments,

$$p_n(0) = p_{n-1}(0) - \frac{8\mu}{\pi a_{n-1}^4} q l_{n-1} \qquad (8.3.13)$$

$$p_n(x_n) = p_{n-1}(l_{n-1}) - \frac{8\mu}{\pi a_n^4} q x_n \qquad (8.3.14)$$

It is convenient to put these expressions in nondimensional, normalized forms, using properties of the first tube segment as reference properties. Thus, the position coordinates in each tube segment are normalized by writing

$$X_0 = \frac{x_0}{l_0} \qquad (8.3.15)$$

$$X_1 = \frac{x_1}{l_1} \qquad (8.3.16)$$

$$\vdots$$

$$X_n = \frac{x_n}{l_n} \qquad (8.3.17)$$

so that, in terms of the new position coordinate X, the normalized length of each tube segment is now 1.0.

The magnitude of the pressure drop in the first tube segment is given by

$$|\Delta p_0| = |p_0(l_0) - p_0(0)| \qquad (8.3.18)$$

$$= \frac{8\mu}{\pi a_0^4} q l_0 \qquad (8.3.19)$$

and this quantity is now used to define a nondimensional form of the pressure in each tube segment, namely

$$P_0(X_0) = \frac{p_0(x_0) - p_0(0)}{|\Delta p_0|} \qquad (8.3.20)$$

$$= -X_0 \qquad (8.3.21)$$

and the nondimensional pressures at the two ends of the first tube segment are then given by

$$P_0(0) = 0 \qquad (8.3.22)$$

$$P_0(1) = -1 \qquad (8.3.23)$$

Similarly, the pressure in the second tube, in nondimensional form, is now defined and given by

$$P_1(X_1) = \frac{p_1(x_1) - p_0(0)}{|\Delta p_0|} \tag{8.3.24}$$

$$= \frac{p_0(l_0) - p_0(0)}{|\Delta p_0|} - \left(\frac{a_0}{a_1}\right)^4 \frac{x_1}{l_0} \tag{8.3.25}$$

$$= P_0(1) - \left(\frac{a_0}{a_1}\right)^4 \left(\frac{l_1}{l_0}\right) X_1 \tag{8.3.26}$$

$$= -1 - \left(\frac{a_0}{a_1}\right)^4 \left(\frac{l_1}{l_0}\right) X_1 \tag{8.3.27}$$

and at the two ends of the tube

$$P_1(0) = -1 \tag{8.3.28}$$

$$P_1(1) = -1 - \left(\frac{a_0}{a_1}\right)^4 \left(\frac{l_1}{l_0}\right) \tag{8.3.29}$$

In the presence of a third tube, we find

$$P_2(X_2) = \frac{p_2(x_2) - p_0(0)}{|\Delta p_0|} \tag{8.3.30}$$

$$= \frac{p_1(l_1) - p_0(0)}{|\Delta p_0|} - \left(\frac{a_0}{a_2}\right)^4 \frac{x_2}{l_0} \tag{8.3.31}$$

$$= P_0(1) - \left(\frac{a_0}{a_1}\right)^4 \left(\frac{l_1}{l_0}\right) X_1 \tag{8.3.32}$$

$$= -1 - \left(\frac{a_0}{a_1}\right)^4 \left(\frac{l_1}{l_0}\right) - \left(\frac{a_0}{a_2}\right)^4 \left(\frac{l_2}{l_0}\right) X_2 \tag{8.3.33}$$

and in general

$$P_n(X_n) = -1 - \left(\frac{a_0}{a_1}\right)^4 \left(\frac{l_1}{l_0}\right) - \left(\frac{a_0}{a_2}\right)^4 \left(\frac{l_2}{l_0}\right) - \ldots$$
$$- \left(\frac{a_0}{a_n}\right)^4 \left(\frac{l_n}{l_0}\right) X_n \tag{8.3.34}$$

The results indicate that in this convenient nondimensional form, the pressure distribution along a sequence of tube segments in series consists of a series of linear pressure drops, with the pressure starting from a normalized value of 0 at entry and dropping linearly to -1 at the end of the first tube. Subsequent values of the pressure depend on the lengths and diameters of subsequent tubes. If, for the purpose of illustration, it is assumed that the tube lengths are proportional to their diameters, the pressure distribution in each tube segment becomes dependent on the ratios of radii only, namely

$$P_1(X_1) = -1 - \left(\frac{a_0}{a_1}\right)^3 X_1 \qquad (8.3.35)$$

$$P_2(X_2) = -1 - \left(\frac{a_0}{a_1}\right)^3 - \left(\frac{a_0}{a_2}\right)^3 X_2 \qquad (8.3.36)$$

$$\vdots$$

$$P_n(X_n) = -1 - \left(\frac{a_0}{a_1}\right)^3 - \left(\frac{a_0}{a_2}\right)^3 - \ldots - \left(\frac{a_0}{a_n}\right)^3 X_n \qquad (8.3.37)$$

If it is assumed further, for the purpose of illustration again, and for reasons to become apparent in the next section, that the radii of successive tube segments are *diminishing* such that

$$\frac{a_0}{a_1} = \frac{a_1}{a_2} = \frac{a_2}{a_3} \ldots = 2^{1/3} \qquad (8.3.38)$$

then these results become

$$P_1(X_1) = -1 - (2)^1 X_1 \qquad (8.3.39)$$
$$P_2(X_2) = -1 - (2)^1 - (2)^2 X_2 \qquad (8.3.40)$$

$$\vdots$$

$$P_n(X_n) = -1 - (2)^1 - (2)^2 - \ldots - (2)^n X_n \qquad (8.3.41)$$

and if, for the purpose of comparison, the radii of successive tube segments are *increasing*, such that

$$\frac{a_0}{a_1} = \frac{a_1}{a_2} = \frac{a_2}{a_3} \ldots = \left(\frac{1}{2}\right)^{1/3} \qquad (8.3.42)$$

we find

$$P_1(X_1) = -1 - \left(\frac{1}{2}\right)^1 X_1 \qquad (8.3.43)$$

$$P_2(X_2) = -1 - \left(\frac{1}{2}\right)^1 - \left(\frac{1}{2}\right)^2 X_2 \qquad (8.3.44)$$

$$\vdots$$

$$P_n(X_n) = -1 - \left(\frac{1}{2}\right)^1 - \left(\frac{1}{2}\right)^2 - \ldots - \left(\frac{1}{2}\right)^n X_n \qquad (8.3.45)$$

In the trivial case where successive tube segments have the same diameters, the results are clearly

$$P_1(X_1) = -1 - X_1 \qquad\qquad (8.3.46)$$

$$P_2(X_2) = -1 - 1 - X_2 \qquad\qquad (8.3.47)$$

$$\vdots$$

$$P_n(X_n) = -1 - 1 - 1 - \ldots - X_n \qquad\qquad (8.3.48)$$

These results are illustrated graphically in Fig. 3.8.2, where it is seen that
the above trivial case serves as a good reference in which the pressure distribution in each tube segment is linear and dropping by the same amount, namely
-1. In the case where the radii of successive tube segments are diminishing,
the drops in pressure in successive segments increase very rapidly, and the
reverse happens when the radii of successive tube segments are increasing.
While these examples are fairly artificial, they serve as useful guides when
considering branching tubes.

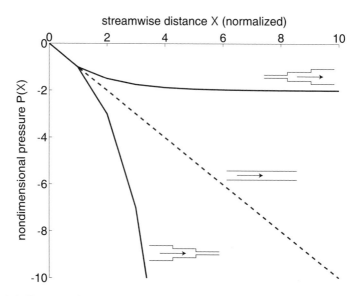

Fig. 8.3.2. Pressure distribution in steady flow along a sequence of tube segments in
series. The streamwise distance X here is a *cumulative* coordinate along the sequence
of tube segments whereby the normalized length of each tube segment is 1.0. Thus,
the first tube segment extends from $X = 0$ to $X = 1.0$, the second extends from
$X = 1.0$ to $X = 2.0$, etc. If the radii of successive tube segments are increasing,
the pressure drops very rapidly, while if the radii are decreasing the pressure drops
very slowly. In the trivial case where the radii of successive tube segments remain
unchanged, the pressure drops by the same amount (-1.0) in each tube segment,
with this case serving as a useful reference for comparison.

8.4 Steady Flow Through a Bifurcation

It is well established that the underlying design of arterial pathways in the cardiovascular system is that of an open tree structure, and the same is true in the coronary circulation. In an open tree structure (Fig. 1.6.1) a root vessel segment divides into branches, then each branch in turn divides into new branches, etc. It has been determined that the number of branches at each division is almost invariably two, that is, the tree structure is formed by repeated *bifurcations* as shown schematically in Fig. 1.6.1. This tree structure is termed "open" in the sense that there are no cross-connections between the branches, so that the path from the root segment to any other vessel segment within this structure is *unique*. The issue of possible cross-connections (collateral vessels) in the coronary circulation was discussed in Section 1.6 and will not be considered further. In this section we consider only open tree structures, in which the building block from which the tree is constructed is an arterial bifurcation.

We begin by considering an arterial bifurcation, being modelled by three tube segments as shown schematically in Fig. 8.4.1. Subscripts 0,1,2 are used to identify the parent and the two branch segments, respectively, as shown in the figure, with the convention that subscript 1 shall always be used to identify the branch with the larger diameter. With the flow being from parent to branches, conservation of mass requires that flow rate q_0 in the parent vessel be equal to the sum of the flow rates in the two branches, that is

$$q_0 = q_1 + q_2 \tag{8.4.1}$$

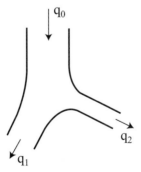

Fig. 8.4.1. Arterial trees in the cardiovascular system are formed largely by repeated bifurcations whereby a vessel segment divides into two branches and then each of the branches in turn divides into two branches, etc. The same is true in the coronary circulation. An arterial bifurcation is shown here schematically, with the parent vessel identified by subscript 0 and the two branches by subscripts 1, 2, with the convention that subscript 1 is always reserved for the branch with the larger radius. Flow rate q_0 in the parent vessel is divided into q_1, q_2 in the branches.

The pressure distribution under conditions of steady flow through the bifurcation can be considered by following two streamwise paths: one from parent to branch-1 and another from parent to branch-2. Along each path the situation is the same as that of two tubes in series, as considered in the previous section. It is important to emphasize again that here too we assume that the idealized conditions of fully developed Poiseuille flow prevail along the full length of each tube segment, ignoring local deviations at the two ends of each segment. The justification for this is that we are interested primarily in the pressure distribution *along* the tubes forming the bifurcation rather than in the local details of the flow field within the bifurcation. The only difference here is that, because of flow division, the flow rates in consecutive tube segments are not the same. Along the path from the root segment to the first branch, we may then return to Eqs. 8.3.5, 6 in the previous section and, using the same notation as in the previous section, write

$$p_0(x_0) = p_0(0) - \frac{8\mu}{\pi a_0^4} q_0 x_0 \tag{8.4.2}$$

$$p_1(x_1) = p_1(0) - \frac{8\mu}{\pi a_1^4} q_1 x_1 \tag{8.4.3}$$

For pressure continuity at the bifurcation point we have, as in the previous section,

$$p_1(0) = p_0(l_0) \tag{8.4.4}$$

thus Eq. 8.4.3 for the pressure distribution along the path to branch-1 becomes

$$p_1(x_1) = p_0(l_0) - \frac{8\mu}{\pi a_1^4} q_1 x_1 \tag{8.4.5}$$

and, similarly, the pressure distribution along the path from the root segment to branch-2 is then given by

$$p_2(x_2) = p_0(l_0) - \frac{8\mu}{\pi a_2^4} q_2 x_2 \tag{8.4.6}$$

The lengths of the three vessel segments can be normalized by defining new normalized coordinates

$$X_0 = \frac{x_0}{l_0} \tag{8.4.7}$$

$$X_1 = \frac{x_1}{l_1} \tag{8.4.8}$$

$$X_2 = \frac{x_2}{l_2} \tag{8.4.9}$$

In terms of these coordinates, the normalized length of each of the three vessel segments is now 1.0, which, as we see shortly, is useful for plotting the pressure distributions along the paths to the two branches using the same scale regardless of their different lengths. Furthermore, these pressure distributions can now be put in nondimensional form by using the properties of the parent tube segment as reference properties. In particular, the magnitude of the pressure drop in the parent tube segment, namely

$$|\Delta p_0| = |p_0(l_0) - p_0(0)| \tag{8.4.10}$$

$$= \frac{8\mu}{\pi a_0^4} q_0 l_0 \tag{8.4.11}$$

is used to put the pressure distributions in nondimensional forms, that is

$$P_0(X_0) = \frac{p_0(x_0) - p_0(0)}{|\Delta p_0|} \tag{8.4.12}$$

$$= -X_0 \tag{8.4.13}$$

$$P_1(X_1) = \frac{p_1(x_1) - p_0(0)}{|\Delta p_0|} \tag{8.4.14}$$

$$= \frac{p_0(l_0) - p_0(0)}{|\Delta p_0|} - \left(\frac{a_0}{a_1}\right)^4 \left(\frac{q_1}{q_0}\right) \frac{x_1}{l_0} \tag{8.4.15}$$

$$= -1 - \left(\frac{a_0}{a_1}\right)^4 \left(\frac{q_1}{q_0}\right) \left(\frac{l_1}{l_0}\right) X_1 \tag{8.4.16}$$

$$P_2(X_2) = \frac{p_2(x_2) - p_0(0)}{|\Delta p_0|} \tag{8.4.17}$$

$$= \frac{p_0(l_0) - p_0(0)}{|\Delta p_0|} - \left(\frac{a_0}{a_2}\right)^4 \left(\frac{q_2}{q_0}\right) \frac{x_2}{l_0} \tag{8.4.18}$$

$$= -1 - \left(\frac{a_0}{a_2}\right)^4 \left(\frac{q_2}{q_0}\right) \left(\frac{l_2}{l_0}\right) X_2 \tag{8.4.19}$$

If it is assumed that vessel lengths are proportional to their radii, the pressure distributions along the two paths become

$$P_1(X_1) = -1 - \left(\frac{a_0}{a_1}\right)^3 \left(\frac{q_1}{q_0}\right) X_1 \tag{8.4.20}$$

$$P_2(X_2) = -1 - \left(\frac{a_0}{a_2}\right)^3 \left(\frac{q_2}{q_0}\right) X_2 \tag{8.4.21}$$

Furthermore, in the theory of vascular branching it is found that a power law relation exists between the radius of a blood vessel and the average flow rate which the vessel is destined to carry, that is

$$q \sim a^{\gamma} \tag{8.4.22}$$

where γ shall be referred to as the "power law index". If this relation is used in Eqs. 8.4.20, 21, the pressure distributions become

$$P_1(X_1) = -1 - \left(\frac{a_0}{a_1}\right)^{3-\gamma} X_1 \tag{8.4.23}$$

$$P_2(X_2) = -1 - \left(\frac{a_0}{a_2}\right)^{3-\gamma} X_2 \tag{8.4.24}$$

If the power law relation between radius and flow rate is used also in Eq. 8.4.1, then the relation between the three flow rates at a bifurcation becomes a relation between the three radii of the vessels involved, namely

$$a_0^{\gamma} = a_1^{\gamma} + a_2^{\gamma} \tag{8.4.25}$$

Essentially, this relation dictates that if one daughter branch at a bifurcation has a comparatively large radius then the other must have a comparatively small one. This is clearly a reflection of the conservation requirement in Eq. 8.4.1, namely that if one branch carries a relatively larger proportion of the flow then the other must carry a correspondingly small proportion. The relation between the radii can be seen more clearly by introducing a "bifurcation index"

$$\alpha = \frac{a_2}{a_1} \tag{8.4.26}$$

Recalling that in our convention branch-1 is always the branch with the larger radius, except when the two radii are equal, this index is a measure of the asymmetry of a bifurcation in terms of the relative radii of its two branches. Its value is 1.0 when the bifurcation is perfectly symmetrical, meaning that its two branches have the same radii, and close to zero when the bifurcation is highly asymmetrical, meaning that one of the two branches has a much larger radius than the other. Thus α has the convenient range of values of 0 to 1.0 for the entire spectrum of possible bifurcations.

The relation between the three radii in Eq. 8.4.25, after division by a_1 or a_2, can now be put in terms of the bifurcation index α, that is

$$\frac{a_0}{a_1} = (1 + \alpha^{\gamma})^{1/\gamma} \tag{8.4.27}$$

$$\frac{a_0}{a_2} = \left(\frac{1 + \alpha^{\gamma}}{\alpha^{\gamma}}\right)^{1/\gamma} \tag{8.4.28}$$

Using these diameter ratios in Eqs. 8.4.23, 24, finally, gives the following expressions for the pressure distributions

$$P_1(X_1) = -1 - (1 + \alpha^\gamma)^{\frac{3}{\gamma} - 1} \tag{8.4.29}$$

$$P_2(X_2) = -1 - \left(\frac{1 + \alpha^\gamma}{\alpha^\gamma}\right)^{\frac{3}{\gamma} - 1} \tag{8.4.30}$$

A considerable volume of work on arterial branching has gone into analysis of the optimal design of arterial bifurcations which, as we see here, depends primarily on the value of the power law index γ in the relation between the radius of a vessel and the flow rate which that vessel is destined to carry (Eq. 8.4.22). Three values in particular were considered on theoretical grounds, namely $\gamma = 2, 3, 4$, while vessel diameters actually measured in the cardiovascular system have produced values of γ highly scattered within and beyond this theoretical range [220]. A key consideration in determining the "optimum" value of γ is the shear stress τ_w which blood flow exerts on endothelial tissue and which under the idealized conditions of Poiseuille flow is given by

$$\tau_w = \mu \left(\frac{du}{dr}\right)_{r=a} \tag{8.4.31}$$

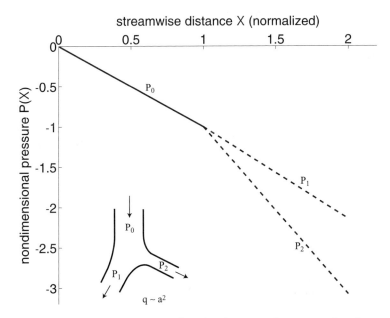

Fig. 8.4.2. Pressure distributions within the three vessel segments forming an arterial bifurcation, under the idealized conditions of steady Poiseuille flow and on the assumption of a power law relation between the radius of each vessel and the flow rate through it. If the power law index is less than 3, as it is here, the pressure drop in the branch with the smaller radius (branch-2) is higher than that in the other branch.

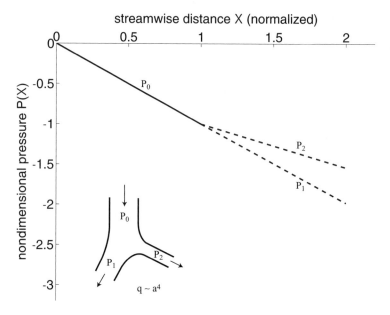

Fig. 8.4.3. Pressure distributions within the three vessel segments forming an arterial bifurcation, under the idealized conditions of steady Poiseuille flow and on the assumption of a power law relation between the radius of each vessel and the flow rate through it. If the power law index is more than 3, as it is here, the pressure drop in the branch with the smaller radius (branch-2) is lower than that in the other branch.

where r is radial coordinate within the vessel and a is its radius. Using the results for flow in a tube and resistance to flow in sections 2.3,4, this equation for the shear stress can be expressed in terms of the flow rate, to give

$$\tau_w = \frac{4\mu}{\pi}\left(\frac{q}{a^3}\right) \tag{8.4.32}$$

or in terms of the pressure drop, to give

$$\tau_w = -\frac{\Delta p}{2\mu}\left(\frac{a}{l}\right) \tag{8.4.33}$$

The first of these results indicates that if a power law relation as in Eq. 8.4.22 exists between the radius of a vessel and the flow rate which it carries, then the shear stress in vessels of different radii varies as

$$\tau_w \sim a^{\gamma-3} \tag{8.4.34}$$

If the value of γ is less than 3, the shear stress will be *higher* in vessels of smaller radii, which clearly cannot be supported on physiological grounds. If the value of γ is more than 3, the shear stress will be lower in vessels of smaller

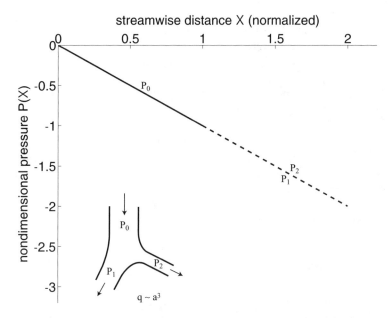

Fig. 8.4.4. Pressure distributions within the three vessel segments forming an arterial bifurcation, under the idealized conditions of steady Poiseuille flow and on the assumption of a power law relation between the radius of each vessel and the flow rate through it. If the power law index is equal to 3, as it is here, the pressure drop is the same along both branches.

radii, which is more plausible on physiological grounds. But if the value of γ is equal to 3, the shear stress in Eq. 8.4.32 will be altogether independent of the radius a, which means that the shear stress will be the same in vessels of different radii. With this value of γ, the flow rate is proportional to the third power of the radius, or conversely, the radius is proportional to the cube root of the flow rate. This relation between the radius and flow rate is widely known as the "cube law" and is of particular interest because it was first derived by showing that it actually provides an optimum compromise between the pumping power required to drive the flow through the bifurcation, which is lower when the vessel radii are large, and the metabolic power required to maintain the volume of blood contained within the three vessels forming the bifurcation, which is lower when the vessel radii are small [147, 212, 213].

Of interest in the present context is the second of the above results, namely that in Eq. 8.4.33, which indicates that if the length of a vessel is assumed to be proportional to its radius then the pressure drop becomes proportional to the shear stress, and hence everything that has been said above about the shear stress now applies equally to the pressure drop. In particular, for the two branches at a bifurcation, if the value of γ is more than 3, the pressure drop along the branch with the smaller radius is lower than that along the branch with the larger radius. This is somewhat unlikely on physiological or fluid dynamic grounds. On the other hand, if the value of γ is less than 3, the reverse is true, which is more plausible on both grounds. If the value of γ is equal to 3, the pressure drop is the same along both branches. These results are shown in Figs. 8.4.2–4.

The interesting conclusion from this discussion is that from the point of view of the shear stress acting on endothelial tissue, the more likely values of γ are 3 or higher, but from the point of view of pressure drop the more likely values are 3 or lower. The only possible compromise between these two conflicting requirements is clearly $\gamma = 3$, and this lends further theoretical support to the cube law. As stated earlier, values of γ based on actual measurements from the cardiovascular system have shown much scatter not only within the range of 2−4 but also outside this range [220]. The scatter, however, is generally found to center around the value $\gamma = 3$.

8.5 Pulsatile Flow in a Rigid Tube

We have seen in Sections 2.3 and 8.3 that in steady flow through a tube the pressure varies linearly along the tube, dropping from a high at entrance to a low at exit, the difference between the two being the pressure drop driving the flow. All other properties of the flow field such as the flow rate, the shear stress at the tube wall, or velocity at any point within the tube are constant in the sense that they do not depend on the streamwise coordinate x. Only the pressure varies along the tube, but the *pressure drop* between the two ends of the tube is constant. These features of the flow are embodied by the equation for Poiseuille flow presented in Section 2.3 which we reproduce here using a subscript s to indicate that the flow properties are for *steady flow* only, as distinct from flow properties in pulsatile flow which we consider below. Thus, from Eq. 2.3.1 we write

$$u_s(r) = \frac{k_s}{4\mu}(r^2 - a^2) \tag{8.5.1}$$

where u_s is the streamwise velocity within the tube, μ is viscosity of the fluid, a is the tube radius, r is radial coordinate measured from the tube axis, and k_s is pressure gradient defined by (Eq. 2.3.2)

$$k_s = \frac{\Delta p_s}{l} \tag{8.5.2}$$

where l is the length of the tube and Δp_s is the pressure difference between the two ends of the tube.

It must be emphasized that the reference here is to *fully developed* flow in a tube as described in Section 2.3. While in practice flow in a tube is rarely fully developed along its full length, we continue to make this assumption here as in the previous three sections because it is necessary in the analysis of flow in a large number of tubes, as in a vascular tree.

Under the assumption of fully developed flow, when the pressure drop between the two ends of a rigid tube is not constant but varies in time, flow properties within the tube vary in time too but, again, they do not vary along the tube. In other words, when the driving pressure drop is a function of time, flow properties within the tube become functions of time too but not functions of the streamwise coordinate x. As the pressure drop changes from one point in time to the next, flow properties change in *the same way at every point along the tube*. This singular behaviour is only possible when (a) the tube is rigid and (b) the fluid is incompressible; both of these conditions are assumed to prevail in this section.

Of particular interest in this section is flow in a tube when the driving pressure drop Δp is a periodic function of time, which may be a simple sinusoidal wave or a composite wave as discussed in Chapter 5. A periodic pressure drop along a tube may be thought to arise from a fixed pressure at the downstream end of the tube and a periodic pressure at the upstream or "input" end. For this reason we often refer to a periodic pressure drop as the "input pressure wave".

If the input pressure waveform has a nonzero mean, as discussed in Section 5.6 (Fig. 5.6.1), then that mean value will serve as a constant pressure drop, producing steady Poiseuille flow within the tube, while the remaining part of the pressure will be purely oscillatory. Denoting the constant part of the waveform by Δp_s, as in Section 5.2, and the oscillatory part by Δp_ϕ, we may then write

$$\Delta p(t) = \Delta p_s + \Delta p_\phi(t) \tag{8.5.3}$$

or, in terms of pressure gradients, as in Eq. 8.5.2,

$$k(t) = k_s + k_\phi(t) \tag{8.5.4}$$

where

$$k(t) = \frac{\Delta p(t)}{l} \tag{8.5.5}$$

$$k_\phi(t) = \frac{\Delta p_\phi(t)}{l} \tag{8.5.6}$$

By definition, Δp_s is the average value of $\Delta p(t)$ over one oscillatory period T as discussed in Section 5.2, Eq. 5.2.15, that is

$$\Delta p_s = \frac{1}{T} \int_0^T \Delta p(t) dt \tag{8.5.7}$$

and, as a consequence, the purely oscillatory part of the waveform has a zero average over one oscillatory period, that is

$$\int_0^T \Delta p_\phi(t)dt = 0 \qquad (8.5.8)$$

Because the equations governing fully developed flow in a tube are *linear*, these two parts of the driving pressure can be dealt with *separately*, each producing a flow field as if it were the only driving pressure, and the two resulting flow fields are then added to produce the complete flow field. The situation is precisely the same as that in which the harmonic components of a composite pressure wave and the flows which they produce can be treated separately and the results are then added, as discussed in Section 6.2. Indeed, here too, the oscillatory part of the pressure, $p_\phi(t)$, may itself be a composite wave which is then decomposed into harmonic components that are dealt with as in Chapter 6.

As a matter of terminology, when $\Delta p(t)$ is a periodic function of time we shall refer to the flow which it produces as "pulsatile flow", to the flow which Δp_s produces as "steady flow", and to the flow which $\Delta p_\phi(t)$ produces as "oscillatory flow". Because the steady and the oscillatory parts of pulsatile flow can be dealt with separately as discussed above, and because steady flow in a tube has already been dealt with in Section 2.3, it remains to deal with only the oscillatory part of the flow.

A classical solution of the equations governing oscillatory flow in a rigid tube exists for an oscillatory pressure gradient of the form

$$k_\phi(t) = k_0 e^{i\omega t} \qquad (8.5.9)$$
$$= k_0(\cos\omega t + i\sin\omega t) \qquad (8.5.10)$$

In other words, the driving pressure consists of a simple sine or cosine wave with amplitude k_0. We recall that, in this complex exponential formulation, the solution obtained is also in complex form where the real part describes the flow when the driving pressure is of the form $k_0\cos\omega t$, and the imaginary part describes the flow when the driving pressure is of the form $k_0\sin\omega t$.

For the oscillatory flow velocity field within the tube, the solution gives [177, 208, 194, 221]

$$u_\phi(r,t) = \frac{ik_0 a^2}{\mu\Omega^2}\left(1 - \frac{J_0(\zeta)}{J_0(\Lambda)}\right)e^{i\omega t} \qquad (8.5.11)$$

where J_0 is Bessel function of order zero of the first kind, and

$$\Omega = \sqrt{\frac{\rho\omega}{\mu}}\, a \qquad (8.5.12)$$

$$\Lambda = \left(\frac{i-1}{\sqrt{2}}\right)\Omega \qquad (8.5.13)$$

$$\zeta(r) = \Lambda\left(\frac{r}{a}\right) \qquad (8.5.14)$$

To interpret $u_\phi(r, t)$ in relation to the corresponding Poiseuille flow velocity in steady flow, namely $u_s(r)$, it is convenient to use as a reference the maximum velocity in Poiseuille flow (Eq. 8.5.1) which occurs on the axis of the tube $(r = 0)$, namely

$$\hat{u}_s = u_s(0) \tag{8.5.15}$$

$$= \frac{-k_s a^2}{4\mu} \tag{8.5.16}$$

Also, comparison between the steady and the oscillatory velocity profiles is made easier by taking the amplitude of the oscillatory pressure gradient to be the same as the pressure gradient in steady flow, that is, we take

$$k_0 = k_s \tag{8.5.17}$$

Nondimensional forms of the steady and the oscillatory velocity profiles are then defined by (Eq. 8.5.1)

$$\frac{u_s(r)}{\hat{u}_s} = 1 - \frac{r^2}{a^2} \tag{8.5.18}$$

$$\frac{u_\phi(r, t)}{\hat{u}_s} = \frac{4}{i\Omega^2} \left(1 - \frac{J_0(\zeta)}{J_0(\Lambda)}\right) e^{i\omega t} \tag{8.5.19}$$

Evaluation of the last expression requires values of the Bessel functions, which are available numerically in math tables and in most computational programs [137, 198]. Because both $u_s(r)$ and $u_\phi(r, t)$ have been normalized in the same way, they can be compared graphically on the same scale, as shown in Fig. 8.5.1. The figure compares the oscillatory velocity profile with the corresponding profile in Poiseuille flow at different times within the oscillatory cycle. The comparison is made when the oscillatory flow is driven by the *real* part of the oscillatory pressure gradient, namely $k_0 \cos \omega t$ $(k_0 = k_s)$. Thus, at $t = 0$ the oscillatory flow and the Poisuille flow are momentarily driven by the same pressure gradient and, in the absence of other factors, we would expect the two velocity profiles to be the same at that point in time within the oscillatory cycle.

The failure of peak velocity in oscillatory flow to reach the same value as peak velocity in Poiseuille flow (normalized at 1.0), is due clearly to inertial effects. More accurately the difference between the two is a function of the nondimensional frequency parameter Ω as defined in Eq. 8.5.12, also known as the Womersley number, which combines the effects of frequency of oscillation, tube radius, and both the density and viscosity of the fluid. The differences observed in Fig. 8.5.1 occur with $\Omega = 2.0$. In particular, the velocity profiles at peak pressure gradient, which occurs at $\omega t = 0$, is short of maximum Poiseuille flow velocity at this value of Ω. Results at higher and lower values of Ω are shown in Fig. 8.5.2.

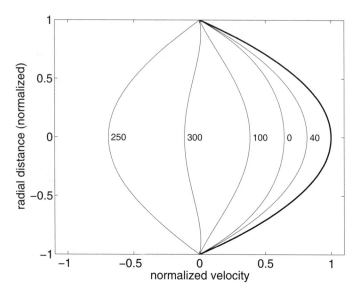

Fig. 8.5.1. Velocity profiles of oscillatory flow in a rigid tube, driven by a pressure gradient of the form $k_s \cos \omega t$ where k_s is a constant, ω is frequency and t is time. The profiles are shown at different times within the oscillatory cycle, as indicated by the value of the phase angle ωt (in degrees) on each curve. For comparison, the bold curve represents the velocity profile in Poiseuille flow driven by a pressure gradient k_s. Velocities are normalized so that their values are comparable on the same scale. The differences between the oscillatory profiles at different points in time within the oscillatory cycle and the Poiseuille flow profile indicate that peak velocity in oscillatory flow is lower than that in Poiseuille flow and occurs at a later time than peak pressure gradient, which occurs at $\omega t = 0$.

Oscillatory flow rate $q_\phi(t)$ is given by

$$q_\phi(t) = \int_0^a 2\pi r u_\phi(r,t)dr \tag{8.5.20}$$

$$= \frac{i\pi k_s a^4}{\mu \Omega^2}\left(1 - \frac{2J_1(\Lambda)}{\Lambda J_0(\Lambda)}\right)e^{i\omega t} \tag{8.5.21}$$

and normalizing in terms of the corresponding flow rate in Poiseuille flow

$$q_s = \int_0^a 2\pi r u_s(r)dr \tag{8.5.22}$$

$$= \frac{-k_s a^2}{8\mu} \tag{8.5.23}$$

we have

$$\frac{q_\phi(t)}{q_s} = \frac{8}{i\Omega^2}\left(1 - \frac{2J_1(\Lambda)}{\Lambda J_0(\Lambda)}\right)e^{i\omega t} \tag{8.5.24}$$

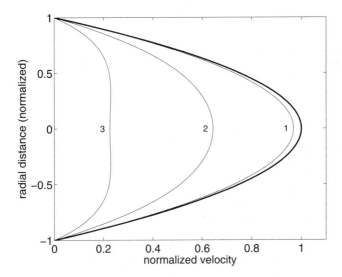

Fig. 8.5.2. Velocity profiles at peak pressure gradient $k_s \cos \omega t$, which occurs at $\omega t = 0$, and different values of the nondimensional frequency parameter Ω as indicated on individual curves. The bold curve represents the velocity profile in Poiseuille flow driven by a pressure gradient k_s.

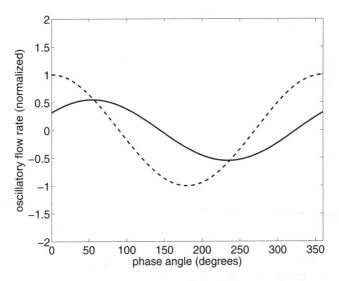

Fig. 8.5.3. Oscillatory flow rate $q_\phi(t)$ (solid) compared with the driving pressure gradient $k_\phi(t)$ (dashed) for flow in a rigid tube with frequency parameter $\Omega = 3.0$. Flow rate lags behind the pressure gradient and peak flow is significantly lower than the corresponding normalized Poiseuille flow value of 1.0.

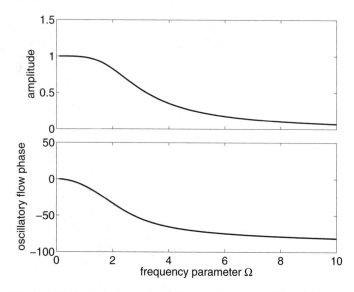

Fig. 8.5.4. Amplitude and phase of oscillatory flow rate $q_\phi(t)$ at different values of the frequency parameter Ω.

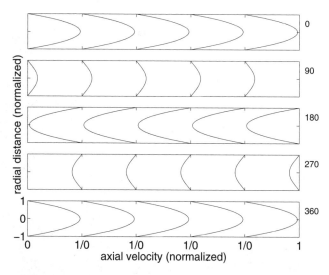

Fig. 8.5.5. Velocity profiles in pulsatile flow in a rigid tube at different times within the oscillatory cycle, indicated in degrees on the right of each panel, and with frequency parameter $\Omega = 1.0$. A characteristic of pulsatile flow in a rigid tube is the absence of any change in the velocity profile as the flow progresses along the tube. The fluid moves in tandem at all positions along the tube.

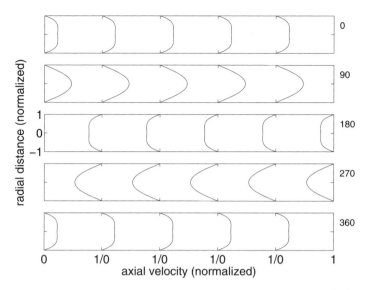

radial distance (normalized)

axial velocity (normalized)

Fig. 8.5.6. Velocity profiles in pulsatile flow in a rigid tube at different times within the oscillatory cycle as in Fig. 8.5.5 but here with the frequency parameter $\Omega = 3.0$.

Variation of the flow rate within the oscillatory cycle, with $\Omega = 3.0$, is shown in Fig. 8.5.3. At this value of Ω peak flow rate is significantly lower than the normalized Poiseuille flow value of 1.0, and it lags behind the driving pressure gradient. At higher values of Ω peak flow diminishes further and the phase lag increases, as shown in Fig. 8.5.4.

One of the characteristics of pulsatile flow in a rigid tube is the absence of any change in the flow field as the flow progresses along the tube. The fluid moves in tandem at all positions along the tube, as shown in Figs. 8.5.5, 6, in other words the flow field is entirely independent of the streamwise coordinate x along the tube. This type of flow is only possible when the tube is strictly rigid. We shall see later that in the presence of any elasticity within the tube wall, pulsatile flow within the tube becomes a propagating wave and the flow field becomes a function of the streamwise coordinate x.

8.6 Pulsatile Flow in an Elastic Tube

A key difference between flow in a rigid tube and that in an elastic tube is that in a rigid tube a local change in pressure is "sensed" instantaneously all along the tube, while in an elastic tube a local change in pressure is first absorbed locally by the elasticity of the tube wall and only then transmitted to other regions of the tube as illustrated schematically in Fig. 8.6.1. In particular, when the input pressure driving the flow in a rigid tube rises and falls in an oscillatory manner, as we saw in Section 8.5, the rise and fall in pressure has its effect simultaneously at every position along the tube. As the input pressure rises, fluid responds by an appropriate increase in flow rate, and while this rise in flow rate is somewhat delayed because of fluid inertia, the delay is the same everywhere along the tube.

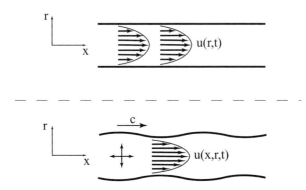

Fig. 8.6.1. Pulsatile flow in a rigid tube, where flow moves in tandem all along the tube (top), compared with pulsatile flow in an elastic tube where flow is in the form of a wave (bottom). In the elastic tube, a rise in pressure is first absorbed by a bulging in the tube wall and then transmitted downstream as the pressure falls and the tube recoils. Thus, in oscillatory flow where the input pressure rises and falls repeatedly, a wave motion is created within the tube. The "wave speed" c is the speed with which the wave progresses downstream.

If the tube is *elastic*, by contrast, as the input pressure rises, the rise is initially absorbed by a local bulge in the tube wall and is therefore not immediately sensed by fluid in other regions of the tube. Only as the input pressure begins to fall does the bulge in the tube wall begin to recoil and the pressure difference driving the flow begins to rise and the flow rate responds accordingly [124, 221]. But by this point in time in the oscillatory cycle the input pressure begins to rise again and the cycle is repeated again and again. The result is a "delayed messaging" of the oscillatory input pressure to the rest of the tube, compared with the instant messaging which occurs in a rigid tube.

The most important result of this difference is that pulsatile flow in an elastic tube produces *wave motion* along the tube (Fig. 8.6.1), which is strictly analogous to the wave motion observed when the calm surface of a lake is disturbed to cause a local change in pressure. In the latter circumstance the local change in pressure is absorbed locally by a rise of a body of fluid against gravity, which then falls back in analogy with the recoiling of the elastic tube, sending the message along the lake surface.

The speed with which a local change in pressure is transmitted downstream in pulsatile flow in an elastic tube is known as the "wave speed", appropriately so because it corresponds to the speed with which the bulges in the tube wall, like the crests on the surface of a lake, move downstream (Fig. 8.6.1). If the material of the tube wall is perfectly elastic, and if it can be assumed further that the tube wall is "thin" compared with the tube radius, then an approximate expression for the wave speed, known as the Moen-Korteweg formula, is given by

$$c_0 = \sqrt{\frac{Eh}{\rho d}} \tag{8.6.1}$$

where E is a measure of elasticity of the tube wall, known as Young's modulus, or modulus of elasticity, h is the thickness of the tube wall, d is tube diameter, and ρ is the density of the fluid which is assumed constant. The measure of elasticity of the tube wall is such that higher values of E represent increased rigidity of the tube wall. The formula thus indicates that the wave speed is higher in a more rigid tube. An important limit is that in which the tube wall is infinitely rigid, that is it lacks any elasticity at all, in which case the value of E is infinite and therefore the wave speed c_0 is infinite. Thus, the "instant messaging" in a rigid tube referred to earlier can now be viewed as messaging occuring at infinite speed, and the difference between pulsatile flow in an elastic tube and that in a rigid tube can now be re-stated by saying that in a rigid tube wave propagation occurs at infinite speed, hence a change in pressure is sensed instantaneously everywhere along the tube. More accurately, wave propagation is actually *absent* in a rigid tube because the wave itself is absent, it simply does not materialize.

The two assumptions on which the Moen-Korteweg formula is based, namely that the tube wall is thin and perfectly elastic, can be dealt with by modified forms of the formula. In most general applications, however, these modifications are not required and the formula provides a reasonable approximation as it stands.

A more important aspect of the Moen-Korteweg formula is that it has been derived under conditions of *inviscid* flow, where the flow does not satisfy the condition of "no-slip" at the tube wall. To satisfy this important physical condition and thereby include the effects of fluid viscosity a full solution of the equations governing the movements of both the fluid and the tube wall is required. The mathematical problem thus becomes considerably more complicated than that in a rigid tube. The problem has been solved, however, and in what follows we use the main elements of the results [145, 208, 8, 46, 125, 221].

In general, while pulsatile flow in a rigid tube is governed critically by the value of the frequency parameter Ω, pulsatile flow in an elastic tube is governed similarly by the value of Ω but also by the value of the wave speed c which in turn depends on Ω as well as on tube properties. We now distinguish between the wave speed c based on a full solution of the problem, including the effects of viscosity, and the inviscid wave speed c_0 provided by the Moen-Korteweg formula in Eq. 8.6.1. The full solution of the problem provides the following relation between the two

$$c = \sqrt{\frac{2}{(1 - \sigma^2)z}} \, c_0 \tag{8.6.2}$$

where σ is Poisson's ratio for the elastic material of the tube wall, and z is a solution of the equation

$$\{(g-1)(\sigma^2 - 1)\}z^2 + \left\{ \frac{\rho_w h}{\rho a}(g-1) + \left(2\sigma - \frac{1}{2}\right)g - 2\right\}z + \frac{2\rho_w h}{\rho a} + g = 0 \tag{8.6.3}$$

where

$$g = \frac{2J_1(\Lambda)}{\Lambda J_0(\Lambda)} \tag{8.6.4}$$

and ρ_w is density of the tube wall.

While the wave speed c_0 in inviscid flow is *real*, as defined by Eq. 8.6.1, the corresponding wave speed c obtained from a full solution of the equations in viscous flow is *complex*, its real and imaginary parts depending on the frequency parameter Ω as shown in Fig. 8.6.2. It is seen that as the value of Ω increases, the real part of c, normalized in terms of c_0, rapidly approaches 1.0 while the imaginary part approaches zero. This means that for values of Ω above 3 or so, c effectively becomes the same as c_0 and the effects of viscosity on the wave speed become negligible.

Other properties of the flow obtained by the full solution of the problem, using the same notation and normalization used in Section 8.5 for flow in a rigid tube, are given by

$$\frac{u_\phi(x, r, t)}{\hat{u}_s} = \frac{4}{i\Omega^2}\left(1 - G\frac{J_0(\zeta)}{J_0(\Lambda)}\right)e^{i\omega(t - x/c)} \tag{8.6.5}$$

$$\frac{v_\phi(x, r, t)}{\hat{u}_s} = \frac{2a\omega}{i\Lambda^2 c}\left(\frac{r}{a} - G\frac{2J_1(\zeta)}{\Lambda J_0(\Lambda)}\right)e^{i\omega(t - x/c)} \tag{8.6.6}$$

$$\frac{q_\phi(x, t)}{q_s} = \frac{8}{i\Omega^2}(1 - Gg)e^{i\omega(t - x/c)} \tag{8.6.7}$$

where

$$G = \frac{2 + z(2\sigma - 1)}{z(2\sigma - g)} \tag{8.6.8}$$

Comparing these expressions with the corresponding expressions for flow in a rigid tube indicates, first, the presence of a new velocity component v_ϕ which represents flow in the radial direction (Fig. 8.6.1). Also, comparing the expressions for the axial velocity component u_ϕ and for the flow rate q_ϕ with

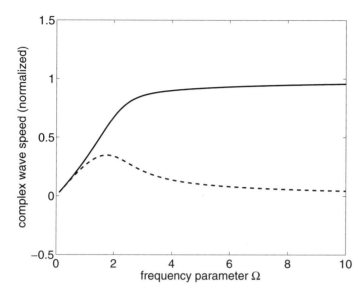

Fig. 8.6.2. The real (solid) and imaginary (dashed) parts of the wave speed c in oscillatory flow in an elastic tube, normalized in terms of the wave speed c_0 in inviscid flow which is purely real. For values of the frequency parameter Ω above 3 or so, the real part of c approaches the value of c_0 and the imaginary part of c approaches zero, thus c effectively becomes the same as c_0.

the corresponding expressions for flow in a rigid tube (Eqs. 8.5.19, 24) indicates that one difference between the two is embodied in what may be referred to as the "elasticity factor" G. Numerical values of G can be found elsewhere [221] and are shown graphically in Fig. 8.6.3, where it is seen that G is a complex quantity whose real part is close to 1.0 and its imaginary part is close to 0.

Another difference between the expressions for pulsatile flow in an elastic tube and those in a rigid tube is the presence of the streamwise coordinate x in the exponential part of the expressions for an elastic tube. This indicates that in the elastic tube oscillations in pressure and flow occur not only in time as they do in a rigid tube, but also in *space*, which is the hallmark of wave motion. In a rigid tube where the fluid everywhere along the tube moves in unison there is no wave motion, the entire bulk of the fluid moves back and forth together. In an elastic tube this is no longer the case because there are now oscillations along the tube as indicated by the presence of the streamwise coodinate x in the exponential part of Eqs. 8.6.5–7, which is now $e^{i\omega(t-x/c)}$ instead of $e^{i\omega t}$.

To see the characteristics of the wave motion more clearly we note that

$$e^{i\omega(t-x/c)} = e^{i\omega t} \times e^{-i\omega x/c} \tag{8.6.9}$$

The first term on the right represents oscillations in time t as for flow in a rigid tube, and the second represents oscillations in x. At a fixed position

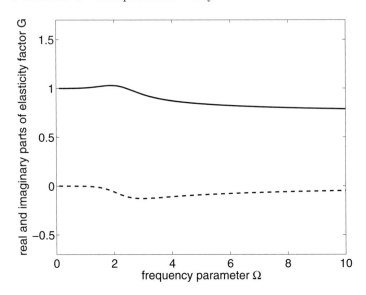

Fig. 8.6.3. Real and imaginary parts of the complex elasticity factor G at different values of the frequency parameter Ω. The real part is close to 1.0 while the imaginary part is close to zero in this range of values of the frequency parameter.

along the tube, which means a fixed value of x, the second term is constant and acts simply as the amplitude of the oscillations in time. At a fixed point in time on the other hand, which means a fixed value of t, the first term is constant and acts simply as the amplitude of the oscillations in x. The wave motion along the tube consists of the *combination* of these two oscillations.

The extent of one complete oscillation along tube is referred to as the "wave length", to be denoted by λ. It is related to the wave speed c and to the frequency of oscillation ω by

$$\lambda = \frac{2\pi c}{\omega} \tag{8.6.10}$$

where λ is in centimeters when ω is in radians per second and c is in centimeters per second. Thus, the flow properties in Eqs. 8.6.5–7 can now be expressed in terms of the wave length λ, that is

$$\frac{u_\phi(x,r,t)}{\hat{u}_s} = \left\{ \frac{4}{i\Omega^2} \left(1 - G\frac{J_0(\zeta)}{J_0(\Lambda)} \right) e^{i\omega t} \right\} e^{-i2\pi x/\lambda} \tag{8.6.11}$$

$$\frac{v_\phi(x,r,t)}{\hat{u}_s} = \left\{ \frac{2a\omega}{i\Lambda^2 c} \left(\frac{r}{a} - G\frac{2J_1(\zeta)}{\Lambda J_0(\Lambda)} \right) e^{i\omega t} \right\} e^{-i2\pi x/\lambda} \tag{8.6.12}$$

$$\frac{q_\phi(x,t)}{q_s} = \left\{ \frac{8}{i\Omega^2} (1 - Gg) e^{i\omega t} \right\} e^{-i2\pi x/\lambda} \tag{8.6.13}$$

In this form of these expressions it is seen that the extent to which space oscillations modify the flow properties in pulsatile flow in an elastic tube depends critically on the ratio x/λ appearing in the exponential terms on the right. In a tube of length l, maximum effect clearly occurs at $x = l$, thus the maximum modifications of the flow depend on the ratio of wave length to tube length, λ/l. To estimate these modifications we may consider a tube in which the wall thickness to diameter ratio h/d is $1/10$, the fluid density ρ is $1.0\,\mathrm{gm/cm^3}$, and Young's modulus E is 10^7 dynes/cm^2. Inserting these values in the Moen-Korteweg formula gives an estimate of the inviscid wave speed

$$c_0 = \sqrt{\frac{Eh}{\rho d}} \tag{8.6.14}$$

$$\sim 10 \text{ m/s} \tag{8.6.15}$$

The corresponding wave length, at a fundamental frequency f_0 of 1 cycle/s, is then

$$\lambda_0 = \frac{c_0}{f_0} \tag{8.6.16}$$

$$\sim 10 \,\mathrm{m} \tag{8.6.17}$$

We use subscript 0 for λ to indicate that here it is based on c_0 and f_0. The values of both c_0 and λ_0 above are high because they are based on inviscid flow but they serve as useful benchmarks. Actual values measured in the cardiovascular system may be closer to one half of these values.

In a coronary artery of length 5 cm, therefore, the ratio of tube length to wave length is approximately $1/100$, and comparison of the flow properties with this value are shown in Figs. 8.6.4, 5.

The radial velocity of the tube wall in pulsatile flow in an elastic tube matches the fluid velocity at the inner surface of the tube, that is at $r = a$, and from Eqs. 8.6.4, 12 and Eq. 8.5.14 we then have

$$\frac{v_\phi(x,a,t)}{\hat{u}_s} = \left\{ \frac{2a\omega}{i\Lambda^2 c}(1 - Gg)e^{i\omega t} \right\} e^{-i2\pi x/\lambda} \tag{8.6.18}$$

and substituting for Λ (Eq. 8.5.13), this simplifies further to

$$\frac{v_\phi(x,a,t)}{\hat{u}_s} = \left\{ \frac{2}{R_c}(1 - Gg)e^{i\omega t} \right\} e^{-i2\pi x/\lambda} \tag{8.6.19}$$

where R_c is a Reynolds number based on the wave speed c, namely

$$R_c = \frac{\rho a c}{\mu} \tag{8.6.20}$$

Plots of the normalized radial velocity scaled in terms of $2/R_c$, that is plots of $(v_\phi(x,a,t)/\hat{u}_s)/(2/R_c)$ are shown in Fig. 8.6.5. Because of the scaling,

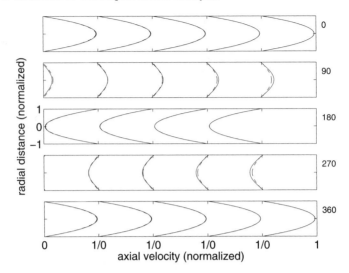

Fig. 8.6.4. Oscillatory velocity profiles in an elastic tube (solid) compared with those in a rigid tube (dashed) when the ratio of tube length to wave length is $1/100$, which would be approximately so in a main coronary artery, and at a moderate value of the frequency parameter, $\Omega = 3.0$. The difference between the two is negligibly small.

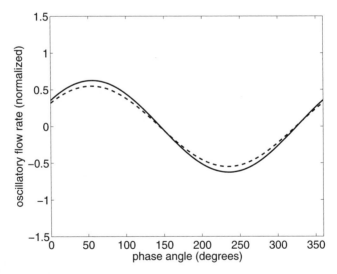

Fig. 8.6.5. Oscillatory flow rate within an oscillatory cycle in an elastic tube (solid) compared with that in a rigid tube (dashed) at a moderate value of the frequency parameter, $\Omega = 3.0$. Oscillatory flow rate reaches higher peaks in an elastic tube than it does in a rigid tube.

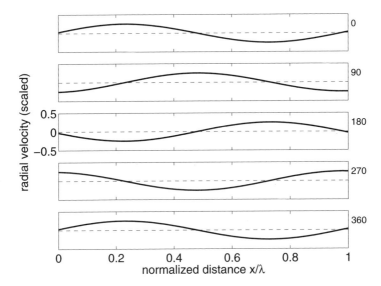

Fig. 8.6.6. Radial velocity of the tube wall in oscillatory flow in an elastic tube at different times within the oscillatory cycle as indicated by the phase angle in degrees on the right. The velocity is scaled in terms of a Reynolds number based on the wave speed c (see text) and is thus exaggerated in magnitude, but the wave pattern of the velocity illustrates the wave motion of the fluid within the tube.

the magnitude of wall velocity is much exaggerated but its pattern illustrates the wave motion of the fluid within the tube since the wall velocity actually matches the fluid velocity at the tube wall.

8.7 Wave Reflections

One of the most important conclusions that can be drawn from the previous section is that pulsatile flow in an elastic tube is associated with wave motion, wave "propagation", along the tube. Wave motion does not arise in pulsatile flow in a rigid tube except when the fluid is compressible, in which case the compressibility of the fluid leads to wave motion within the tube, in much the same way as the elasticity of a tube. This scenario is not of particular interest in coronary blood flow or in blood flow in general where the fluid is very nearly incompressible while blood vessels are decidedly not rigid. Thus, our working model here is that of wave motion of an incompressible fluid in an elastic tube.

One of the most important characteristics of a propagating wave is that it can be *reflected* in the presence of an obstacle, that is, in the presence of a change in the conditions under which the wave is propagating. The analogy of a wave travelling on the calm surface of a lake is again useful here. It is a very common observation that when the wave reaches the shore or other obstacle such as a boat, it is partially or totally reflected, producing a wave moving in the opposite direction. This backward moving wave then combines with the forward moving wave to produce a complex pattern of wave motion.

The same scenario occurs in an elastic tube, though it is not as clearly visible. In this case a propagating wave may be reflected because of an occlusion or narrowing within or at the end of the tube, or a local change in its elasticity. Any of these will act as an obstacle in the way of a propagating wave and thus act as a *reflection site*. The most important reflection sites in coronary blood flow and in blood flow in general are vascular junctions. They are important because of their very large number. Vascular trees typically consist of many millions of vascular junctions, and any path for blood flow within the tree typically consists of only short tube segments, each terminating at a vascular junction (Figs. 8.5.1, 2). The propagating wave rarely enjoys any significant length of tube free from obstacles, thus wave reflections are ubiquitous in the coronary circulation as they are in the cardiovascular system in general. The effect of wave reflections from a single vascular junction may be very small, but the *cumulative* effect from many thousands or millions of such junctions can be very large. In the lake analogy this is equivalent to a wave being reflected from many boats of different sizes and at different positions on the lake surface, leading to a highly complex wave pattern. In a vascular tree the result of this complex pattern of forward and backward moving waves is *a change in the pressure distribution within the tree*, which in turn affects blood flow within the tree. Wave reflection effects in the coronary circulation have received little attention in the past despite the significant role they may play in the dynamics of coronary blood flow [5, 219, 164]

The effects of wave reflections on the pressure distribution within a vascular tree is therefore a key element in the analysis of pulsatile blood flow. Because these effects are cumulative, coming from many reflection sites within the tree, and because each of these sites is typically positioned at the end of a short tube segment, there are two essential steps in the computation of these effects. In the first, one considers the effects of wave reflections in a single tube, and, in the second step, the results are applied to the hierarchy of tube segments in a branching tree structure to calculate the cumulative effect. We consider the first of these steps in the present section, and the second in the next chapter.

Since the ultimate aim in this work is to deal with a large number of tube segments in a tree structure, the analysis of wave reflections is essentially one dimensional. The detailed analysis of pulsatile flow in an elastic tube considered in the previous section cannot be carried in full to each of many thousands of tube segments in a tree structure. Instead, we take only the main conclusions from that analysis, namely that an oscillatory input pressure applied at the entrance of an elastic tube produces a travelling wave along the tube. This conclusion can in fact be reached alternatively by considering solutions of wave equations instead of the equations on which the results of pulsatile flow in an elastic tube discussed in the previous section are based. In either case, our starting point is that an input oscillatory pressure of the form [221]

$$p_{in}(t) = p_0 e^{i\omega t} \tag{8.7.1}$$

applied at the entry to an elastic tube, produces a travelling wave within the tube, of the form

$$P(x,t) = p_0 e^{i\omega(t-x/c)} \tag{8.7.2}$$

where p_0 is the constant amplitude of the input wave, ω is the frequency of oscillation, c is the wave speed, t is time, and x is distance along the tube measured from the entrance.

As discussed in the previous section, this travelling wave consists of two oscillations, one in time and one in space. To separate these two oscillations, and to put the pressure in normalized form, we write

$$\overline{P}(x,t) = \frac{P(x,t)}{p_0} \tag{8.7.3}$$

$$= e^{-i\omega x/c} \times e^{i\omega t} \tag{8.7.4}$$

$$= p(x)e^{i\omega t} \tag{8.7.5}$$

where

$$p(x) = e^{-i\omega x/c} \tag{8.7.6}$$

$$= \cos\left(\frac{\omega x}{c}\right) - i\sin\left(\frac{\omega x}{c}\right) \tag{8.7.7}$$

Thus, at different points in time within the oscillatory cycle and at different positions along the tube, the normalized pressure is given by

$$\overline{P}(x,t) = \cos\omega(t - x/c) + i\sin\omega(t - x/c) \tag{8.7.8}$$

$$= p(x)e^{i\omega t} \tag{8.7.9}$$

which is a simple cosine or sine function of x, depending on whether we use the real or the imaginary part of the solution. Using the real part, we have a simple cosine wave which progresses along the tube at increasing values of t, as illustrated in Fig. 8.7.1.

The amplitude of this travelling wave represents the peak values of pressure reached at different points along the tube and at different points in time and is given by

$$|\overline{P}(x,t)| = |e^{i\omega(t-x/c)}| \tag{8.7.10}$$

$$= |p(x)| \times |e^{i\omega t}| \tag{8.7.11}$$

$$= |p(x)| \tag{8.7.12}$$

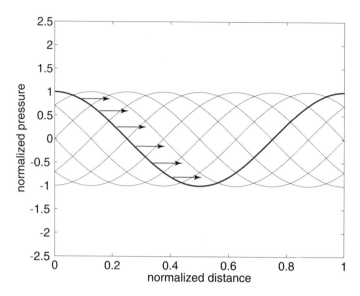

Fig. 8.7.1. Wave propagation in a tube can be thought of graphically as a simple sine or cosine wave moving along the tube as time goes on.

Thus, because $|e^{i\omega t}| = 1.0$, the amplitude of the travelling wave is always equal to the amplitude of the pressure oscillations in space within the tube, namely $|p(x)|$, which is a function of x representing the distribution of peak pressure along the tube and we shall refer to it simply as the "pressure distribution" within the tube. It is a key element in the analysis of wave reflections within a tube because it provides a measure of the amount by which the pressure distribution along the tube is modified by wave reflections and hence a measure of the amount by which the flow is being affected by the moving wave. The reason for this is that *in the absence of wave reflections* we have, from Eq. 8.7.7 above,

$$|p(x)| = \left| \cos\left(\frac{\omega x}{c}\right) - i\sin\left(\frac{\omega x}{c}\right) \right| \tag{8.7.13}$$

$$= 1.0 \tag{8.7.14}$$

This means that in the present case the pressure peaks reached at different points along the tube are equal, as shown graphically in Fig. 8.7.2. That is, the pressure distribution is uniform at a normalized value of 1.0. We shall find this result for the pressure distribution is a singular result in the sense that it can only occur in the absence of wave reflections. Because of this, any deviations from $|p(x)| = 1.0$ in other cases can be attributed immediately to the effects of wave reflections.

To see the way in which the pressure distribution along a tube is actually modified by wave reflections, consider a tube of length l in which the following

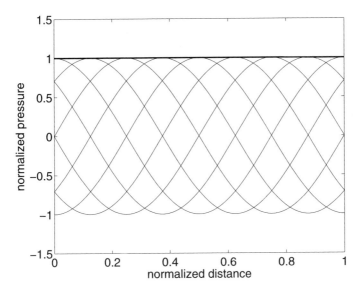

Fig. 8.7.2. The peaks of pressure oscillations at different positions along the tube, represented by the bold line, shall be referred to as the "pressure distribution" along the tube. Only in the absence of wave reflections is this distribution uniform at a normalized value of 1.0 as it is in this case.

forward moving pressure wave, to be identified by subscript f, is moving towards the end of the tube

$$\overline{P}_f(x,t) = e^{i\omega(t-x/c)} \qquad (8.7.15)$$

If a reflection site at the end of the tube ($x = l$) causes a fraction R of the wave to be reflected, then the following backward moving wave, to be identified by subscript b, will arise at that point [221]

$$\overline{P}_b(x,t) = Re^{i\omega(t-(2l-x)/c)} \qquad (8.7.16)$$

where R is known as the "reflection coefficient" and is defined by

$$R = \frac{\overline{P}_b(l,t)}{\overline{P}_f(l,t)} \qquad (8.7.17)$$

The time and space oscillations of pressure within the tube now consists of the combination of the two waves, namely

$$\overline{P}(x,t) = \overline{P}_f(x,t) + \overline{P}_b(x,t) \qquad (8.7.18)$$

and the form of the combined wave depends on how the forward and the backward moving waves add up. They add up differently at different points in time within the oscillatory cycle, as illustrated in Fig. 8.7.3, because the

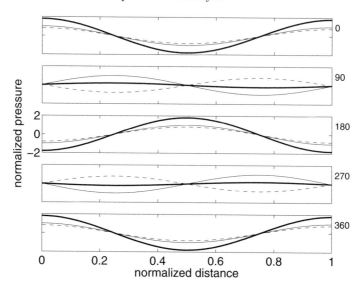

Fig. 8.7.3. A reflection site at the end of a tube causes a fraction R (here $R = 0.8$) of the forward moving wave (thin solid line) to be reflected as a backward moving wave (dashed line). The oscillatory pressure within the tube now consists of the *sum* of the two waves (bold line), and the result depends on how they add up. They add up differently at different points in time within the oscillatory cycle because the forward wave being reflected is different at different points in time within the oscillatory cycle as illustrated in the different panels. The panels represent different times within the oscillatory cycle in terms of the phase angle indicated in degrees on the right.

forward wave being reflected is different at different points in time within the oscillatory cycle.

To find an expression for the pressure distribution $|p(x)|$, as modified by wave reflections, we note first that

$$\overline{P}(x,t) = e^{i\omega(t-x/c)} + Re^{i\omega(t-(2l-x)/c)} \tag{8.7.19}$$

$$= p(x)e^{i\omega t} \tag{8.7.20}$$

where now

$$p(x) = e^{-i\omega x/c} + Re^{-i\omega(2l-x)/c} \tag{8.7.21}$$

and the amplitude of $p(x)$ is clearly no longer equal to 1.0. To find an expression for $|p(x)|$ we note first that if the wave length is denoted by λ as before, then the following relation can be used to eliminate the wave speed

$$c = \frac{\omega\lambda}{2\pi} \tag{8.7.22}$$

If both x and λ are normalized in terms of the tube length l, using the notation

$$\overline{x} = \frac{x}{l} \qquad (8.7.23)$$

$$\overline{\lambda} = \frac{\lambda}{l} \qquad (8.7.24)$$

then in terms of these normalized parameters we have, finally

$$|p(x)| = |e^{-i\omega x/c} + Re^{-i\omega(2l-x)/c}| \qquad (8.7.25)$$

$$= |e^{-i2\pi\overline{x}/\overline{\lambda}} + Re^{-i2\pi(2-\overline{x})/\overline{\lambda}}| \qquad (8.7.26)$$

$$= \sqrt{1 + R^2 + 2R\cos\left(\frac{4\pi}{\overline{\lambda}}[\overline{x} - 1]\right)} \qquad (8.7.27)$$

with some algebra required in the last step.

Eq. 8.7.27 indicates that the pressure distribution along the tube $|p(x)|$, as modified by wave reflections, depends on two key parameters: the reflection coefficient R and the wave-length-to-tube-length ratio $\overline{\lambda}$. If $R = 0$, the equation reduces to $|p(x)| = 1.0$ as we found earlier in the absence of wave reflections. For other values of R the pressure distribution is no longer uniform, as illustrated in Figs. 8.7.4–9. In all cases, the value of $|p(x)|$ at the reflection site, namely at $\overline{x} = 1.0$, is determined by the value of the reflection coefficient such that

$$|p(l)| = \sqrt{1 + R^2 + 2R\cos\left(\frac{4\pi}{\overline{\lambda}}[1 - 1]\right)} \qquad (8.7.28)$$

$$= \sqrt{1 + R^2 + 2R} \qquad (8.7.29)$$

$$= 1 + R \qquad (8.7.30)$$

Also, as $\overline{\lambda}$ becomes large, Eq. 8.7.27 gives

$$\lim_{\overline{\lambda}\to\infty} |p(x)| = \lim_{\overline{\lambda}\to\infty} \sqrt{1 + R^2 + 2R\cos\left(\frac{4\pi}{\overline{\lambda}}[\overline{x} - 1]\right)} \qquad (8.7.31)$$

$$= \sqrt{1 + R^2 + 2R} \qquad (8.7.32)$$

$$= 1 + R \qquad (8.7.33)$$

At the benchmark case in which the wave length and the tube length are equal, hence $\overline{\lambda} = 1.0$, there are four distinct points along the tube where the forward and the backward moving waves combine to produce maxima and minima in the oscillatory pressure. The result is a pressure distribution in which the four points (spaced at quarter tube lengths apart, and sometimes referred to as "node points") are prominent, as shown in Figs. 8.7.4, 5.

As the ratio of wave length to tube length increases, the pressure distribution changes considerably. Examples are shown in Figs. 8.7.6–9 where

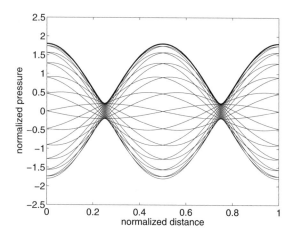

Fig. 8.7.4. Wave reflections in a tube when the wave length and the tube length are equal, i.e., $\overline{\lambda} = 1.0$, and the reflection coefficient $R = 0.8$. The thin lines represent the oscillatory pressure at different points in time within the oscillatory cycle, each line representing the *sum* of the forward and the backward moving waves which are not shown here. The bold line represents the "pressure distribution" along the tube, which is the distribution of peak pressures reached at different points along the tube. The distribution in this case is to be compared with that in Fig. 8.7.2 where wave reflections are absent.

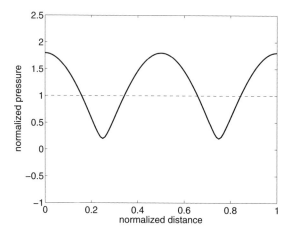

Fig. 8.7.5. Solid line: Pressure distribution along a tube (from Fig. 8.7.4) as modified by wave reflections when the wave length and the tube length are equal ($\overline{\lambda} = 1.0$) and the reflection coefficient $R = 0.8$. The points of maxima and minima are sometimes referred to as "node points" and in this case are spaced one quarter tube lengths apart. Dashed line: Pressure distribution in the absence of wave reflections.

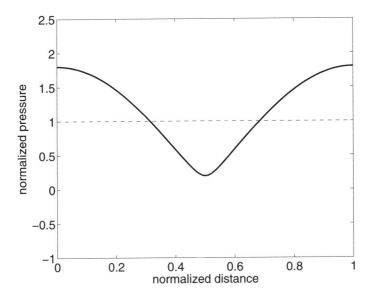

Fig. 8.7.6. Solid line: Pressure distribution along a tube as modified by wave reflections when the wave-length-to-tube-length ratio $\overline{\lambda} = 2$ and the reflection coefficient $R = 0.8$. Dashed line: Pressure distribution in the absence of wave reflections.

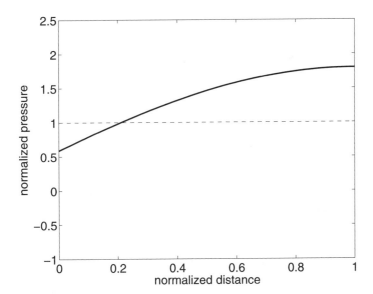

Fig. 8.7.7. Pressure distributions in the presence (solid line) and absence (dashed line) of wave reflections as in Fig. 8.7.6 but here with $\overline{\lambda} = 5$ and $R = 0.8$.

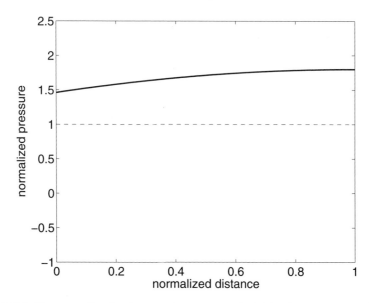

Fig. 8.7.8. Pressure distributions in the presence (solid line) and absence (dashed line) of wave reflections as in Fig. 8.7.6 but here with $\overline{\lambda} = 10$ and $R = 0.8$.

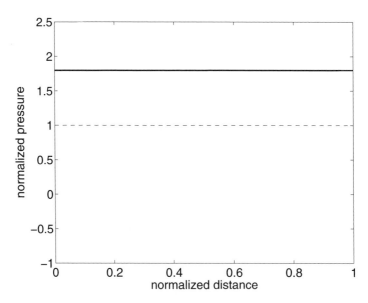

Fig. 8.7.9. Pressure distributions in the presence (solid line) and absence (dashed line) of wave reflections as in Fig. 8.7.6 but here with $\overline{\lambda} = 100$ and $R = 0.8$.

$\overline{\lambda} = 2, 5, 10, 100$. It is seen that the number of node points is reduced to two at $\overline{\lambda} = 2$ and there are no node points at all for $\overline{\lambda} > 4$. In all cases, the value of the normalized pressure amplitude at the terminal end of the tube ($\overline{x} = 1.0$) is $1 + R$ as prescribed by Eq. 8.7.30. The characteristic pattern of the pressure distribution at higher values of $\overline{\lambda}$ is described by a monotonically increasing function, edging closer and closer towards a uniform value of $1 + R$ as prescribed by Eq. 8.7.33. These higher values of $\overline{\lambda}$ are particularly relevant in coronary blood flow where the lengths of main coronary arteries and their branches are of the order of centimeters while the wave length is of the order of meters.

8.8 Summary

In unlumped-model analysis internal structures and events *within* the coronary circulation are considered, thus introducing a space dimension which does not exist in lumped models. In particular, the intricate vascular structure within the system and flow events within this structure, including wave propagation and wave reflections, provide the main grounds for unlumped-model analysis.

A space dimension is an essential part of unlumped-model analysis, whether for flow in a single vessel segment or flow along the tree structure of coronary vasculature. In both cases flow properties are considered to be functions of space and time. In lumped-model analysis, by contrast, they are considered to be functions of time only.

In a sequence of tube segments in series, assuming idealized Poiseuille flow in each segment, the streamwise distribution of pressure along the tube segments depends critically on their successive diameters. The pressure drops linearly if successive diameters are unchanged, it drops more steeply if successive diameters are decreasing and less steeply if they are increasing.

Arterial bifurcations are the building block from which the tree structure of coronary vasculature is constructed. The pressure distributions in steady Poiseuille flow along the two paths in a bifurcation depend on the power law relation between the diameter of a vessel segment and the flow rate which the vessel is destined to convey. Under the "cube law", whereby the diameter is proportional to the cube root of the flow rate, the pressure drops linearly and equally along both paths.

Pulsatile flow in a rigid tube may be divided into a steady part and an oscillatory part and, because the equations governing the flow are linear, the two parts can be dealt with separately and the results then simply added. A characteristic feature of oscillatory flow in a rigid tube is that velocity profiles at a particular moment in time are the same at every streamwise position along the tube. In other words, the velocity field is a function of time only, not a function of position along the tube. This is only possible when the tube is rigid, hence the results cannot be applied to coronary vasculature because of the elasticity of the vessels, but the results and the solutions on which they are based form the basis for a solution for pulsatile flow in an elastic tube.

Pulsatile flow in an elastic tube consists of wave motion along the tube. The velocity field within the tube is a function both of time and of streamwise position along the tube. At a fixed point in time the pressure distribution along the tube is periodic in position, while at a fixed position along the tube the pressure is periodic in time. A key parameter of the flow is the wave speed, which represents the speed with which the wave motion progresses along the tube. Pulsatile flow within coronary vasculature has the characteristics of pulsatile flow in an elastic tube.

One of the most important consequences of wave propagation in an elastic tube is the possibility of wave reflection. Reflections arise when a forward moving wave meets an "obstacle", a change in the conditions under which it is moving. Part of the wave is reflected, thus giving rise to a backward moving wave and the combination of the two waves changes the pressure distribution within the tube. In coronary blood flow the most common obstacle is a vascular junction, and since there are so many of these within the coronary vasculature, the cumulative effect of wave reflections can be very large.

Basic Unlumped Models

9.1 Introduction

There are as yet no well defined or complete unlumped models of the coronary circulation. Work towards this goal has been going on for some time but the subject is still in its infancy because the first few decades of that work were spent largely on attempts to unravel the functional anatomy of coronary vasculature [94, 69, 16, 127, 61, 73, 196, 133, 182, 81, 227, 228, 223, 224, 214, 229, 216, 106, 218, 175, 233, 39, 103, 181]. Also, studies of flow within this vasculature, whether theoretical or experimental, have been rather sporadic because of the enormous difficulties involved, and have so far been able to consider only flow in specific parts of coronary vasculature or only specific aspects of coronary blood flow [108, 38, 11, 12, 160, 101, 143, 88, 104, 159, 170, 219, 105, 109, 18, 144, 148, 99, 23, 231].

The ultimate *unlumped* model of the coronary circulation would be one in which the branching architecture of coronary vasculature is represented in accurate details, the elasticities and other properties of all vessel segments within the coronary network are specified, intramyocardial pressure is mapped out in space within the myocardium and in time within the oscillatory cycle so that its effect on flow in each vessel segment at each moment during the oscillatory cycle is determined, regulatory feedback loops and mechanisms are integrated into the dynamics of the system, and the form of the input pressure wave is specified.

It is an unattainable model, and likely an unnecessary luxury even if it could be attained. The representation of every vessel segment within the coronary network in a model of the coronary circulation is not only an (almost) impossible task but also a wasteful one because the detailed architecture of the coronary network is highly variable from one heart to another [228, 216]. Two coronary networks are never the same in terms of the location and properties of every vessel segment. Therefore, to strive for the validity of such details in a model of the coronary circulation is not highly meaningful, even if it were possible.

The challenge for unlumped model analysis of the coronary circulation is to (1) extract salient features of the coronary network that are more global in nature, such as the scale of the network, its general branching structure, and the flow rate that it is required to carry compared with that in the systemic circulation and to use combinations and variations of such features to demonstrate the type of dynamics that they can or cannot produce; (2) show how the propagating pressure wave driving the flow is likely to evolve under different models and properties of the coronary arterial tree and under different circumstances; and (3) determine under what circumstances wave reflections are likely to affect the flow and therefore to what extent they may play a significant role in the dynamics of the coronary circulation. These issues are considered in the present chapter.

9.2 Steady Flow in Branching Tubes

Arterial trees are generally found to consist of a succession of bifurcations whereby a root vessel segment divides into two branches, then each of the branches in turn divides into two branches, etc. [212]. The "symmetry" of each bifurcation, that is the relative radii of the two branches, is measured by the bifurcation index $\alpha = a_2/a_1$ introduced in the previous section, where a_1, a_2 are radii of the two branches, subscript 1 being reserved by convention to the branch with the larger radius. The branching process is illustrated schematically in Fig. 9.2.1 where $\alpha = 1.0$ and in Fig. 9.2.2 where $\alpha = 0.7$. In each case, the value of α is the same at every bifurcation within the tree structure, thus producing a degree of uniformity which is *not* characteristic of arterial trees in the cardiovascular system where it is found that the value of α varies widely throughout the tree [224, 215, 226, 106, 218, 220]. Nevertheless, these theoretical structures are useful for the purpose of the present section, which is to analyse the pressure distribution in steady flow along such a vascular tree structure by generalizing the results of the previous chapter.

The strategy we follow is based on that used for a single bifurcation in the previous chapter. Here too the pressure distribution within the tree structure is determined simply by following all possible paths from the root segment of the tree to the terminal branch segments. Since each of these paths is unique and consists of a simple succession of tube segments in series, the results of Section 8.3 for tubes in series can then be used, noting only that the flow rate in this succession of tube segments is not the same but varies according to the bifurcation rules discussed in the previous chapter.

To carry out this strategy, the notation of the previous chapter must clearly be generalized to cater for a much larger number of branches. The term "branches" used in the previous chapter to identify the two branch vessel segments at a bifurcation is no longer adequate here because most vessel segments within the tree structure are both parents and branches. The only vessel segments that can be identified by name here are the root segment and

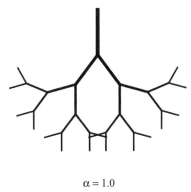

$$\alpha = 1.0$$

Fig. 9.2.1. A 5-level branching tree structure in which the value of the bifurcation index α is 1.0, which means that the two branches at each bifurcation along the tree structure have the same radius.

the terminal branch segments, and we shall continue to use that terminology for these segments. For general notation throughout the tree structure, however, this descriptive scheme is rather inadequate and a more analytic scheme is required. For this purpose we note that each vessel segment has a unique position within the tree structure in terms of the generation or "level" of the tree in which it is located and in terms of its sequential position among other vessel segments at that level. Thus each vessel segment within the tree structure can be identified uniquely by a double subscript notation j, k whereby the first denotes the "level coordinate" of that segment and the second denotes its "sequential coordinate", as illustrated schematically in Fig. 9.2.3.

The convention used in the previous chapter to designate branch-1 as the branch with the larger radius at a bifurcation can also be extended. Here we

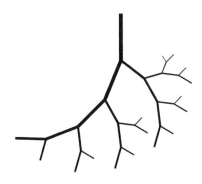

$$\alpha = 0.7$$

Fig. 9.2.2. A 5-level branching tree as in Fig. 9.2.1 but with $\alpha = 0.7$.

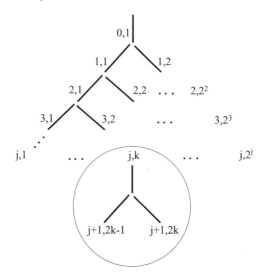

Fig. 9.2.3. A notation scheme for identifying the position of vessel segments in a branching tree structure. Each segment is assigned a coordinate pair j, k in which the first identifies the level of the tree at which the segment is located, and the second identifies the sequential position of that segment among other segments at that level of the tree. The insert shows that in general at each bifurcation one of the two branch segments has an odd sequential number and the other has an even sequential number. We use the convention of reserving the odd sequential number for the branch with the larger radius at each bifurcation.

note that a vessel segment with position coordinates of segment j, k in general has two branches with position coordinates $j + 1, 2k - 1$ and $j + 1, 2k$. The sequential coordinate of the first of these $(2k - 1)$ is an *odd* number while that of the second $(2k)$ is *even*. Thus, to generalize the convention of the previous chapter we reserve the odd sequential number at each bifurcation for the branch segment with the larger radius and the even sequential number for the branch with the smaller radius. An application of this scheme to the 5-level tree structure is illustrated in Fig. 9.2.4.

Using these notation schemes, together with results from the previous two sections, we may now evaluate the pressure distribution along any vessel segment within the tree structure by simply following the unique path from the root segment of the tree to that particular segment. Along this path, as we proceed from one level of the tree to the next, the ratio of parent to branch radius will be given by Eq. 8.4.27 or Eq. 8.4.28, depending on whether the branch has the larger or smaller radius at that particular bifurcation. In the first case the ratio is given by Eq. 8.4.27 and we shall denote it by λ_1, so that

$$\lambda_1 = \frac{a_{j,k}}{a_{j+1,2k-1}} \tag{9.2.1}$$

$$= (1 + \alpha^\gamma)^{1/\gamma} \tag{9.2.2}$$

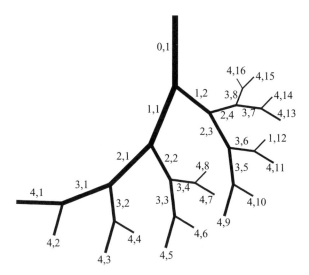

Fig. 9.2.4. Position coordinates of vessel segments along the 5-level tree structure shown in Fig. 9.2.2, illustrating the notation scheme used in the text and in Fig. 9.2.4. At each bifurcation, the ratio of the radii of the two branch segments is 0.7, which means that at each bifurcation one of the two branches has a larger diameter than the other. The path from the root segment to any other segment within the tree structure is unique. One path of particular significance is that of following the branch with the larger radius at each bifurcation, another is that of following the branch with the smaller diameter. These two singular paths are referred to as "bounding paths" in the text, and here they are seen to be bounding in the sense of lying on the two boundaries of the tree structure.

where α is the bifurcation index ($= a_{j+1,2k}/a_{j+1,2k-1}$) and γ is the power law index. In the second case the ratio is given by Eq. 8.4.28 and we shall denote it by λ_2, that is

$$\lambda_2 = \frac{a_{j,k}}{a_{j+1,2k}} \tag{9.2.3}$$

$$= \left(\frac{1+\alpha^\gamma}{\alpha^\gamma}\right)^{1/\gamma} \tag{9.2.4}$$

As an example, for the 5-level tree shown in Fig. 9.2.4, using Eqs.8.4.23,24, the pressure distribution in the terminal branch segment with position coordinates $4, 1$ is given by

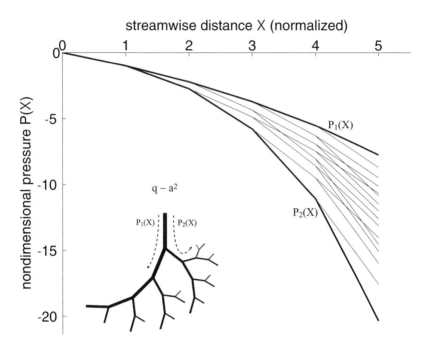

Fig. 9.2.5. Pressure distributions along paths from the root segment to all terminal segments of the 5-level tree structure shown in the inset. The two bold curves
represent the pressure distributions along the two "bounding paths", the pressure
distributions along all other paths fall in between those two. The pressure falls linearly in each segment in accordance with the pressure drop in Poiseuille flow, but
the magnitude of the drop depends on the radius of each vessel segment and on the
amont of flow. Results in this figure are based on the assumption that the flow rate
in a vessel segment is proportional to the square of its radius.

$$P_{4,1}(X_{4,1})$$

$$= -1 - \left(\frac{a_{0,1}}{a_{1,1}}\right)^{3-\gamma} - \left(\frac{a_{0,1}}{a_{2,1}}\right)^{3-\gamma} - \left(\frac{a_{0,1}}{a_{3,1}}\right)^{3-\gamma} - \left(\frac{a_{0,1}}{a_{4,1}}\right)^{3-\gamma} X_{4,1}$$

$$(9.2.5)$$

$$= -1 - \left(\frac{a_{0,1}}{a_{1,1}}\right)^{3-\gamma} - \left(\frac{a_{0,1}}{a_{1,1}} \times \frac{a_{1,1}}{a_{2,1}}\right)^{3-\gamma} - \left(\frac{a_{0,1}}{a_{1,1}} \times \frac{a_{1,1}}{a_{2,1}} \times \frac{a_{2,1}}{a_{3,1}}\right)^{3-\gamma}$$

$$- \left(\frac{a_{0,1}}{a_{1,1}} \times \frac{a_{1,1}}{a_{2,1}} \times \frac{a_{2,1}}{a_{3,1}} \times \frac{a_{3,1}}{a_{4,1}}\right)^{3-\gamma} X_{4,1}$$

$$(9.2.6)$$

$$= -1 - (\lambda_1)^{3-\gamma} - (\lambda_1^2)^{3-\gamma} - (\lambda_1^3)^{3-\gamma} - (\lambda_1^4)^{3-\gamma} X_{4,1}$$

$$(9.2.7)$$

Similarly, the pressure distribution in the terminal branch segment with position coordinates 4, 16 is given by

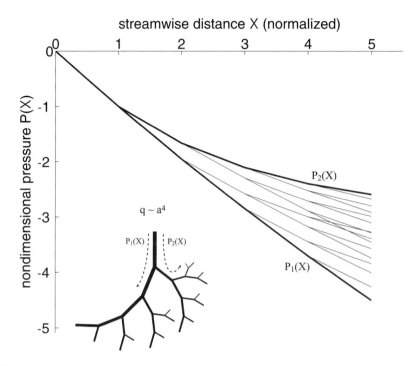

Fig. 9.2.6. Pressure distributions in the 5-level tree structure as in Fig. 9.2.5, but here the results are based on the assumption that the flow rate in a vessel segment is proportional to the fourth power of its radius. It is seen that under this assumption the pressure drops more steeply along the path of branches with the larger radii than it does along the path of branches with the smaller radii, which is somewhat unlikely on physiological or fluid dynamic grounds.

$$P_{4,16}(X_{4,16})$$

$$= -1 - \left(\frac{a_{0,1}}{a_{1,2}}\right)^{3-\gamma} - \left(\frac{a_{0,1}}{a_{2,4}}\right)^{3-\gamma} - \left(\frac{a_{0,1}}{a_{3,8}}\right)^{3-\gamma} - \left(\frac{a_{0,1}}{a_{4,16}}\right)^{3-\gamma} X_{4,16}$$

$$(9.2.8)$$

$$= -1 - \left(\frac{a_{0,1}}{a_{1,2}}\right)^{3-\gamma} - \left(\frac{a_{0,1}}{a_{1,2}} \times \frac{a_{1,2}}{a_{2,4}}\right)^{3-\gamma} - \left(\frac{a_{0,1}}{a_{1,2}} \times \frac{a_{1,2}}{a_{2,4}} \times \frac{a_{2,4}}{a_{3,8}}\right)^{3-\gamma}$$

$$- \left(\frac{a_{0,1}}{a_{1,2}} \times \frac{a_{1,2}}{a_{2,4}} \times \frac{a_{2,4}}{a_{3,8}} \times \frac{a_{3,8}}{a_{4,16}}\right)^{3-\gamma} X_{4,16} \qquad (9.2.9)$$

$$= -1 - (\lambda_2)^{3-\gamma} - (\lambda_2^2)^{3-\gamma} - (\lambda_2^3)^{3-\gamma} - (\lambda_2^4)^{3-\gamma} X_{4,16} \qquad (9.2.10)$$

These two particular segments and the paths leading to them from the root segment of the tree are clearly singular in the sense that the path to segment

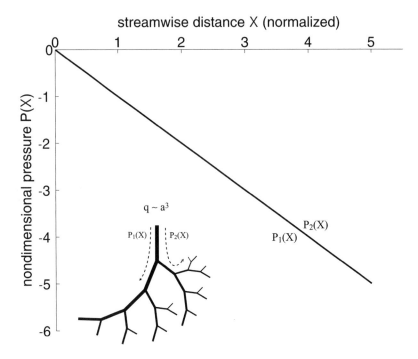

Fig. 9.2.7. Pressure distributions in the 5-level tree structure as in Fig. 9.2.5, but here the results are based on the assumption that the flow rate in a vessel segment is proportional to the third power of its radius, which is widely known as the "cube law". Under this assumption the pressure distributions are identical along all paths from the root segment of the tree to the terminal branches, which lends strong theoretical support to the validity of the cube law.

4, 1 is followed by selecting the branch with the larger radius and hence using λ_1 at every bifurcation, while the path to segment 4, 16 is followed by selecting the branch with the smaller radius and hence using λ_2 at every bifurcation. We may thus refer to these as "bounding paths" in the sense that along all other paths along the tree structure a combination of λ_1 and λ_2 must be used.

The pressure distributions along the two bounding paths and along all other paths are shown in Figs. 9.2.6–8 where the singular nature of the two bounding paths is seen again in terms of the pressure distributions within the entire tree structure. The figures also show the critical dependence of the pressure distribution on the value of the power law index γ in the relation between flow rate and vessel radius (Eq. 8.4.22). Again, the cube law ($\gamma = 3$) appears to present the ideal compromise as concluded in the previous chapter. However, pressure distributions based on vessel radii actually measured in the cardiovascular system exhibit characteristics of Fig. 9.2.6 rather than Fig. 9.2.7. This indicates clearly that the assumptions on which the results in Fig. 9.2.7 are based, namely the cube law relation between flow rate and vessel radius, and the linear relation between vessel length and radius, are not met exactly in the cardiovascular system but are met with considerable scatter as stated in the previous chapter and as data from the cardiovascular system indeed indicate [224, 215, 226, 106, 218, 220].

9.3 Pulsatile Flow in Rigid Branching Tubes

While pulsatile flow in a rigid tube is an idealized model of pulsatile flow in an elastic blood vessel as remarked at the end of the previous section, the results which the model produces provide important benchmarks for pulsatile flow in elastic tubes. Before we consider that problem, therefore, we continue to use the rigid tube model in this section to examine pulsatile flow in a tree structure consisting of rigid tube segments such as that considered for steady flow in Section 9.2.

From the previous chapter the key parameter that determines the properties of pulsatile flow in a rigid tube is the frequency parameter, also known as the Womersley number,

$$\Omega = \sqrt{\frac{\rho\omega}{\mu}}\, a \qquad (9.3.1)$$

If the flow in a tree structure made up of many rigid tube segments is driven by an oscillatory input pressure of the form

$$k_\phi(t) = k_0 e^{i\omega t} \qquad (9.3.2)$$

then the frequency of oscillation ω in Eq. 9.3.1 for tube segments throughout the tree will be determined by the frequency of that input pressure. Assuming, also, that the fluid density ρ and viscosity μ in that equation remain constant throughout the tree, then the value of the frequency parameter Ω will change only with the radius a of tube segments within the tree.

To illustrate the variation of Ω along the 5-level tree structure considered in Section 9.2, taking the following property values

$$\rho = 1.0 \quad \text{g/cm}^3 \qquad (9.3.3)$$
$$\mu = 0.04 \quad \text{g/(cm.s)} \qquad (9.3.4)$$
$$\omega = 1.0 \quad \text{cycles/s} \qquad (9.3.5)$$
$$= 2\pi \quad \text{radians/s} \qquad (9.3.6)$$

the value of the frequency parameter in Eq. 9.3.1 is then given by

$$\Omega = \sqrt{50\pi}\, a \qquad (9.3.7)$$

where a is the radius of the tube *in centimeters*. This expression can be used to map out the values of Ω along the 5-level tree structure in which the radii

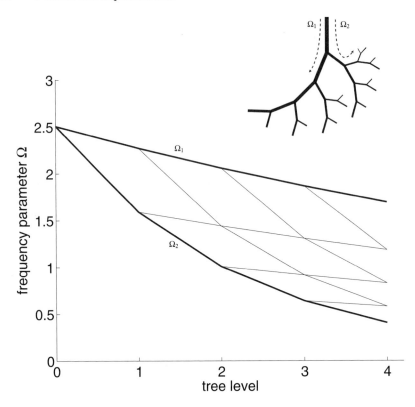

Fig. 9.3.1. Values of the frequency parameter Ω at different segments of the 5-level tree shown in the inset. The tree is based on a power law relation between flow rate and vessel radius, with power law index $\gamma = 3.0$ and bifurcation index $\alpha = 0.7$. The values of Ω decrease most rapidly along the bounding path marked Ω_2 consisting of branch segments with the smaller radii at each bifurcation, and most slowly along the other bounding path, marked Ω_1. Values of Ω at other branch segments fall in between these two extremes.

of branch segments are determined by the power law relation between flow and radii used in Section 8.4.

Starting out with $a = 0.2$ cm as the radius of the root segment of the tree, which is representative of the radius of a main human coronary artery, the radii of subsequent branch segments are then given by Eqs.8.4.27,28 assuming the power law relation on which these equations are based. Values of the frequency parameter Ω obtained with these radii, and with the parameter values in Eqs.9.3.3-6 are shown in Fig. 9.3.1. In general the value of Ω decreases along any streamwise path from the root segment of the tree. It decreases most rapidly along the bounding path consisting of branch segments with the smaller radii at each bifurcation, and most slowly along the other bounding path. In a tree structure with a larger number of levels the values of Ω continue

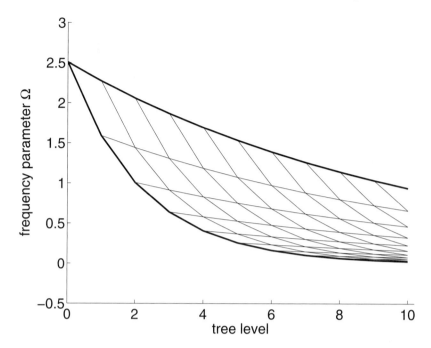

Fig. 9.3.2. Values of the frequency parameter Ω in a tree with the same parameters as that in Fig. 9.3.1 but here the tree has 11 levels (marked 0 to 10). Values of Ω continue to decrease, ultimately reaching towards zero.

to decrease, reaching ultimately towards zero, as illustrated in Fig. 9.3.2 for an 11-level tree.

Based on values of the frequency parameter, the maximum flow rate reached within the oscillatory cycle in each tube segment within the tree, which we shall refer to simply as "peak flow", can be calculated using Eq. 8.5.24 for the oscillatory flow rate $q_\phi(t)$. In normalized form, this peak flow is given by

$$\frac{|q_\phi(t)|}{q_s} = \left| \frac{8}{i\Omega^2} \left(1 - \frac{2J_1(\Lambda)}{\Lambda J_0(\Lambda)} \right) \right| \qquad (9.3.8)$$

The distribution of peak flow within the 5-level vascular tree is shown in Fig. 9.3.3. Because it is normalized in terms of the steady flow rate q_s in Poiseuille flow (Eq. 9.3.8), this peak flow is a measure of how close the oscillatory flow at each point in time is to a Poiseuille flow driven by a pressure gradient equal to the value of the oscillatory pressure gradient at that point in time. Thus, a normalized peak flow of 1.0 represents an oscillatory flow in which the velocity profile at each point in time is a Poiseuille flow profile, while values less than 1.0 represent oscillatory flows in which peak flow does not quite reach the corresponding Poiseuille flow value. The results in

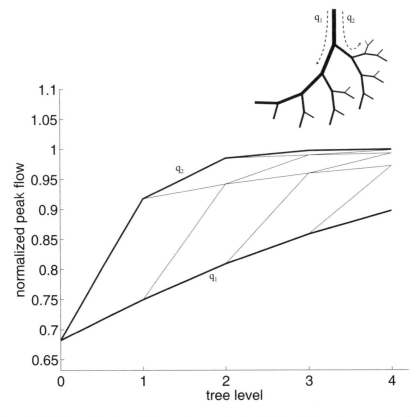

Fig. 9.3.3. Normalized peak flow rates reached at different branch segments of the 5-level tree shown in the inset. A value of 1.0 represents a peak flow equal to that in steady flow. This value is reached more rapidly along the bounding path marked q_2 consisting of branch segments with the smaller radii, and more slowly along the other path, marked q_1. Other lines represent values of the peak flow along other paths within the tree structure.

Fig. 9.3.3 indicate that peak flows reach the Poiseuille flow values more rapidly along the bounding path with the smaller branch segments than they do along the path with the larger branch segments. The reason for this is that the oscillatory flow profile is a more complete Poiseuille profile at smaller values of the frequency parameter Ω, which are reached more rapidly along the path with the smaller branch segments.

Another flow property of particular interest is the oscillatory shear stress τ_ϕ exerted by the fluid on the tube wall and defined by

$$\tau_\phi(t) = -\mu \left(\frac{\partial u_\phi(r,t)}{\partial r} \right)_{r=a} \tag{9.3.9}$$

Using the solution for $u_\phi(r,t)$ in Eq. 8.5.11, and noting that

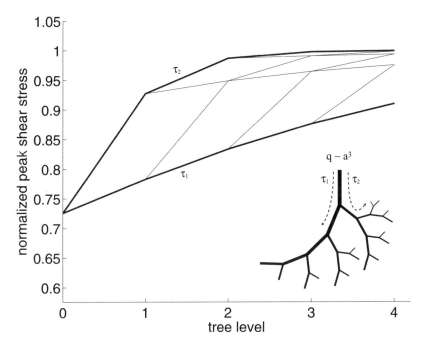

Fig. 9.3.4. Normalized peak shear stress reached at different branch segments of the 5-level tree shown in the inset. A value of 1.0 represents a peak shear stress equal to that in steady flow. This value is reached more rapidly along the bounding path marked τ_2 consisting of branch segments with the smaller radii, and more slowly along the other path, marked τ_1. Other lines represent values of the peak flow along other paths within the tree structure. The results are based on the cube law relation between flow rate and vessel radius, $q \sim a^\gamma$, $\gamma = 3.0$, as indicated in the inset.

$$\frac{dJ_0(\zeta)}{d\zeta} = -J_1(\zeta) \tag{9.3.10}$$

we find

$$\frac{\tau_\phi(t)}{\tau_s} = \frac{2}{\Lambda}\left(\frac{J_1(\Lambda)}{J_0(\Lambda)}\right)e^{i\omega t} \tag{9.3.11}$$

where τ_s is the constant shear stress in steady Poiseuille flow in a tube of radius a under a pressure gradient k_s, which is given by

$$\tau_s = \frac{-k_s a}{2} \tag{9.3.12}$$

Within each oscillatory cycle the normalized shear stress reaches a peak value given by

$$\frac{|\tau_\phi(t)|}{\tau_s} = \left|\frac{2}{\Lambda}\left(\frac{J_1(\Lambda)}{J_0(\Lambda)}\right)\right| \tag{9.3.13}$$

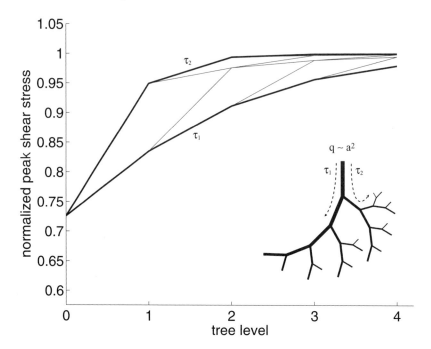

Fig. 9.3.5. Normalized peak shear stress reached at different branch segments of the 5-level tree shown in the inset, as in Fig. 9.3.4, but here the results are based on a power law index $\gamma = 2.0$.

Because of the way it is normalized the value of the peak shear stress is expressed as a fraction of the constant shear stress in steady Poiseuille flow. Thus, a normalized value of 1.0 represents peak oscillatory shear stress equal to that in Poiseuille flow. The distribution of peak shear stress within the 5-level tree is shown in Fig. 9.3.4. It is similar to that of peak flow rate, as would be expected, because shear stress is high at high flow rates and low at low flow rates.

We recall that in *steady* flow, a cube law relation between vessel radius and flow rate, namely $q \sim a^3$, ensures a constant shear stress throughout the tree structure, as illustrated in Section 9.2. This is not the case in *pulsatile* flow, as we see in Fig. 9.3.4 where the results are based on the cube law. Other values of the power law index, namely $\gamma = 2.0$ and $\gamma = 4.0$, produce similar results as shown in Figs. 9.3.5, 6. The reason for this is that in steady flow the shear stress depends on the ratio of flow rate over the third power of the radius (Eq. 8.4.32), while in pulsatile flow the corresponding relation is not as simple [221].

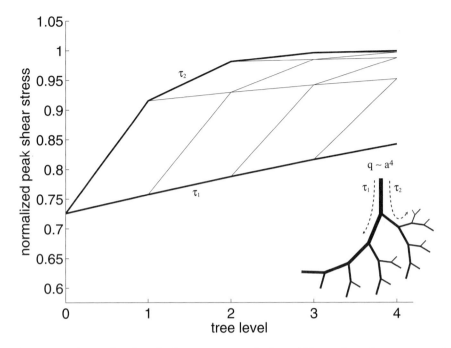

Fig. 9.3.6. Normalized peak shear stress reached at different branch segments of the 5-level tree shown in the inset, as in Fig. 9.3.4, but here the results are based on a power law index $\gamma = 4.0$.

9.4 Elastic Branching Tubes

Results for pulsatile flow in an elastic tube discussed in Section 8.6 indicate that when the wave-length-to-tube-length ratio, $\overline{\lambda} = \lambda/L$, is of the order of 100 or greater, the basic characteristics of the flow are nearly the same as those of pulsatile flow in a rigid tube. In particular, the oscillatory velocity profiles in the elastic tube (Fig. 8.6.4) and the wave form of the oscillatory flow rate (Fig. 8.6.5) are very close to those in a rigid tube. In the human coronary circulation the length of a main coronary artery may be between 5 and 10 cm, while the length of the propagating wave is between 5 and 10 m. Thus, $\overline{\lambda} \sim 100$ is a good estimate of the order of magnitude of the wave-length-to-tube-length ratio in the coronary circulation. However, the length λ of a propagating wave is related to the wave speed c and the angular frequency ω by

$$\lambda = \frac{2\pi c}{\omega} \tag{9.4.1}$$

Thus, at a fundamental frequency of 1 Hz, or $\omega = 2\pi$ radians/s, the wave length is directly related to the wave speed. If the wave speed can be assumed

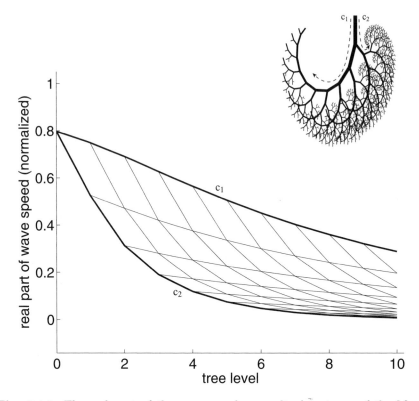

Fig. 9.4.1. The real part of the wave speed, normalized in terms of the Moen-Korteweg wave speed c_0, in a vascular tree model in which the root segment has approximately the same diameter (4 mm) as a main coronary artery in the human heart and in which subsequent branching follows the cube law with power law index $\gamma = 3.0$ as described in Section 8.4 and bifurcation index $\alpha = 0.7$. The two bounding paths marked c_1, c_2 in the tree model are singular paths along which the branch with the larger diameter is followed at each junction in one case (c_1), and the branch with the smaller diameter is followed in the second (c_2). They represent two paths along which the real part of the wave speed decreases most slowly (c_1), or most rapidly (c_2), as indicated on the graph. Everywhere else within the tree structure the value of the real part of the wave speed is bound by these two curves. Thus, since the normalized values are everywhere less than 1.0, the figure indicates that the wave speed is everywhere lower than the Moen-Korteweg wave speed.

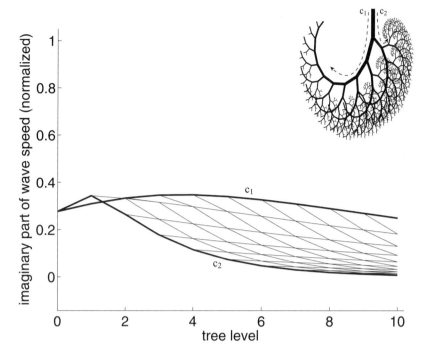

Fig. 9.4.2. The imaginary part of the wave speed associated with the real part shown in Fig. 9.4.1. Since the Moen-Korteweg wave speed is purely real, it follows that the wave speed here is different from the Moen-Korteweg wave speed everywhere along the tree structure, consistent with the results in Fig. 9.4.1. Remaining caption is the same as in that figure.

constant throughout the coronary circulation, the same assumption can be made about the wave length, and the above estimate of $\overline{\lambda} \sim 100$ can be used throughout. However, in Section 8.6 we saw that the value of the wave speed actually depends on the frequency parameter Ω such that when $\Omega > 3$ the wave speed can indeed be assumed constant and equal to the Moen-Korteweg wave speed c_0 (Eq. 8.6.1), but when $\Omega < 3$ the wave speed departs significantly from this value. In fact, its value becomes complex, with real and imaginary parts depending on the value of Ω as shown in Fig. 8.6.2. Now, values of Ω in the coronary arterial tree were estimated in Section 9.3 and are shown in Figs. 9.3.1, 2. It is seen in these two figures and in Fig. 8.6.2 that the coronary circulation lies entirely in the region $\Omega < 3$ where the wave speed is certainly not constant and not equal to the Moen-Korteweg wave speed c_0. Therefore, in a model of the coronary circulation consisting of elastic tubes, it is necessary to map out the values of c within the tree structure on which the model is based. To do so, the value of Ω for each vessel segment in that model must be used to calculate the corresponding value of c according to the solution for pulsatile flow in an elastic tube as was done in Section 8.6. Results, using the 11-level tree model as an example, are shown in Figs.9.4.1,2 where the real and the imaginary parts of the wave speed are shown normalized in terms of the Moen-Korteweg wave speed c_0.

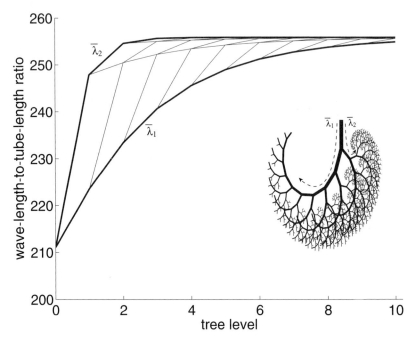

Fig. 9.4.3. The ratio of wave length to tube length $(\overline{\lambda} = \lambda/L)$ for vessel segments along the same 11-level tree model used in Fig. 9.4.1 and using values of the wave speed shown in that figure. The two bounding paths marked $\overline{\lambda}_1$ and $\overline{\lambda}_2$ correspond to those marked c_1, c_2 in Fig. 9.4.1 and have the same interpretation. The figure indicates that the value of $\overline{\lambda}$ is significantly above 100 everywhere along the tree structure, which means that the effects of wave propagation on flow within individual vessel segments is minimal *if wave reflections are absent*. Because of the large number of vessel junctions, however, wave reflections are ubiquitous and their effects on pressure and flow within the tree structure must be calculated.

At the root segment of this tree model, in which the root segment has approximately the same diameter (4 mm) as a main coronary artery in the human heart, the real part of the wave speed is below the normalized value of 1.0, which means that it is below the Moen-Korteweg value. Thereafter, at smaller and smaller branch segments, the wave speed continues to decrease in value, more rapidly along the bounding path with the smaller branches. Similarly, the imaginary part of the wave speed, is above zero everywhere along the tree structure, which means that the wave speed is complex and hence different from c_0 everywhere along the tree structure.

Based on these values of c, the wave-length-to-tube-length ratio can be calculated for each vessel segment using the radii and lengths of vessel segments prescribed in the model. In particular, the model assumes that the hierarchy of radii at different levels of the tree follows the cube law as described earlier (Section 8.4), and that the length-to-diameter ratio for each vessel segment is 10. Thus, at a fundamental frequency of 10 Hz, or $\omega = 2\pi$ radians/s, from Eq. 9.4.1 we have

$$\overline{\lambda} = \frac{\lambda}{L} \tag{9.4.2}$$

$$= \frac{c}{10 \times d} \tag{9.4.3}$$

Results are shown in Fig. 9.4.3 where it is seen that, based on this model, values of $\overline{\lambda}$ would be well above 100 throughout the coronary circulation. From these results, therefore, it would seem that the model of pulsatile flow in branching *rigid* tubes considered in Section 9.3 should be a good model for pulsatile flow in the coronary circulation. This is decidedly *not* the case, however, because pulsatile flow in an elastic tube is characterized by *wave propagation*, a phenomenon which does not (cannot) arise in a rigid tube. The only exception to this, as mentioned previously, is when the fluid is compressible, which is not under consideration here. Wave propagation is a critically important element in any model of pulsatile flow in an elastic tube not only because of the different patterns of flow that it brings with it, namely those described in Sections 8.5 and 8.6, but because wave propagation brings the prospect for wave reflections.

9.5 Effective Impedance, Admittance

From a functional standpoint, the effects of wave reflections in a tube or vascular tree can be thought of in terms of the way they affect the opposition to flow. The term "opposition" is used here deliberately because the opposition to pulsatile flow in the presence of wave reflections is neither pure "resistance" as it is in steady flow, nor pure "impedance" as it is in oscillatory flow in a rigid tube. In fact, the opposition to pulsatile flow in the presence of wave reflections is best described by a modified impedance usually referred to as "effective" impedance. The pure impedance in oscillatory flow in a rigid tube is then renamed "characteristic impedance" to differentiate between the two. Thus, the difference between the characteristic impedance and the effective impedance in a tube or vascular tree is a direct and functionally meaningful measure of the effects of wave reflections.

In general, opposition to flow is defined in terms of the amount of flow produced by a given amount of driving pressure. More specifically, in steady

Poiseuille flow the ratio of pressure difference driving the flow to flow rate is termed the "resistance" to flow, as discussed in Section 2.4. In oscillatory flow, if the driving pressure and the flow rate are simple harmonic (sine or cosine) functions, the ratio of the *amplitudes* of pressure over flow is termed the "impedance" as discussed in Section 4.9.

The resistance in steady Poiseuille flow in a tube depends on only *static* properties of the tube and of the fluid (Eq. 2.4.4). The impedance in oscillatory flow through a rigid tube (hence in the absence of wave propagation and the potential for wave reflections) depends on static properties of the tube and of the fluid as well as on the *frequency* of oscillation (Eqs. 4.9.8,11). The impedance in oscillatory flow through an elastic tube, where flow is characterized by wave propagation and the potential exists for wave reflections, depends on static properties of the tube and of the fluid as well as on the frequency of oscillation, as in a rigid tube, but now also on the extent of *wave reflection effects* if any. Now, wave reflection effects depend on the properties of a reflection site at the downstream end of a tube or many reflection sites far downstream in a vascular tree structure, and in both cases the effects depend critically on the frequency of oscillation or more accurately on the ratios of wave lengths to tube lengths as we saw in Section 8.7. As mentioned earlier, in order to distinguish between impedance to flow when these added wave reflection effects are absent or when they are present, the term "characteristic impedance" is used for the first and "effective impedance" is used for the second.

From a solution of the wave equations for pressure and flow, an input oscillatory pressure of the form

$$p_{in}(t) = p_0 e^{i\omega t} \tag{9.5.1}$$

in the absence of wave reflections, leads to a pressure wave and a flow wave within the tube, of the form

$$P(x,t) = p_0 e^{i\omega(t-x/c)} \tag{9.5.2}$$

$$Q(x,t) = q_0 e^{i\omega(t-x/c)} \tag{9.5.3}$$

$$q_0 = Y_0 p_0 \tag{9.5.4}$$

$$Y_0 = \frac{\pi a^2}{\rho c} \tag{9.5.5}$$

where a is tube radius, ρ is fluid density and c is wave speed which is assumed constant within the tube. It is seen that

$$Y_0 = \frac{q_0}{p_0} \tag{9.5.6}$$

$$= \frac{Q(0,t)}{P(0,t)} \tag{9.5.7}$$

The name "admittance" for Y_0 is thus appropriate as it represents a measure of the amount of oscillatory flow which the tube "admits" for a given oscillatory pressure. By the same logic, the reciprocal of Y_0, which represents the extent to which the tube "impedes" the flow, is then given the name "impedance" and is defined by

$$Z_0 = \frac{1}{Y_0} \tag{9.5.8}$$

$$= \frac{q_0}{p_0} \tag{9.5.9}$$

$$= \frac{P(0,t)}{Q(0,t)} \tag{9.5.10}$$

$$= \frac{\rho c}{\pi a^2} \tag{9.5.11}$$

In the way they are defined here, Y_0 and Z_0 are referred to as "input" admittance and "input" impedance, respectively, because they are based on pressure and flow at the input end of the tube. A number of other definitions are possible.

It is important to note that Y_0, Z_0 represent the admittance and impedance in a given tube not only in the absence of wave reflections, but also on the assumption of constant wave speed within the tube. As we saw in Section 8.6, in the coronary circulation this constant wave speed is typically not equal to the Moen-Korteweg wave speed but must be determined from the solution for pulsatile flow in an elastic tube *for each vessel segment*. As seen in Fig. 8.6.2, the wave speed then depends on the frequency parameter ω and therefore Y_0, Z_0 will also depend on ω.

In the presence of wave reflections, the pressure and flow waves are no longer given by Eqs.9.5.2,3 because they then consist of both forward and backward moving waves as discussed in Section 8.7. As determined for the pressure in detail in that section, and using similar analysis for the flow wave, the two waves are now given by

$$P(x,t) = p_0 e^{i\omega(t-x/c)} + R p_0 e^{i\omega(t-(2l-x)/c)} \tag{9.5.12}$$
$$Q(x,t) = q_0 e^{i\omega(t-x/c)} - R q_0 e^{i\omega(t-(2l-x)/c)} \tag{9.5.13}$$

from which it is clear that admittance or impedance are no longer determined by p_0, q_0 only. For this reason they are now referred to as the "effective" admittance and "effective" impedance and denoted by Y_e, Z_e to distinguish these from the characteristic admittance and impedance. They are defined to have the same meaning as before, however, in terms of the pressure and flow at input to the tube, namely, as in Eq. 9.5.10

$$Y_e = \frac{Q(0,t)}{P(0,t)} \tag{9.5.14}$$

$$= \frac{q_0 - Rq_0 e^{-i\omega(2l/c)}}{p_0 + Rp_0 e^{-i\omega(2l/c)}} \tag{9.5.15}$$

$$= Y_0 \left(\frac{1 - Re^{-i\omega(2l/c)}}{1 + Re^{-i\omega(2l/c)}} \right) \tag{9.5.16}$$

If wave reflections arise at a junction between two tubes denoted by A and B whose characteristic admittances are Y_{0A} and Y_{0B}, and if there are no further reflections at the downstream end of the second tube, as shown schematically in Fig. 9.5.1, then the reflection coefficient R_A at the junction can be expressed in terms of the two characteristic admittances

$$R_A = \frac{Y_{0A} - Y_{0B}}{Y_{0A} + Y_{0B}} \tag{9.5.17}$$

But if there *are* wave reflections at the downstream end of the second tube then its admittance is now its effective admittance Y_{eA}, and the expression for the reflection coefficient becomes

$$R_A = \frac{Y_{0A} - Y_{eB}}{Y_{0A} + Y_{eB}} \tag{9.5.18}$$

Similarly, at a bifurcation, since the combined admittance of the two branches is simply the sum of the two (hence it is convenient to use admittance

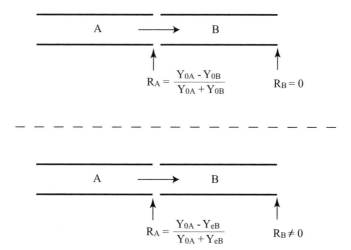

Fig. 9.5.1. The reflection coefficient R_A at the junction between two tubes depends on the difference between the characteristic admittances Y_{0A} of the first tube and Y_{0B} of the second if wave reflections are absent (top), or the difference between the characteristic admittance Y_{0A} of the first tube and the effective admittance Y_{eB} of the second if wave reflections are present (bottom).

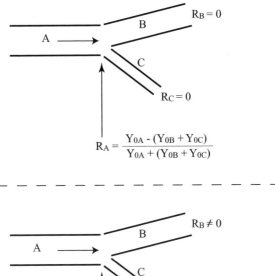

Fig. 9.5.2. The reflection coefficient R_A at an arterial bifurcation depends on the difference between the characteristic admittance Y_{0A} of the parent vessel segment and the sum of the characteristic admittances Y_{0B}, Y_{0C} of the two branch segments if wave reflections are absent (top), or the difference between the characteristic admittance Y_{0A} of the parent vessel segment and the sum of the effective admittance Y_{eB}, Y_{eC} of the two branch segments if wave reflections are present (bottom).

rather than impedance in the analysis of branching trees), then, as illustrated schematically in Fig. 9.5.2, in the absence of wave reflections at the downstream ends of the two branches we have

$$R_A = \frac{Y_{0A} - (Y_{0B} + Y_{0C})}{Y_{0A} + (Y_{0B} + Y_{0C})} \qquad (9.5.19)$$

and in the presence of wave reflections

$$R_A = \frac{Y_{0A} - (Y_{eB} + Y_{eC})}{Y_{0A} + (Y_{eB} + Y_{eC})} \qquad (9.5.20)$$

In a tree structure the effective admittance of each vessel segment depends on wave reflections from all junction sites downstream of that segment. To calculate these, we may use Eq. 9.5.16 for the effective admittance of the parent vessel segment, with the reflection coefficient as given by Eq. 9.5.20, that is,

$$Y_{eA} = Y_{0A} \left\{ \frac{1 - R_A e^{-i\omega(2l_A/c_A)}}{1 + R_A e^{-i\omega(2l_A/c_A)}} \right\} \tag{9.5.21}$$

It is then convenient to eliminate the reflection coefficient R_A from this expression by using Eq. 9.5.20 to substitute for R_A, which requires some algebra, to get

$$Y_{eA} = Y_{0A} \left(\frac{(Y_{eB} + Y_{eC}) + iY_{0A} \tan\theta}{Y_{0A} + i(Y_{eB} + Y_{eC}) \tan\theta} \right) \tag{9.5.22}$$

$$\theta = \frac{\omega l_A}{c_A} \tag{9.5.23}$$

In this form we see that the effective admittance of the parent vessel segment in a bifurcation is determined by its characteristic admittance and the effective admittance of the two branches. If there are no wave reflections at the downstream ends of the two branches, then their effective admittances are the same as their characteristic admittances.

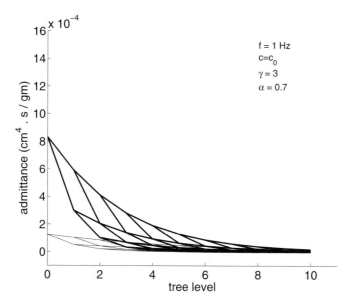

Fig. 9.5.3. Absolute values of the effective (bold) and characteristic (thin) admittances at different levels of the 11-level tree model, with parameter values as shown in the figure. The difference between the two distributions is due entirely to wave reflections.

An important measure of the effects of wave reflections in a vascular tree structure is therefore the difference between the characteristic and the effective admittances of all vessel segments within the tree. The characteristic impedances can be calculated from the prescribed properties of these segments, using Eq. 9.5.5. The calculation of effective impedances, since it involves the effects of wave reflections from all junctions up to and including the upstream ends of the peripheral terminal segments, must therefore begin at this end of the tree. From known or prescribed reflection coefficients at these ends, the effective admittances of the terminal vessel segments are determined, using Eqs.9.5.22,23. If there are no wave reflections at these ends, then the reflection coefficients are zero and the effective admittances of the terminal branches are the same as their characteristic admittances. In either case, the calculation can then progress to the next level of the tree in which each vessel segment is a parent segment in a bifurcation in which the two branches are two of the peripheral segments, thus the effective admittance of the parent segment is determined using Eq. 9.5.21. This process then continues to the next upstream level of the tree, and so on, until the root segment is reached. The results of such calculations are illustrated for the 11-level tree model in Fig. 9.5.3.

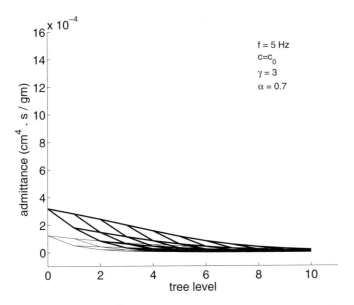

Fig. 9.5.4. Absolute values of the effective (bold) and characteristic (thin) admittances at different levels of the 11-level tree model, as in Fig. 9.5.3 but here with a frequency of 5 Hz.

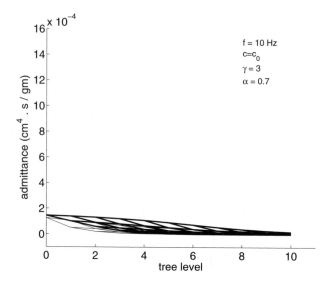

Fig. 9.5.5. Absolute values of the effective (bold) and characteristic (thin) admittances at different levels of the 11-level tree model, as in Fig. 9.5.3 but here with a frequency of 10 Hz.

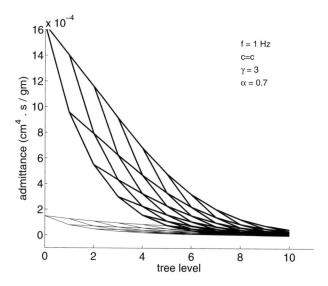

Fig. 9.5.6. Absolute values of the effective (bold) and characteristic (thin) admittances at different levels of the 11-level tree model at a frequency of 1 Hz, as in Fig. 9.5.3, but here using a value of the wave speed c obtained from a solution of the pulsatile flow in an elastic tube for each vessel segment.

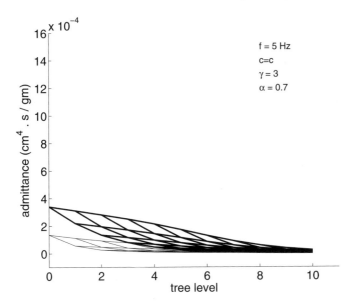

Fig. 9.5.7. Absolute values of the effective (bold) and characteristic (thin) admittances at different levels of the 11-level tree model at a frequency of 5 Hz, as in Fig. 9.5.4, but here using a value of the wave speed c obtained from a solution of the pulsatile flow in an elastic tube for each vessel segment.

The difference between the distribution of characteristic admittances and that of effective admittances, which is due entirely to the effects of wave reflections, is seen to be fairly large. Furthermore, because these effects are *cumulative*, they reach their highest value at the root of the tree as observed in the figure.

The most important aspect of the results in Fig. 9.5.3 is that the effects of wave reflections in this tree model are seen to produce *higher* admittance, and hence lower impedance, within the tree. This is particularly significant because the physical dimensions of the model are of the same order of magnitude as those in the coronary circulation, with wave-length-to-tube-length ratios well above 100 everywhere along the tree as illustrated in Fig. 9.4.3. *This means that the direct effects of wave propagation, due to the difference between pulsatile flow in an elastic tube and that in a rigid tube, are negligibly small as shown in Figs. 8.6.4, 5, but the indirect effects, due to wave reflections, are very large and far from being negligible.*

Indeed, while the direct effects of wave propagation diminish at higher values of $\overline{\lambda}$ as observed in Figs. 8.6.4, 5, the reverse is true for the indirect effects, namely the effects of wave reflections. This is because the relation between the wave length λ and the frequency f (in cycles per second) is

$$\lambda = \frac{c}{f} \qquad (9.5.24)$$

where c is the wave speed. Thus, all else being unchanged, higher frequencies are associated with lower wave lengths and hence lower values of wave-length-to-tube-length ratios $\bar{\lambda}$. In Section 8.6 we saw that this produces more significant *direct* effects of wave propagation, in terms of direct changes in the flow field, than those in a rigid tube. By contrast, the *indirect* effects of wave propagation due to wave reflections, are less significant at higher frequencies as shown in Figs. 9.5.4, 5 where the calculations for the 11-level tree model are repeated at frequencies of 5 and 10 Hz. Compared with the results at 1 Hz in Fig. 9.5.3, it is seen that the difference between effective and characteristic admittances is much higher at 1 Hz. This is particularly significant in the physiological system because the fundamental frequency of the composite pressure wave (in humans), which represents the frequency of the largest harmonic component of the composite wave, is very close to 1 Hz. The higher frequencies are associated with the much smaller harmonics of the composite wave.

The situation is somewhat more complicated, however, because the wave speed c in Eq. 9.5.23 is also a function of frequency. In Figs. 9.5.3-5 this was simplified by taking $c = c_0$ where c_0 is the constant Moen-Korteweg wave speed. In order to account for this effect, the calculations of characteristic admittances in these figures must be repeated by using a value of c obtained from the solution for pulsatile flow in an elastic tube for each vessel segment.

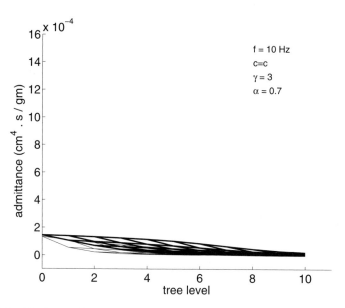

Fig. 9.5.8. Absolute values of the effective (bold) and characteristic (thin) admittances at different levels of the 11-level tree model at a frequency of 10 Hz, as in Fig. 9.5.5, but here using a value of the wave speed c obtained from a solution of the pulsatile flow in an elastic tube for each vessel segment.

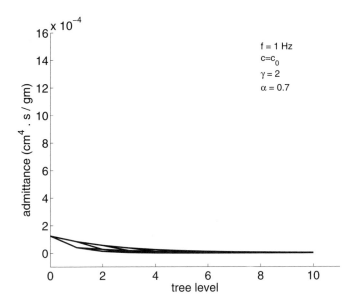

Fig. 9.5.9. Absolute values of the effective (bold) and characteristic (thin) admittances (which in this case coincide) at different levels of the 11-level tree model at a frequency of 1 Hz and $c = c_0$, as in Fig. 9.5.3, but here using a square law ($\gamma = 2$) for the hierarchy of radii at different levels of the tree. The law produces what is generally referred to as "impedance matching" at arterial bifurcations because it implies that the impedance of the parent vessel is equal to the combined impedance of the two branches. Under these conditions the effective admittances are everywhere the same as the corresponding characteristic admittances, which means that the effects of wave reflections are entirely eliminated. It is important to note that impedance matching requires not only $\gamma = 2$ but also $c = c_0$.

The results are shown in Figs.9.5.6-8. This "refinement" produces essentially the same results, though with even *higher* effects of wave reflections.

Finally, the results in Figs.9.5.3-8 are all based on a tree model in which the hierarchy of vessel radii follows the cube law discussed in Section 8.4, namely $\gamma = 3.0$. An interesting case to consider is a model in which $\gamma = 2.0$ *and* $c = c_0$. The results are shown in Fig. 9.5.9, where the effects of wave reflections are seen to be totally absent and the distributions of characterisic and of effective admittances are identical. The result shown is for a frequency $f = 1$ Hz, but the same results are obtained for other frequencies.

The reason for these rather singular results is that when the power law index $\gamma = 2.0$, the sum of cross-sectional areas of the two branches at a bifurcation is equal to the cross-sectional area of the parent vessel, that is (Eq. 8.4.25)

$$a_A^2 = a_B^2 + a_C^2 \qquad (9.5.25)$$

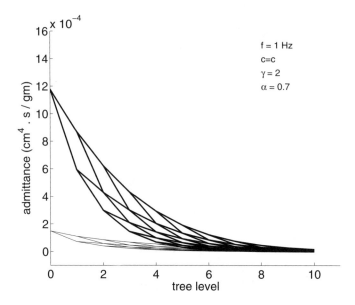

Fig. 9.5.10. Absolute values of the effective (bold) and characteristic (thin) admittances at different levels of the 11-level tree model, as in Fig. 9.5.9, but here using a value of the wave speed c obtained from a solution of the pulsatile flow in an elastic tube for each vessel segment, instead of $c = c_0$ on which the results in Fig. 9.5.9 are based. The results show that impedance matching does not occur in this case because it requires $\gamma = 2$ *and* $c = c_0$.

Now, using Eq. 9.5.5, the characteristic admittances of the three vessels are given by

$$Y_{0A} = \frac{\pi a_A^2}{\rho c_A} \tag{9.5.26}$$

$$Y_{0B} = \frac{\pi a_B^2}{\rho c_B} \tag{9.5.27}$$

$$Y_{0C} = \frac{\pi a_C^2}{\rho c_C} \tag{9.5.28}$$

and the corresponding characteristic impedances are given by

$$Z_{0A} = \frac{\rho c_A}{\pi a_A^2} \tag{9.5.29}$$

$$Z_{0B} = \frac{\rho c_B}{\pi a_B^2} \tag{9.5.30}$$

$$Z_{0C} = \frac{\rho c_C}{\pi a_C^2} \tag{9.5.31}$$

If $c = c_0$, which is assumed to be the case in Fig. 9.5.9, it is clear that in view
of Eq. 9.5.25 the three admittances and three impedances are related by

$$Y_{0A} = Y_{0B} + Y_{0C} \tag{9.5.32}$$

and

$$\frac{1}{Z_{0A}} = \frac{1}{Z_{0B}} + \frac{1}{Z_{0C}} \tag{9.5.33}$$

which constitute what is generally referred to as "impedance matching" at
the bifurcation, meaning that the propagating wave does not encounter any
change of impedance (or admittance) as it crosses the bifurcation and hence
there are no wave reflections at the junction. Since this is true at all bifurcations of the 11-level tree model, with $\gamma = 2$, it follows that no wave
reflections arise throughout the tree and hence the effective admittances are
everywhere the same as the corresponding characteristic admittances as observed in Fig. 9.5.9.

It has been suggested that "square law" conditions ($\gamma = 2.0$) prevail in the
first branching levels of the aorta [230, 201], and that this is a deliberate design
feature of the cardiovascular system which has the advantage of avoiding wave
reflection effects at these levels of the vascular tree. This may indeed be the
case in the larger vessels of the cardiovascular system where values of the
frequency parameter Ω are sufficiently high that the wave speed is close to
the constant Moen-Korteweg wave speed (Fig. 8.6.2) which the conditions for
impedance matching require. However, in the coronary circulation where the
values of Ω are typically much lower (Fig. 9.3.2) these conditions are not met,
and as we saw in Section 9.4, the wave speed in fact is significantly different
from c_0 (Fig. 9.4.1). Thus, strictly, for application to the coronary circulation
the calculations on which the results of impedance matching in Fig. 9.5.9 are
based must be repeated using the actual value of c obtained from a solution for
pulsatile flow in an elastic tube for each vessel segment. The results are shown
in Fig. 9.5.10 where it is seen clearly that impedance matching does not occur
at the junctions, and effective admittances are significantly different from the
corresponding characteristic admittances in most of the tree structure.

9.6 Pulsatile Flow in Elastic Branching Tubes

A principal tool in the analysis of pulsatile flow in branching elastic tubes is
the "pressure distribution" along a tube segment within the branching tree
structure, which we defined earlier as the amplitude of time oscillations in
pressure at different points along that segment. In the absence of wave reflections the pressure distribution is uniform at a normalized value of 1.0 along
the entire length of the tube segment (Fig. 8.7.2), but in the presence of wave
reflections this is no longer the case, as illustrated in Figs.8.7.5-9. Thus, in

a tube segment within a tree structure, any departure of the pressure distribution from the "benchmark" of a uniform normalized value of 1.0 provides a direct and useful measure of the effects of wave reflections in that structure. It is also a highly *meaningful* measure because the pressure distribution produced by wave reflections in a vessel segment within a vascular tree is superimposed on a pre-existing pressure distribution associated with steady flow within that vessel segment, and the net result is a modified distribution of pressure gradients driving the flow. For the same reasons, the pressure distribution along a vascular tree as a whole has the same utility and functional significance. The focus in this section, therefore, is on determining the pressure distribution within a vascular tree structure under different conditions.

We recall from Section 8.7 that an oscillatory input pressure applied at entry to an elastic tube produces a travelling wave within the tube. If this scenario occurs in a tube segment within a vascular tree structure, the travelling wave will progress to the next tube segment within the hierarchy of the tree structure, with the junction point between the two segments usually acting as a wave reflection site because of any difference in the admittance properties of the two tubes. Typically, only part of the wave will be reflected. The remaining part, usually referred to as the "transmitted" part of the wave, will continue on to the next junction point, and so on. The net result is that the pressure distribution within the tree structure will be modified because of the many reflected waves within the system.

In order to determine the pressure distribution in a branching tree structure, it is necessary to keep track of the "net" pressure wave as it progresses from the root segment of the tree to the periphery, the net pressure wave in each tube segment being the sum of the transmitted wave and any reflected waves. In each tube segment along the way, we use essentially the same analysis as that in Section 8.7 for wave reflections in a single tube, but now the tube position within the tree structure must be identified so that the pressure and flow within it can be mapped along with the pressure and flow within the tree as a whole. For this purpose we use the j, k notation introduced in Section 9.2, whereby j denotes the level of the tree in which a given tube segment is located and k denotes the sequential position of that segment within other segments at that level, as illustrated in Figs. 9.2.3, 4.

Using that notation, the results of Section 8.7 for wave reflections in a single tube can now be placed in the context of a tree structure. In particular, the net pressure wave within a general tube segment at position j, k in a tree structure, using Eqs.8.7.3,19, is now given by

$$P_{j,k}(x_{j,k}, t) = p_{0,j,k}\left(e^{i\omega(t - x_{j,k}/c_{j,k})} + R_{j,k}e^{i\omega(t - (2l_{j,k} - x_{j,k})/c_{j,k})}\right) \quad (9.6.1)$$

where $l_{j,k}$ is the length of the tube segment and $x_{j,k}$ is a position coordinate within that length such that the tube segment extends from $x_{j,k} = 0$ to $x_{j,k} = l_{j,k}$, $R_{j,k}$ is the reflection coefficient at the downstream end of the vessel segment, $c_{j,k}$ is the wave speed and $p_{0,j,k}$ is the amplitude of the input oscillatory pressure at entry to that segment, namely

$$p_{in,j,k}(t) = p_{0,j,k}e^{i\omega t} \tag{9.6.2}$$

As was done in Section 8.7, the net pressure wave in Eq. 9.6.1 can now be broken into its oscillatory components in time and space such that

$$P_{j,k}(x_{j,k}, t) = p_{j,k}(x) \times e^{i\omega t} \tag{9.6.3}$$

where

$$p_{j,k}(x) = p_{0,j,k}\left(e^{-i\omega x_{j,k}/c_{j,k}} + R_{j,k}e^{-i\omega(2l_{j,k}-x_{j,k})/c_{j,k}}\right) \tag{9.6.4}$$

The absolute value of this complex function of x, namely $|p_{j,k}(x)|$, is what we refer to as the "pressure distribution" in a tube segment within a tree structure, and the map of all such distributions together make up the pressure distribution in the tree as a whole. It is a key property of the propagating pressure wave within each tube segment. It represents the net pressure at a fixed point in time at each point along the segment, including forward and backward moving waves. Also, in view of Eq. 9.6.3, it represents the amplitude of time oscillations in pressure at a fixed position within the tube segment.

In the context of a tree structure, it is seen from Eq. 9.6.4 that the pressure distribution in a vessel segment at position j, k within the tree depends on the wave speed $c_{j,k}$ within that segment, on the reflection coefficient $R_{j,k}$ at the downstream end of the segment, and on the input pressure amplitude $p_{0,j,k}$ at entry to the tube segment. The way in which each of these is determined is outlined below.

Depending on the desired accuracy, the wave speed in a vessel segment within a tree structure may be taken as the constant Moen-Korteweg wave speed c_0 as defined by Eq. 8.6.1 and which depends on properties of that segment only, or it may be taken as the wave speed c as defined by Eq. 8.6.2 based on a solution of pulsatile flow in an elastic tube which depends on properties of the tube as well as on the frequency parameter Ω. For comparison purposes, results based on these two alternatives to be presented below shall be identified by $c = c_0$ and $c = c$ respectively.

The reflection coefficient at the downstream end of a tube segment in a tree structure is determined by the characteristic admittance of that segment and by the *effective* admittances of the two branch segments forming the bifurcation at that end, as discussed in Section 9.5. If the position of the vessel segment under consideration is j, k, then the positions of the two branch segments at its downstream end $(x_{j,k} = l_{j,k})$ are $j + 1, 2k - 1$ and $j + 1, 2k$, as illustrated in Figs. 9.2.3, 4. Using the results of Section 9.5, therefore, Eq. 9.5.20 expressed in the notation of the present section gives for the reflection coefficient

$$R_{j,k} = \frac{Y_{0,j,k} - (Y_{e,j+1,2k-1} + Y_{e,j+1,2k})}{Y_{0,j,k} + (Y_{e,j+1,2k-1} + Y_{e,j+1,2k})} \tag{9.6.5}$$

The characteristic admittance Y_0 is defined by Eq. 9.5.5 and is determined by the radius of the tube segment and by the wave speed which again may be taken as c or c_0 as discussed above, depending on the desired accuracy. The effective admittance Y_e, using Eq. 9.5.22 in present notation, is given by

$$Y_{e,j,k} = Y_{0,j,k} \times \left(\frac{(Y_{e,j+1,2k-1} + Y_{e,j+1,2k}) + iY_{0,j,k} \tan\theta_{j,k}}{Y_{0,j,k} + i(Y_{e,j+1,2k-1} + Y_{e,j+1,2k}) \tan\theta_{j,k}} \right)$$

$$\theta_{j,k} = \frac{\omega l_{j,k}}{c_{j,k}} \tag{9.6.6}$$

Finally, the input pressure amplitude $p_{0,j,k}$ at entry to the tube segment at position j, k is obtained by equating the local pressure in all three vessel segments that meet at that point $(x_{j,k} = 0)$, recalling that in a branching tree structure based on repeated bifurcations, three vessel segments meet at entry to and at exit from each interior tube segment, as illustrated in Figs. 9.2.3, 4. For junctions beyond the root level of the tree, that is for $j > 0$, this gives

$$p_{0,j,k}(x) = p_{0,j-1,n} \left(\frac{(1 + R_{j-1,n})e^{-i\omega l_{j-1,n}/c_{j-1,n}}}{1 + R_{j,k}e^{-i2\omega l_{j,k}/c_{j,k}}} \right), \qquad j > 0 \tag{9.6.7}$$

where

$$n = \frac{k}{2} \quad \text{if k is even} \tag{9.6.8}$$

$$= \frac{k+1}{2} \quad \text{if k is odd} \tag{9.6.9}$$

For the root segment of the tree where $j = 0$ the input pressure amplitude $p_{0,0,1}$ is clearly equal to the amplitude of input pressure to the tree as a whole. That is, if the latter is given by

$$p(t) = p_0 e^{i\omega t} \tag{9.6.10}$$

then

$$p_{0,0,1} = p_0 \tag{9.6.11}$$

Note, however, that the oscillatory pressure that actually *prevails* at the root of the tree is given by

$$P_{0,1}(0,t) = p_{0,0,1} \left\{ 1 + R_{0,1}e^{-i2\theta_{0,1}} \right\} e^{i\omega t} \tag{9.6.12}$$

$$= \left\{ 1 + R_{0,1}e^{-i2\theta_{0,1}} \right\} p_0 e^{i\omega t} \tag{9.6.13}$$

$$\neq p(t) \quad \text{unless } R_{0,1} = 0 \tag{9.6.14}$$

and

$$|P_{0,1}(0,t)| \neq |p_0(t)| \tag{9.6.15}$$

$$\neq p_0 \quad \text{unless } R_{0,1} = 0 \tag{9.6.16}$$

In other words, the oscillatory pressure that actually prevails at entry to the tree is in general not equal to the input oscillatory pressure, and the amplitude of the oscillatory pressure at entry to the tree is in general not equal to the amplitude of the input pressure. The difference in each case is due to wave reflections from downstream junctions.

To illustrate these results, examples of pressure distributions are presented below for the 5-level and the 11-level tree models used earlier. In each case the two bounding paths are identified as path-1 and path-2, where the first path follows the branch with the larger diameter at each bifurcation while the second path follows the branch with the smaller diameter at each bifurcation, as illustrated in Fig. 9.6.1.

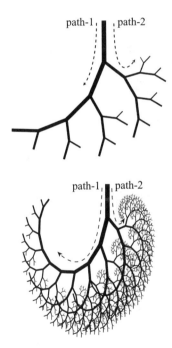

Fig. 9.6.1. The 5-level and the 11-level tree models along which the pressure distributions are calculated to illstrate the effects of wave reflections. The bounding paths identified as path-1 and path-2 are of particular interest because they represent the paths along which the diameter of vessel segments diminish most slowly (path-1) or most rapidly (path-2).

Pressure distributions along the 5-level tree are shown in Figs.9.6.2-4 at high, moderate, and low frequencies, respectively. We recall that the pressure

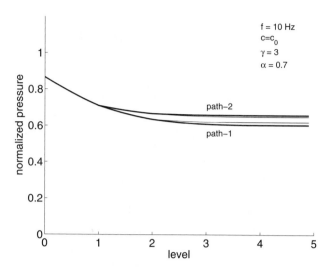

Fig. 9.6.2. Pressure distribution along different paths within the 5-level tree model, at a frequency of 10 Hz. In the absence of wave reflections the pressure distribution would be uniform at a normalized constant value of 1.0, thus the observed deviations from this benchmark are due entirely to the effects of wave reflections. Pressure distributions along the two bounding paths defined in Fig. 9.6.1 are shown in bold. They are seen to be "bounding" here too, in the sense that they represent bounding values for the pressure distribution in the tree as a whole.

shown is normalized such that in the absence of wave reflections the pressure distribution would be uniform at a normalized value of 1.0 (Eq. 8.7.14). The figures show that in the presence of wave reflections the pressure distributions lie below this value at all three frequencies. At the highest frequency, which corresponds to the lowest wave length, the pressure decreases from the root segment of the tree to the periphery, while at the lowest frequency the pressure is uniformly below the normalized benchmark value. This is consistent with effects of wave reflection at high and low wave-length-to-tube-length ratio discussed in Section 8.7.

Corresponding results for the 11-level tree model are shown in Figs.9.6.5-7. The results indicate that the effects of wave reflection are even more pronounced here than in the 5-level tree model, as indicated by larger deviations of the pressure distribution from the benchmark of uniform distribution at a normalized value of 1.0. This indicates clearly that the effects of wave reflections in a vascular tree structure are *cumulative*, thus producing larger overall effects in the 11-level tree where there are a larger number of wave reflection sites.

The dimensions of the 11-level tree model on which the results in Figs.9.6.2-4 are based have been chosen to be of the same order of magnitude as those of the human coronary arterial tree, with the root segment of the tree having a diameter of 4.0 mm and the diameters of its subsequent branches diminishing hierarchically according to the cube law thereafter. Results in Section 9.4 showed that these dimensions produce fairly high wave-length-to-tube-length ratios, in excess of 200 throughout the 11-level tree. It has been thought in the past that because such high ratios lead to what seem to be "inconspicuous" effects in a *single tube* (Fig. 8.7.9 compared with Fig. 8.7.4), the same would likely be true in the coronary circulation. The results in Figs.9.6.5-7 indicate clearly that this is not the case. The reasons for this are twofold. First, while the effects of wave reflections in a single tube at high wave-length-to-tube-length ratio are inconspicuous, they are not insignificant. As seen clearly in Fig. 8.7.9, while the pressure distribution within the tube remains uniform under these circumstances, it is so at a much higher level than the benchmark level of 1.0 when wave reflections are absent. Second, in a branching tree consisting of many tube segments these effects are *cumulative* because of the large number of reflection sites.

To pursue these issues further, the dimensions of the 11-level tree can now be changed to the effect that they become of the order of magnitude of the human *systemic* arterial tree rather than the coronary arterial tree. To do so we give the root segment of the tree a diameter of 25 mm, being representative of the human aorta, and let subsequent branches assume diminishing diameters in accordance with the cube law as before. The results are shown in Figs.9.6.7-9. The effects of wave reflections on the pressure distribution are indeed more conspicuous here and have the characteristic highs and lows associated with low wave-length-to-tube-length ratios (as in Figs. 8.7.5, 6). The difference between these results and those based on the smaller coronary scale

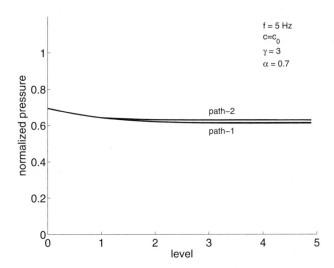

Fig. 9.6.3. Pressure distribution along different paths within the 5-level tree model, as in Fig. 9.6.2, but here at a frequency of 5 Hz.

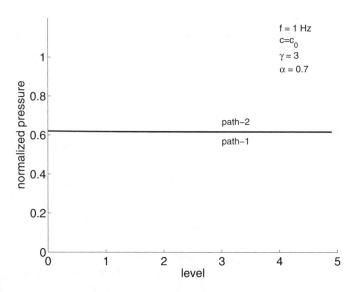

Fig. 9.6.4. Pressure distribution along different paths within the 5-level tree model, as in Fig. 9.6.2, but here at a frequency of 1 Hz.

(Figs. 9.6.5–7) is a difference in the *pattern* of pressure distributions which they produce rather than a difference of significance. While under the smaller scale of the coronary vasculature the pressure distributions have a monotonically decreasing pattern at all three frequencies, under the larger scale of the systemic vasculature there are regions of *increasing* pressure at the higher frequencies and at levels of the tree closer to the root segment. It is under these conditions that the well known "peaking" phenomenon occurs in the systemic circulation, but not in the coronary circulation where these conditions are absent.

All the results above are based on the wave speed approximation $c = c_0$ as indicated in the figures, where c_0 is the constant Moen-Korteweg wave speed. Results based on the more accurate $c = c$ model, where c is the wave speed obtained from a solution of pulsatile flow in an elastic tube, are shown in Figs. 9.6.11-13. Comparison of these with results in Figs. 9.6.5-7 based on $c = c_0$ indicates that the difference between the two is not highly significant.

A more significant difference arises if the hierarchy of branch diameters within the tree structure is changed from one based on the cube law ($\gamma = 3$) to one based on the square law ($\gamma = 2$). Under the square law the sum of the cross-sectional areas of the two branch segments is equal to the cross-sectional area of the parent segment at all bifurcations within the tree structure. If this is combined with the $c = c_0$ wave speed assumption, then there is no change of admittance as the flow crosses each bifurcation within the entire tree structure, which means that the bifurcations are no longer reflection sites

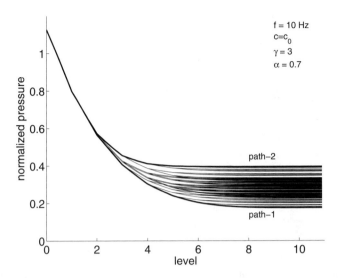

Fig. 9.6.5. Pressure distributions along different paths within the 11-level tree model at a frequency of 10 Hz. Comparison with the corresponding results for the 5-level tree model shown in Fig. 9.6.2 indicates clearly that the effects of wave reflections are larger here, as marked by larger departures from the benchmark of uniform distribution at a normalized value of 1.0 which occurs when wave reflections are absent.

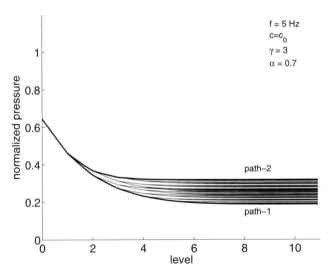

Fig. 9.6.6. Pressure distributions along different paths within the 11-level tree model as in Fig. 9.6.5 but here at a frequency of 5 Hz.

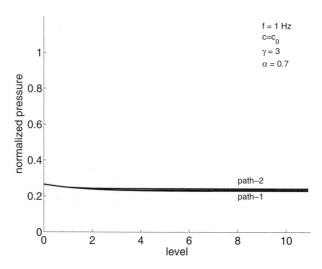

Fig. 9.6.7. Pressure distributions along different paths within the 11-level tree model as in Fig. 9.6.5 but here at a frequency of 1 Hz.

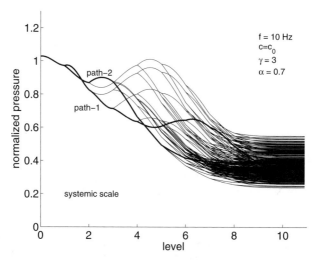

Fig. 9.6.8. Pressure distributions along different paths within the 11-level tree model at a frequency of 10 Hz as in Fig. 9.6.5 but here the root segment of the tree has a diameter of 25 mm compared with 4 mm in that figure. Thus the scale of the tree model in this case is representative of the scale of the human systemic arterial tree while that in Fig. 9.6.5 is representative of the coronary arterial tree. Wave-length-to-tube-length ratios are much lower in this case, thus the characteristic highs and lows in the pressure distributions.

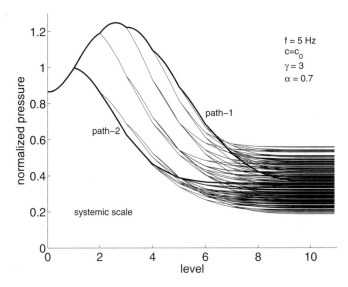

Fig. 9.6.9. Pressure distributions along different paths within the 11-level tree model as in Fig. 9.6.8, but here at a frequency of 5 Hz which means higher wave lengths and thus higher wave-length-to-tube-length ratios than in that figure.

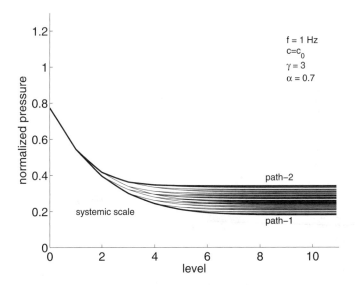

Fig. 9.6.10. Pressure distributions along different paths within the 11-level tree model as in Fig. 9.6.8, but here at a frequency of 1 Hz, which means considerably higher wave lengths and thus considerably higher wave-length-to-tube-length ratios than in that figure.

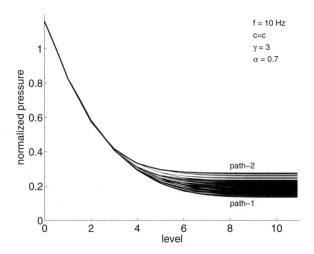

Fig. 9.6.11. Pressure distributions along different paths within the 11-level tree model at a frequency of 10 Hz as in Fig. 9.6.5, but here the results are based on the wave speed c obtained from a solution of pulsatile flow in an elastic tube rather than on the constant Moen-Korteweg wave speed c_0 on which the results in that figure are based. The difference between the two is quantitative rather than qualitative.

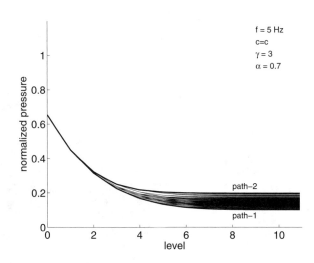

Fig. 9.6.12. Pressure distributions along different paths within the 11-level tree model at a frequency of 5 Hz as in Fig. 9.6.6, but here the results are based on the wave speed c obtained from a solution of pulsatile flow in an elastic tube rather than on the constant Moen-Korteweg wave speed c_0 on which the results in that figure are based.

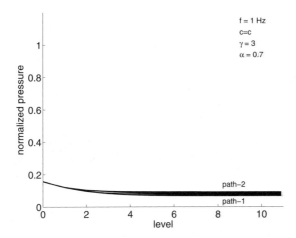

Fig. 9.6.13. Pressure distributions along different paths within the 11-level tree model at a frequency of 1 Hz as in Fig. 9.6.7, but here the results are based on the wave speed c obtained from a solution of pulsatile flow in an elastic tube rather than on the constant Moen-Korteweg wave speed c_0 on which the results in that figure are based.

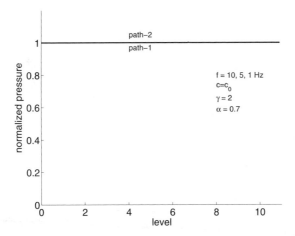

Fig. 9.6.14. Pressure distributions along all paths within the 11-level tree model at frequencies of 10 Hz, 5 Hz, and 1 Hz, as in Figs.9.6.5-7, but here the results are based on a square power law index ($\gamma = 2$) compared with the cube law ($\gamma = 3$) in those figures. The square law produces a unique set of circumstances known as "impedance matching" whereby there is no change of admittance across bifurcations, and hence no wave reflections arise. Pressure distributions along any path within the tree, therefore, are identical with the benchmark of uniform distribution at a normalized value of 1.0.

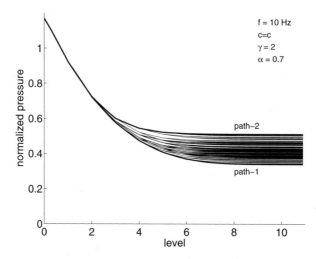

Fig. 9.6.15. Pressure distributions along different paths within the 11-level tree model at a frequency of 10 Hz and based on a square power law index ($\gamma = 2$) as in Fig. 9.6.14, but here the results are based on the calculated wave speed $c = c$ rather than on $c = c_0$ in that figure. Wave reflection effects are clearly evident here, thus the unique conditions of "impedance matching" observed in Fig. 9.6.14 require *both* $\gamma = 2$ and $c = c_0$.

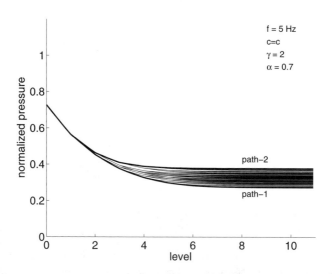

Fig. 9.6.16. Pressure distributions along different paths within the 11-level tree model based on $\gamma = 2$ and $c = c$ as in Fig. 9.6.15, but here at a frequency of 5 Hz where wave reflection effects are still evident.

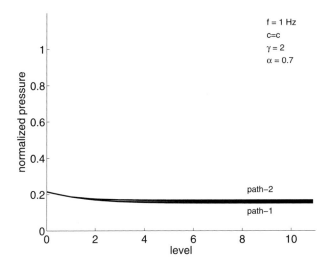

Fig. 9.6.17. Pressure distributions along different paths within the 11-level tree model based on $\gamma = 2$ and $c = c$ as in Fig. 9.6.15, but here at the much lower frequency of 1 Hz where significant wave reflection effects are still evident. In the absence of such effects the pressure distributions would be uniform at a normalized value of 1.0, as in Fig. 9.6.14.

and wave reflections do not arise, a condition known as "impedance matching". Results based on these circumstances are shown in Fig. 9.6.14 where it is seen that the pressure distribution under these circumstances is identical with the benchmark distribution at all three frequencies.

It is important to note, however, that this only occurs under the $c = c_0$ wave speed assumption. Under the more accurate $c = c$ model the special condition of impedance matching at arterial bifurcations is broken and wave reflections arise and lead to significant effects as shown in Figs.9.6.15-17.

9.7 Cardiac Pressure Wave in Elastic Branching Tubes

What is of ultimate interest in an unlumped model of the coronary circulation is to determine the way in which a *composite* pressure wave is modified by wave reflections as it travels down a branching tree structure consisting of elastic tube segments. For this purpose we use the cardiac pressure wave considered in previous chapters which represents the oscillatory pressure measured in the ascending aorta and is produced by the pumping action of the left ventricle. Because the main left and the main right coronary arteries have their origin at the ascending aorta (or more accurately in the sinus of valsalva), as discussed in Section 1.3, the cardiac pressure wave, sometimes referred to as "aortic pressure", may be considered reasonably accurately as the oscillatory input pressure for the coronary circulation. In other words, the cardiac pressure wave may be taken as the "input pressure" $p_{in}(t)$ at the root segments of the two main coronary arterial trees, with the role and significance of $p_{in}(t)$ as described in Section 8.7 and used in the previous two sections. However, in those sections $p_{in}(t)$ was taken as a single harmonic function, meaning that it has the form of a simple sine or cosine function. In the present section we consider $p_{in}(t)$ to be a composite wave that has the form of the cardiac pressure wave shown in Fig. 9.7.1.

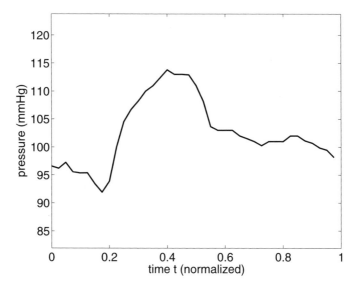

Fig. 9.7.1. The composite pressure wave considered as the input pressure driving coronary blood flow. Of ultimate interest is to determine the way in which the form of the wave is modified as it travels down the coronary vasculature. To do so requires that the wave be decomposed into its harmonic components and then recomposed in terms of its modified components.

As discussed in Section 8.7, when the oscillatory input pressure $p_{in}(t)$ at entry to an elastic tube is a single harmonic, the pressure wave which it produces within the tube is also a single harmonic function of time but with amplitude which depends on position x within the tube. The way in which the amplitude varies with x is what we have called the "pressure distribution" along the tube. When the oscillatory input pressure $p_{in}(t)$ is a *composite wave*, these concepts cannot be applied to the composite wave as a whole because the notion of an amplitude does not apply to a composite wave. But the concepts of amplitude and pressure distribution do apply to the *individual harmonics* of the composite wave.

Thus, if the cardiac wave in Fig. 9.7.1 is decomposed into its individual harmonics as discussed in Chapter 5, the pressure distribution associated with each harmonic can be obtained as in the previous section. This pressure distribution represents the amplitude of the propagating wave associated with that particular harmonic at differrent positions x along each tube segment and along the tree as a whole. At a particular position x, therefore, the collective amplitudes associated with all the harmonics of the composite cardiac wave represent the amplitudes of the individual harmonics of the *modified* form of the composite wave at that position. The modified form of the wave is thus determined from the collective amplitudes at that position by simply recomposing the harmonics associated with these amplitudes, using the Fourier analysis techniques discussed in Chapter 5.

In fact, since the pressure distributions obtained in the previous section were all normalized so that the input pressure at entry has an amplitude of 1.0, then the same analysis and results can be applied to *all* the harmonics of the composite cardiac pressure wave if each harmonic is first normalized so that the input pressure at entry associated with it has a normalized value of 1.0, and then "denormalized" when the modified harmonics are put together to recompose the modified composite wave.

The end results of this exercise provide a picture of the evolution of the composite pressure wave as it travels along different paths within the tree structure. These results are strongly linked to results obtained in the previous section because the evolution of the composite wave along a given path indeed consists of the aggregate pressure distributions associated with its individual harmonics along that path. In a graphical presentation of the results, however, only one path can be considered at a time because the full form of the wave must be presented at different points along the path. For this reason we shall focus on the evolution of the wave along the two bounding paths defined in Fig. 9.6.1. Since the pressure distributions along these two paths represent the bounds on the pressure distribution within the tree as a whole, the evolution of the composite pressure wave along these two paths will represent similar bounds on the evolution of the composite pressure wave along the tree as a whole.

Computed evolutions of the composite pressure wave along the two bounding paths of the 5-level tree model at three different frequencies are shown in Figs.9.7.2-7. At each frequency, the evolution of the composite wave along a given path can be interpreted in terms of the corresponding pressure distribution along that path obtained in the previous section. Thus, the pressure distribution along path-1 of the 5-level tree model in Fig. 9.6.2 of the previous section indicates that at a frequency of 10 Hz the amplitude of a harmonic wave starting with a normalized amplitude of 1.0 is modified by wave reflections so that its values at different points along the path are as indicated by that curve. Since the values of the modified amplitude are everywhere below the initial value of 1.0 and continue to decrease along the path, it follows that the same will be true of the individual harmonics of a composite wave travelling down the same path. Thus, because the amplitudes of its harmonic components are diminishing as it travels, the composite wave will be "flattened", or "blunted", as it progresses along the path, as is indeed observed in Figs. 9.7.2, 3 where the frequency is 10 Hz as it is in Fig. 9.6.2.

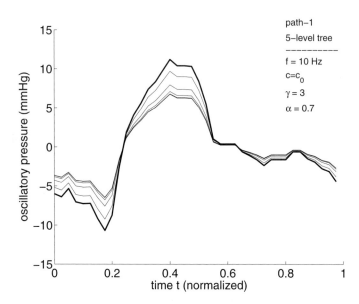

Fig. 9.7.2. Evolution of the composite cardiac pressure wave as it travels down path-1 of the 5-level tree model. The top bold curve represents the input pressure wave at entry while the thin curve next to it in sequence represents the form of the pressure wave that actually prevails at entry, which illustrates graphically that the two are different because of wave reflections returning from downstream as discussed in Section 9.6 (Eqs.9.6.13,14). Thin curves further down in sequence represent modified forms of the wave at higher levels of the tree. There is thus a general "flattening", or "blunting", of the wave as it progresses, consistent with the corresponding pressure distribution along the same path shown in Fig. 9.6.2.

At the lower frequencies of 5 Hz and 1 Hz, the evolutions of the composite pressure wave are shown in Figs. 9.7.4–7 and they can be interpreted in the same way in relation to the corresponding pressure distributions in Figs.9.6.3,4. In particular, at the lowest frequency of 1 Hz, the pressure distribution in Fig. 9.6.4 shows not a gradual but a uniform drop in the amplitude of harmonic components, therefore leading to the abrupt change in the form of the composite wave observed in Figs. 9.7.6, 7.

Evolutions of the composite cardiac pressure wave along the 11-level tree model are shown in Figs.9.7.8-13, to be interpreted in terms of the corresponding pressure distributions in Figs.9.6.5-7. In particular, results in Figs. 9.7.8, 9 indicate that at the highest frequency of 10 Hz some "peaking" of the wave-

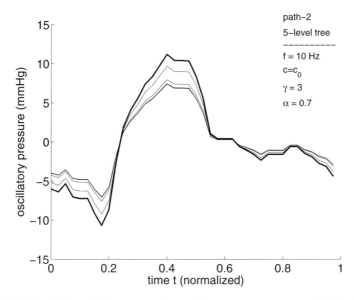

Fig. 9.7.3. Evolution of the composite cardiac pressure wave as in Fig. 9.7.2 but here as it travels down path-2 of the 5-level tree model.

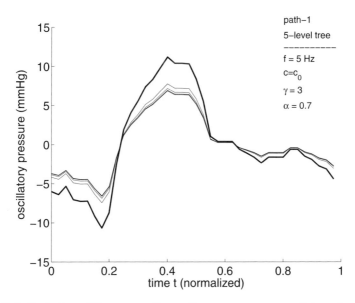

Fig. 9.7.4. Evolution of the composite cardiac pressure wave as it travels along the 5-level tree model as in Figs. 9.7.2, 3 but here at a frequency of 5 Hz, consistent with the corresponding pressure distribution along the same path shown in Fig. 9.6.3.

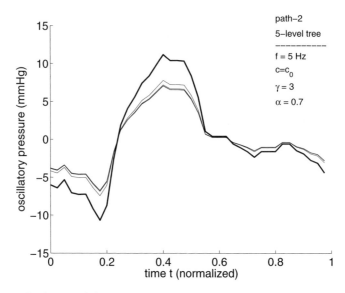

Fig. 9.7.5. Evolution of the composite cardiac pressure wave as it travels along the
5-level tree model as in Figs. 9.7.2, 3 but here at a frequency of 5 Hz, consistent with
the corresponding pressure distribution along the same path shown in Fig. 9.6.3.

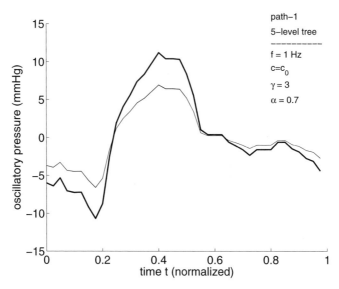

Fig. 9.7.6. Evolution of the composite cardiac pressure wave as it travels along the
5-level tree model as in Figs. 9.7.2, 3 but here at a frequency of 1 Hz, consistent with
the corresponding pressure distribution along the same path shown in Fig. 9.6.4.

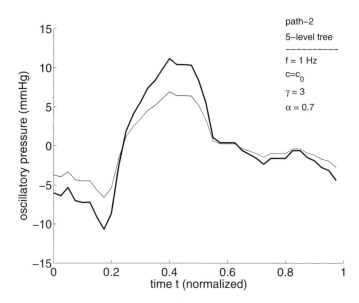

Fig. 9.7.7. Evolution of the composite cardiac pressure wave as it travels along the 5-level tree model as in Figs. 9.7.2, 3 but here at a frequency of 1 Hz, consistent with the corresponding pressure distribution along the same path shown in Fig. 9.6.4.

form occurs initially. This is consistent with the pressure distributions in Fig. 9.6.5 where it is seen that at the root segment of the tree the normalized pressure amplitude is higher than the benchmark of 1.0. Peaking of the pressure waveform due to wave reflections is known to occur in the systemic circulation as the cardiac pressure wave travels down the abdominal aorta. Results obtained in this section so far indicate that in the coronary circulation, because of the much higher wave-length-to-tube-length ratios, wave reflections lead mostly to a *flattening* of the pressure waveform, which is the reverse of peaking. But the results in Figs.9.7.8,9 demonstrate that peaking may occur in the coronary circulation too, though only locally and under limited circumstances. Indeed, the phenomenon is no longer present at higher levels of the tree at 10 Hz, and not present at all at the lower frequencies of 5 Hz and 1 Hz as seen in Figs.9.7.10-13.

To pursue the peaking phenomenon a little further, and to compare the evolution of the pressure waveform on the scale of the coronary circulation with that on the scale of the *systemic* circulation, the dimensions of the 11-level tree model are increased accordingly as before, with the root segment of the tree being given a diameter of 25 mm. Results are shown in Figs.9.7.14-16. Compared with the corresponding results on the scale of the coronary circulation (Figs. 9.7.8, 10, 12), the peaking phenomenon is clearly more prominent here than it is on the scale of the coronary circulation. Also, here the evolu-

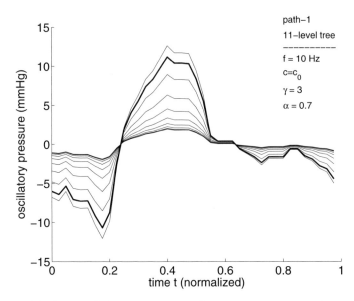

Fig. 9.7.8. Evolution of the composite cardiac pressure wave as it travels down path-1 of the 11-level tree model at a frequency of 10 Hz. The bold curve at the top represents the initial form of the wave as applied at entry while the thin curve above it represents the form of the wave that actually prevails at entry, the two being different because of wave reflections returning from downstream. This demonstrates graphically that some local "peaking" occurs initially before the waveform begins to flatten as it progresses further along the hierarchy of the tree and as indicated by subsequent curves in the sequence. Both the local peaking and subsequent flattening is consistent with the corresponding pressure distributions shown in Fig. 9.6.5.

tion of the wave is more gradual at all three frequencies, consistent with the corresponding pressure distributions in Figs. 9.6.8–10.

Effects of using the wave speed c obtained from a solution of pulsatile flow in an elastic tube instead of the constant Moen-Korteweg wave speed on which all the above results are based, are shown in Figs. 9.7.17-19. Comparing these with the results in Figs. 9.7.8, 10, 12 which are based on $c = c_0$ indicates that the difference between the two is unremarkable at all three frequencies.

As for the pressure distribution obtained in the previous section, however, it is important to note that the unique condition of "impedance matching" occurs only when $\gamma = 2$ *and* $c = c_0$ (Fig. 9.6.14). When the more accurate wave speed c is used instead of the Moen-Korteweg wave speed c_0, matching is no longer attained and significant wave reflection effects arise as illustrated in Figs. 9.7.21-23.

However, as for the pressure distributions discussed in the previous section, a more significant difference arises if the hierarchy of branch diameters within the tree is changed from one based on the cube law ($\gamma = 3$) to one based on the square law ($\gamma = 2$). Under the square law the sum of the cross-sectional areas

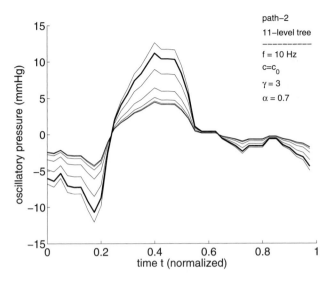

Fig. 9.7.9. Evolution of the composite cardiac pressure wave as it travels down path-2 of the 11-level tree model at a frequency of 10 Hz, which again indicates some initial peaking as in Fig. 9.7.8.

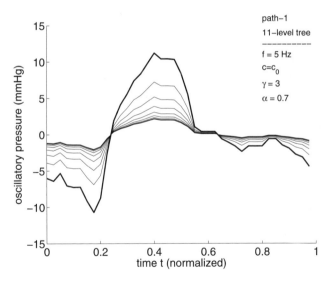

Fig. 9.7.10. Evolution of the composite cardiac pressure wave as it travels down path-1 of the 11-level tree model, as in Fig. 9.7.8 but here at a frequency of 5 Hz, which means higher wave-length-to-tube-length ratios, and hence no peaking in the waveform is observed, consistent with the corresponding pressure distributions in Fig. 9.6.6.

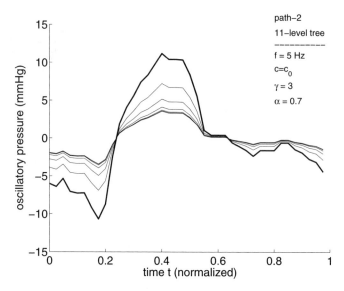

Fig. 9.7.11. Evolution of the composite cardiac pressure wave as in Fig. 9.7.10, but here as it travels down path-2 of the 11-level tree model.

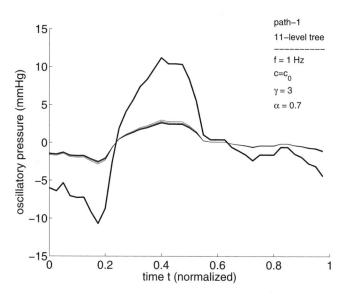

Fig. 9.7.12. Evolution of the composite cardiac pressure wave as it travels down path-1 of the 11-level tree model at the lowest frequency of 1 Hz, which means much higher wave-length-to-tube-length ratios. The abrupt flattening of the waveform is consistent with the pressure distributions in Fig. 9.6.7.

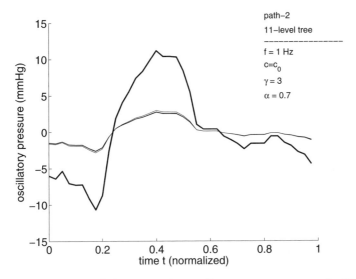

Fig. 9.7.13. Evolution of the composite cardiac pressure wave, as in Fig. 9.7.12, but here as it travels down path-2 of the 11-level tree model.

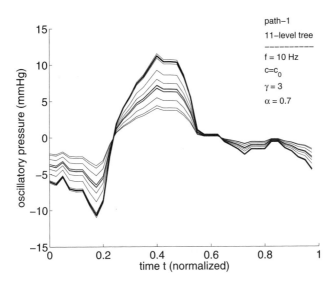

Fig. 9.7.14. Evolution of the composite cardiac pressure wave as it travels down path-1 of the 11-level tree model at a frequency of 10 Hz as in Fig. 9.7.8, but here the dimensions of the tree are increased to the scale of the systemic circulation, with the root segment of the tree given a diameter of 25 mm compared with 4 mm in that figure. Some initial peaking is observed, consistent with the corresponding pressure distribution in Fig. 9.6.8.

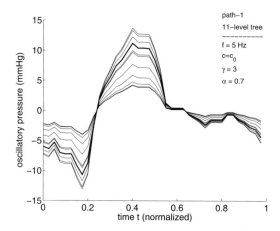

Fig. 9.7.15. Evolution of the composite cardiac pressure wave as it travels down path-1 of the 11-level tree model at a frequency of 5 Hz as in Fig. 9.7.10, but here the dimensions of the tree are increased to the scale of the systemic circulation, with the root segment of the tree given a diameter of 25 mm compared with 4 mm in that figure. Considerably more peaking is observed here while it is completely absent on the scale of the coronary circulation, consistent with the corresponding pressure distribution in Fig. 9.6.9.

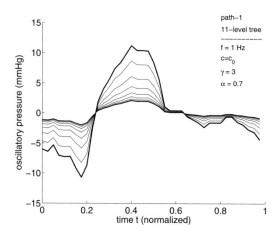

Fig. 9.7.16. Evolution of the composite cardiac pressure wave as it travels down path-1 of the 11-level tree model at a frequency of 1 Hz as in Fig. 9.7.12, but here the dimensions of the tree are increased to the scale of the systemic circulation, with the root segment of the tree given a diameter of 25 mm compared with 4 mm in that figure. Peaking is absent in both cases, but the evolution of the wave is considerably more *gradual* on the systemic scale, clearly because of the lower wave-length-to-tube-length ratios on that scale.

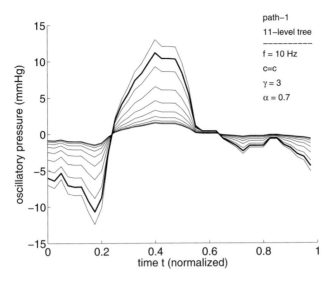

Fig. 9.7.17. Evolution of the composite cardiac pressure wave as it travels down path-1 of the 11-level tree model at a frequency of 10 Hz as in Fig. 9.7.8, but here using the actual wave speed c instead of the Moen-Korteweg wave speed c_0 on which the results in that figure are based. The difference between the two is unremarkable.

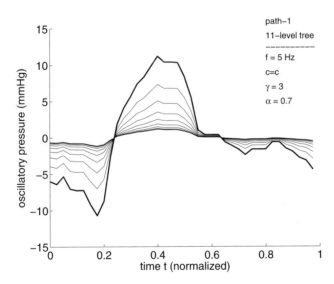

Fig. 9.7.18. Evolution of the composite cardiac pressure wave as it travels down path-1 of the 11-level tree model at a frequency of 5 Hz as in Fig. 9.7.10, but here using the actual wave speed c instead of the Moen-Korteweg wave speed c_0 on which the results in that figure are based. The difference between the two is unremarkable.

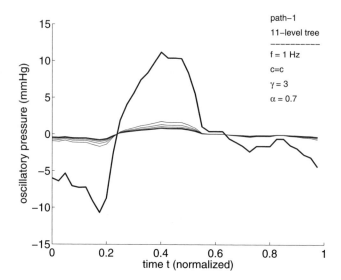

Fig. 9.7.19. Evolution of the composite cardiac pressure wave as it travels down path-1 of the 11-level tree model at a frequency of 1 Hz as in Fig. 9.7.12, but here using the actual wave speed c instead of the Moen-Korteweg wave speed c_0 on which the results in that figure are based. The difference between the two is unremarkable.

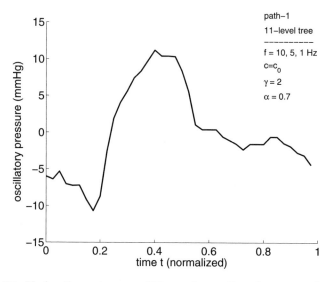

Fig. 9.7.20. Under the unique conditions of $\gamma = 2$ *and* $c = c_0$, the composite cardiac pressure waveform remains unchanged at all levels of the tree and at all three frequencies, a condition known as "impedance matching".

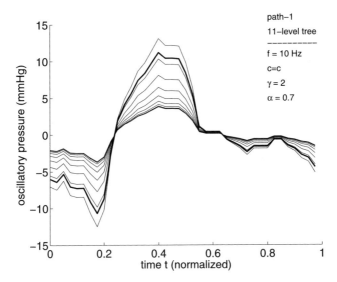

Fig. 9.7.21. Evolution of the composite cardiac pressure waveform as it travels down path-1 of the 11-level tree model at a frequency of 10 Hz. Although $\gamma = 2$ here as in Fig. 9.7.20, impedance matching is not attained here because the more accurate wave speed is used here instead of the Moen-Korteweg wave speed c_0.

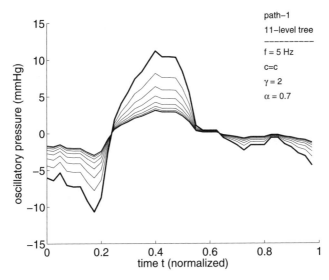

Fig. 9.7.22. Evolution of the composite cardiac pressure waveform as it travels down path-1 of the 11-level tree model as in Fig. 9.7.21, but here at a frequency of 5 Hz. Impedance matching is not attained at this frequency.

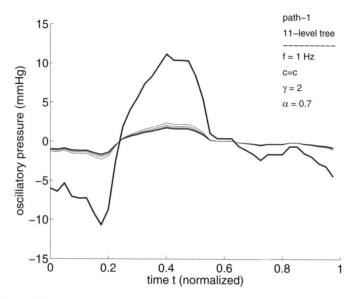

Fig. 9.7.23. Evolution of the composite cardiac pressure waveform as it travels down path-1 of the 11-level tree model as in Fig. 9.7.21, but here at a frequency of 1 Hz. Impedance matching is not attained at this frequency.

of the two branch segments is equal to the cross-sectional area of the parent segment at all bifurcations within the tree structure. If this is combined with the $c = c_0$ wave speed assumption, then there is no change of admittance as the flow crosses each bifurcation within the entire tree structure, which means that the bifurcations are no longer reflection sites and wave reflections do not arise. Results obtained in the previous section (Fig. 9.6.14) indicated that the pressure distributions obtained under these circumstances are identical with the benchmark of uniform distribution at 1.0, which means, therefore, that the composite pressure waveform would be unchanged as it travels down the tree, a condition known as "admittance matching". This is indeed the case as illustrated in Fig. 9.7.20. The waveform is unchanged at all levels of the tree and at all three frequencies, consistent with the corresponding pressure distribution in Fig. 9.6.14

9.8 Summary

The ultimate unlumped model of the coronary circulation is unattainable in practice because of the overwhelming details of coronary vasculature and the high degree of variablility in these details from one heart to another, and because of the enormous difficulties involved in a study of all aspects of flow in this vasculature. A more modest approach is that of constructing unlumped models that have only the broad features of scale and branching pattern of coronary vasculature, and to use combinations and variations of these features to probe into the type of dynamics that they can or cannot give rise to, and under what type of conditions.

In order to track the flow in an arterial tree structure it is necessary to identify the position of each vessel segment within that structure. A simple j, k coordinate system can be used for that purpose, whereby the first coordinate identifies the level of the tree in which a vessel segment is located and the second identifies the sequential position of that segment among other segments at that level of the tree. Any path within the tree structure can then be specified in terms of the vessel segments that make up that path, and the pressure distribution along the path can be pieced together from the pressure distributions along each vessel segment. A theoretical tree structure based on a power law with three different values of the power law index is used to illustrate this scheme and to find the pressure distributions under steady flow conditions. Because of their uniformity, these theoretical structures do not represent the characteristic heterogenous structure of coronary vasculature, but they serve to illustrate the way in which the pressure varies along the hierarchy of a tree structure.

In pulsatile flow through a rigid tube flow properties depend on the value of the frequency parameter Ω which in turn depends on the tube radius, amongst other things. In a tree structure, therefore, the distribution of the values of Ω within the tree structure is an important determinant of the properties of pulsatile flow along the tree. In general, the value of Ω has its highest value at the root segment of the tree, then decreases from there, along the hierarchy of the tree as the diameters of vessel segments decrease. Peak flow rate and peak shear stress, the maximum values reached within the oscillatory cycle, actually increase from the root segment of the tree towards the periphery because of the decrease in the value of Ω in that direction.

Pulsatile flow in a tree structure consisting of elastic tube segments gives rise to wave propagation and widespread wave reflections because of the large number of vascular junctions. With the ratio of wave length to tube length $\overline{\lambda} \sim 100$ or greater in the coronary circulation as indicated in Fig. 9.4.3, effects of wave propagation on the flow field are minimal, but the cumulative effects of wave reflections can be enormous. The aim of the elastic branching tubes model is therefore to establish a method of tracking these reflections and calculating the effects which they produce in a vascular tree structure consisting of many vascular tube segments and hence many vascular junctions which act as reflection sites.

Pulsatile flow in an elastic tube gives rise to wave propagation within the tube. Impedance is the total opposition to pulsatile flow in an elastic tube, which consists of normal resistance to the steady part of the flow plus opposition to wave propagation. Admittance is the reciprocal of impedance and it represents the extent to which pulsatile flow in an elastic tube is "admitted", rather than opposed. In a vascular tree structure consisting of elastic tube segments, the total opposition to pulsatile flow registered at entry into the tree is referred to as "input impedance". One part of this opposition depends on the elasticity and other geometric properties of the tube segments as well as on the frequency of the oscillatory flow and is referred to as "characteristic" impedance. Another part is due to wave reflections and it depends on the succession of local impedances at vascular junctions as well as on the frequency of the oscillatory flow. When this part is included in the total opposition to pulsatile flow it is referred to as "effective" impedance. Similar terminology is used for admittance. Only under very special circumstances is the effective impedance in a vascular tree the same as the characteristic impedance. These circumstances require that the characteristic impedance of the parent vessel at every vascular junction be the same as the combined characteristic impedance of the two daughter segments, which is generally referred to as "impedance matching", and that the wave speed be constant throughout the tree and equal to the Moen-Korteweg wave speed. These circumstances do not occur in the coronary circulation.

In a tree model consisting of elastic tube segments and having dimensions comparable with those of the human coronary arterial tree, pulsatile flow is associated with significant wave reflection effects that produce a negative pressure gradient along the tree that actually aids rather than impedes the flow. If the dimensions of the tree are made comparable with those of the systemic arterial tree, this trend is reversed initially to produce a positive pressure gradient, consistent with the well known "peaking" phenomenon in the descending aorta, and then becomes negative again towards the periphery.

A cardiac pressure wave representing input pressure into a tree model consisting of elastic tube segments and having dimensions comparable with those of the human coronary arterial tree, is generally "blunted" or "flattened" as it progresses from the root segment of the tree to the periphery. This is the reverse of the "peaking" phenomenon observed in the descending aorta, the difference being due to the different scales of the coronary and systemic circulations, but the change in the form of the wave in both cases is due to wave reflections. In the systemic circulation the wave would be blunted by viscous dissipation as it travels along the descending aorta, but instead it peaks because of wave reflections coming from downstream. In the coronary circulation model being considered here the wave is blunted with or without the effects of viscosity (by using c or c_0) thus the blunting is due largely to wave reflections.

10

Dynamic Pathologies

10.1 Introduction

The dynamics of the coronary circulation are under strict regulation, mediated by several neural and humoral control mechanisms [83, 128]. While a great deal has been learned about the isolated effects of certain humoral agents or neural stimuli, a full understanding of the feedback loops associated with these mechanisms does not exist at present. These aspects of the coronary circulation have yet to be integrated into a unified model of the dynamics of the system.

In the absence of an integrated model, it is not unreasonable to assume that the regulatory mechanisms of the system will aim to keep it in a mode of operation in which its dynamics are "optimal" in some way. And if the parameter values associated with these optimal dynamics are changed, whether by disease or clinical intervention, the dynamics of the system will then be less than optimal. They will be "pathological" in the sense of being away from the normal dynamics of the system, and it is in this sense that we use the term "dynamic pathology" in this chapter.

A dynamic pathology is not unlike a structural or a functional pathology in a tissue or organ, but it deals specifically with *dynamics*. Furthermore, it deals with the dynamics of an integrated system as a whole, such as the dynamics of the coronary circulation as a whole. Thus, an atherosclerotic lesion in a coronary artery is a vascular pathology but the ultimate significance of this lesion lies in the dynamic pathology which it produces. The ultimate significance of the lesion lies in the way it affects the dynamics of the coronary circulation as a whole, and in particular in the way it affects the ultimate goal of the coronary circulation to provide the heart with sufficient blood supply. If the occlusive progression of a lesion is sufficiently slow, so that the occlusion is compensated for by vascular restructuring, then the dynamic pathology may be minimal or nonexistent. Thus, the presence of vascular pathology does not necessarily imply the presence of dynamic pathology. *The tenet of this book is that the reverse of this statement is also true, namely that the absence of vascular pathology in the coronary circulation does not necessarily imply the absence of dynamic pathology.*

Arrhythmia, in all its forms, is a familiar example of a dynamic pathology. It affects the harmony between the pumping rhythm of the heart and the compliance of the aorta, thereby affecting cardiac output and hence the dynamics of the *systemic* circulation. Similarly, a disruption in the harmony between the pulsatile rhythm of the input pressure driving the flow into the coronary circulation, the pulsatile rhythm of the contracting cardiac muscle and its effect on coronary vessels imbedded within it, and the combination of wave propagation and wave reflections within the coronary vascular tree, will lead to dynamic pathology in the coronary circulation. Anyone who has pushed a child on a swing in the park will know the exquisite harmony required between the timing of the push and the rhythmic momentum of the swing. Any disruption in that harmony will cause the swing to lose rather than gain momentum, thus producing a dynamic pathology. The analogy is quite pertinent and we shall use it again later. Any disruption in the harmony between the several factors involved in the dynamics of the coronary circulation will reduce the efficiency of coronary blood flow and thereby produce dynamic pathology.

While an integrated model of the coronary circulation that can deal with these dynamic issues and with dynamic pathologies in the system does not exist at present, the lumped and the unlumped models introduced in previous chapters, though incomplete, can provide tools for a useful prelude to dealing with these issues. The models can be used to assess the consequences of a change in diameters or mechanical properties of the vessels involved or in the frequency of oscillation, to then speculate on the possible role such changes might play in an integrated model of the coronary circulation. Alternatively, results of lumped and unlumped models, though incomplete, can be used in combination with known dynamic features of the coronary circulation to speculate on possible dynamic factors in coronary heart disease. We follow some of these lines in this concluding chapter of the book, which is therefore largely speculative in scope.

10.2 Magic Norms?

While the goal of uncovering the full range of dynamics of the coronary circulation may be out of reach, a more modest goal is that of finding "ideal" or "normal" modes of operation of the system, modes of operation in which the dynamics of the system appear to be particularly simplified or optimal in some sense. We shall refer to these loosely as "magic norms" of the system in the sense that they may represent modes of operation for which the system is designed (or towards which the system has evolved).

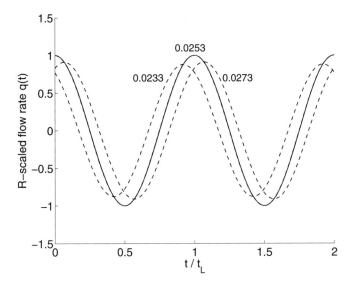

Fig. 10.2.1. Oscillatory pressure (solid) and flow (dashed) in an RLC system in series. The flow rate is "R-scaled" such that the pressure and flow curves become identical under pure resistance, that is when L and C are absent. In the presence of L and C, however, the system has a magic norm in which it behaves *as if L and C are absent*. If the value of the inertial time constant t_L is taken as $0.1\,s$, the magic norm conditions occur when the value of the capacitive time constant t_C is $0.0253\,s$. Any deviation from this value moves the system away from this mode, as indicated by the departure of the flow curve away from the pressure curve.

Thus, a lumped or unlumped model of the coronary circulation may not embody the full range of dynamics of the system, but it may embody some of its magic norms. While the model may be incomplete in terms of the range of parameters on which it is based, it may point to modes of operation of the system in which a unique combination of the values of these parameters produce remarkably simplified dynamics.

An example from another area of application may illustrate the point. The dynamics of a pendulum depend on several parameters such as length, weight, and any driving forces. The differential equations governing these dynamics, as well as their solutions, are fully known in even the most complicated cases. When a child sits on a swing in the park for the first time, however, the child does not know that the swing is in fact a pendulum with fully known dynamics, yet he or she very quickly discovers the magic norm of the system. The child soon discovers how to use his or her body weight to drive the swing to higher and higher altitudes. He or she soon learns that the amount and *exquisite timing* of the applied body force is crucial to attaining the magic norm, and reaching the ideal conditions under which the system operates.

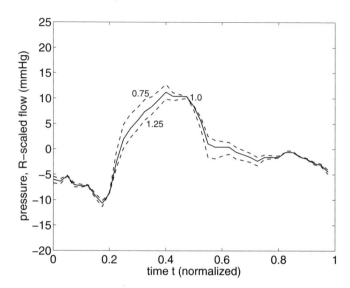

Fig. 10.2.2. Oscillatory pressure (solid) and flow (dashed) in Lumped Model 2 of Section 7.4 whose composition in the notation of that section is $\{\{R_1+L\},\{R_2+C\}\}$. A magic norm occurs when $\lambda = R_2/R_1 = 1.0$ and $t_L = t_C = 0.2\,s$ whereby the system behaves as if L and C were absent, and the pressure and R-scaled flow curves again become identical. Other values of λ (shown) move the system away from the magic norm.

Any deviation from the precise timing or the precise force moves the system away from its magic norm.

While the dynamics of the coronary circulation may be several orders of magnitude more complicated than the dynamics of a swing, the example is pertinent in the sense that we are somewhat in the position of the child who does not know the equations governing the full dynamics of the system, yet we may recognize conditions under which the dynamics of the system appear to be ideal in some sense.

The first example of a magic norm was actually found in Section 4.8, where under certain circumstances the inertial and the capacitive effects of the RLC system were found to "cancel" each other and produce zero reactance. These conditions were found to occur when the value of the capacitive time constant t_C $(= CR)$ and the value of the inertial time constant t_L $(= L/R)$ were such that (Eq. 4.8.22)

$$\frac{1}{t_C \times t_L} = \omega^2 \tag{10.2.1}$$

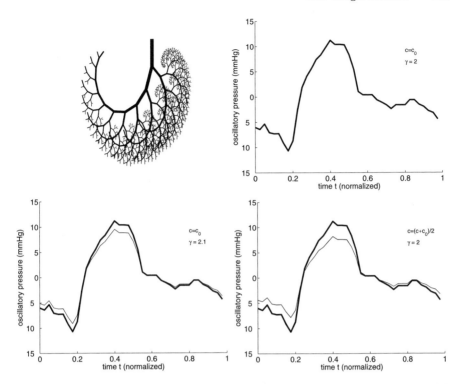

Fig. 10.2.3. Propagation of the cardiac pressure wave along the 11-level tree model. The input wave, shown in bold, would normally be modified as it progresses along the tree structure, but a "magic norm" occurs when the wave speed c is equal to the Moen-Korteweg wave speed c_0 and the value of the power law index γ is 2.0 whereby the wave remains unchanged. Departure from any of these values moves the system away from this magic norm. The thin curves indicate the changed form of the wave as it travels in that case.

where ω is the frequency of oscillation of the driving pressure, R is resistance, L is inductance, and C is capacitance. Under this unique combination of values the RLC system behaves as if L and C do not exist, thus the relation between pressure and flow is particulary simple and is the same as that in the presence of resistance R only, namely

$$q(t) = \frac{\Delta p(t)}{R} \tag{10.2.2}$$

where $q(t)$ is the oscillatory flow rate and $\Delta p(t)$ is the oscillatory driving pressure. Thus, if the flow rate is multiplied by R, the resulting "R-scaled" flow rate becomes identical with the driving pressure, as illustrated in Fig. 10.2.1.

For the purpose of illustration, in Fig. 10.2.1 the value of t_L is taken as $1\,s$ and the frequency of oscillation is taken as 1 Hz, thus $\omega = 2\pi\,rad/s$. This means that the magic norm conditions occur when $t_C = 1/\omega^2 = 0.0253\,s$

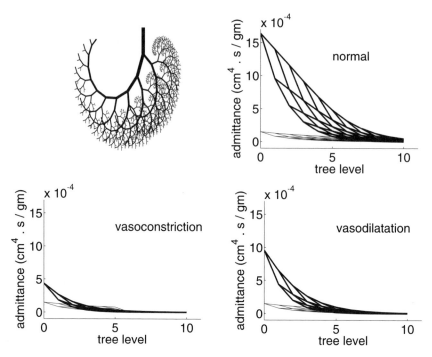

Fig. 10.2.4. Effects of "vasoconstriction" and "vasodilatation", as defined in the text, on the values of admittance within the 11-level tree model, compared with the "normal" case, at a frequency of 1 Hz. The two conditions produce significant changes in the distribution of admittance and hence in the dynamics of pressure and flow within the tree model.

as required by Eq. 10.2.1. The "magic" in this magic norm is illustrated in Fig. 10.2.1 where it is seen that any small departure from this particular value of t_C leads to a departure from the ideal relation between pressure and flow.

As discussed in Section 4.7, an appropriate value of t_L for the coronary circulation is not known, if indeed a single such value is appropriate at all. Nevertheless, the results in Fig. 10.2.1 are valuable in providing a hint at the *relative values* of the two time constants, as provided by Eq. 10.2.1. Because of the great simplicity of the pressure-flow relations under these conditions, it is highly tempting to speculate that the coronary circulation may be designed to operate with values of the two time constants that are close to those provided by Eq. 10.2.1. It would then follow that any departure from these ideal values would be functionally undesirable in the sense that it would disrupt the ideal relation between pressure and flow. As in the case of the child on a swing, any departure from the perfect timing of the force which the child applies with his or her body leads to disruption in the ideal operation of the swing.

While the magic norm in Fig. 10.2.1 is based on a driving pressure consisting of a single harmonic through a simple RLC system in series, similar

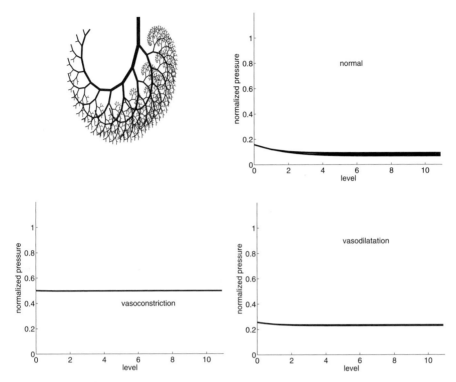

Fig. 10.2.5. Effects of "vasoconstriction" and "vasodilatation", as defined in the text, on the pressure distribution within the 11-level tree model, compared with the "normal" case, at a frequency of 1 Hz. While the *form* of the distribution is not greatly changed by the two conditions, there is considerable change in the *level* of the pressure, moving higher in both cases.

results were found in Section 7.4 (Fig. 7.4.4) where the driving pressure is a composite wave and the RLC system is somewhat more complicated. Results in that section for $LM2$, whose composition in the notation of Chapter 7 is $\{\{R_1+L\}, \{R_2+C\}\}$, show that when $\lambda = R_2/R_1 = 1.0$ and $t_L = t_C = 0.2\,s$ a magic norm occurs whereby this RLC system behaves as if L and C were absent. As illustrated in Fig. 10.2.2, with this combination of values of the time constants and the ratio λ of the two resistances, the pressure and R-scaled flow curves again become identical, but any departure from these values moves the system away from this magic norm.

The concept of magic norms extends also to unlumped model analysis where the architecture of the vascular tree plays a key role. Here, the *propagation* of the pressure and flow waves along the tree structure is an important issue as discussed in Section 9.7. Results in that section (Fig. 9.7.20) showed that under the ideal conditions of the so-called "impedance matching", the pressure wave remains unchanged as it travels along the tree structure. This

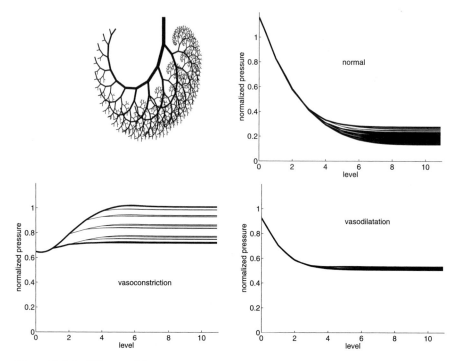

Fig. 10.2.6. Effects of "vasoconstriction" and "vasodilatation", as defined in the text, on the pressure distribution within the 11-level tree model, compared with the "normal" case as in Fig. 10.2.5 but here at a frequency of 10 Hz. The effects of the two conditions on the pressure distribution are clearly more dramatic at the higher frequency.

magic norm occurs when the value of the power law index γ is 2.0 and the wave speed c is taken equal to the Moen-Korteweg wave speed c_0. Any departure from these conditions leads to departure from the magic norm, as illustrated in Fig. 10.2.3.

It must be remembered, of course, that the coronary circulation is not a simple RLC system nor an idealized tree model, thus the parameter values for the magic norms described here cannot be applied directly to the coronary circulation. The ultimate value of these results lies in pointing out that the coronary circulation very likely has its own magic norm or norms and that any disruption in the conditions required for these norms will move the system away from its optimal dynamics. Such disruptions may occur as a result of vascular disease or aging, which would change the mechanical properties of the vessel wall or constrict the lumen available for the flow, or as a result of clinical intervention, whether by drugs or surgery. To illustrate this more succinctly, we finish this section by considering the 11-level tree model with architecture on the scale of the coronary circulation as was done in Chapter

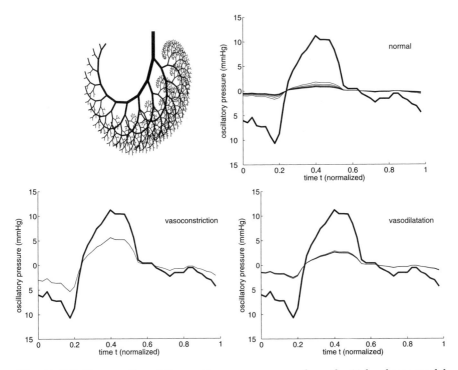

Fig. 10.2.7. Propagation of the cardiac pressure wave along the 11-level tree model, the input waveform being shown in bold, while the thin curves indicate the changed form of the wave as it travels in that case. The three panels show the effects of "vasoconstriction" and "vasodilatation", as defined in the text, compared with the "normal" case at a frequency of 1 Hz.

9. Then we consider two scenarios in which the peripheral 6 levels of the tree are either constricted, with vessel diameters reduced by 50%, or dilated, with vessel diameters increased by 50%, while the elasticity of the same vessels is reduced to the effect that the vessels become more rigid, with Young's Modulus increased from 10^7 to $10^9 \, dyn/cm^2$. While the two scenarios are highly idealized, we shall refer to them loosely as conditions of "vasoconstriction" and "vasodilatation", respectively, suggesting not that they replicate the corresponding physiological conditions but that they may be relevant to them. Effects of these scenarios on admittance or pressure distribution or on the form of the propagating wave within the tree structure are shown in Figs.10.2.4-9.

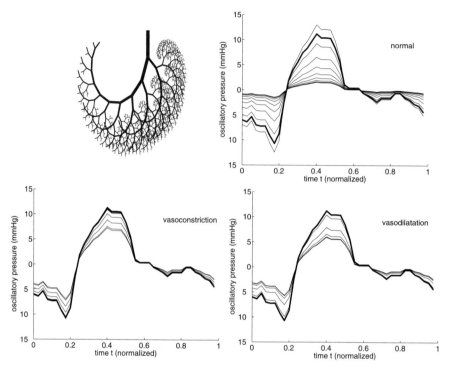

Fig. 10.2.8. Propagation of the cardiac pressure wave along the 11-level tree model, the input waveform being shown in bold, while the thin curves indicate the changed form of the wave as it travels in that case. The three panels show the effects of "vasoconstriction" and "vasodilatation", as defined in the text, compared with the "normal" case as in Fig. 10.2.7, but here at a frequency of 10 Hz.

10.3 Coronary Heart Disease, Physical Exercise, and the Conundrum of Coronary Flow Reserve

One of the characteristic features of the coronary circulation that distinguishes it from other circulations is its capacity to increase blood flow to the heart "on demand" by as much as five or six times normal flow rate, a feature generally referred to as "coronary flow reserve" [83, 42, 66, 64, 129, 128, 100, 183, 26, 112]. Elements of this subject were introduced earlier in Sections 1.8,9. In the present section we attempt to unify these elements into a useful whole and place the subject of coronary flow reserve in the context of coronary heart disease and physical exercise.

It was shown in Section 1.8 that coronary flow reserve is *not* provided by coronary arteries that are "oversized" compared with other arteries, or oversized in relation to the flow rate which they carry under normal conditions. It was shown in fact that the coronary arteries are of fairly normal caliber compared with arteries elsewhere in the body. It is widely accepted that coronary flow reserve is provided simply by an increase in the caliber of the so called "resistance vessels" within the coronary vascular tree, thus leading to a higher flow rate through the system [83, 42, 66, 64, 129, 128, 100, 183, 26, 112]. The conventional wisdom is that the five- to six-fold increase in flow is achieved simply by decreasing the resistance to flow. While this wisdom is to a large extent correct, it is not entirely so because it is based on a model of the coronary circulation in which flow is assumed to be *steady* and flow rate is assumed to be determined entirely by resistance to flow.

Coronary blood flow, on the other hand, is *pulsatile*, and in pulsatile flow, as we have seen throughout this book, many more factors are involved in determining the flow rate through a vascular system. In particular, the effects of capacitance, inductance, and wave reflections, as discussed in previous chapters, make it clear that the coronary circulation has more tools at its disposal for changing the flow rate than that of a simple change in pure resistance. An increase in the caliber of the so-called resistance vessels will indeed reduce resistance to the steady part of the flow, but as we have seen in previous sections it will also change the admittance to the pulsatile part of the flow, change the dynamics of wave reflections, reduce the elasticity of the vessels involved and hence increase the wave speed, all of which pointing to the number of variables that may be involved in the facility and function of coronary flow reserve. The concept of magic norms discussed in the previous section and the examples of vasoconstriction and vasodilatation discussed in that section make it clear that a change in one variable may lead to more than the effect of that particular variable *in isolation*. What is ultimately important is whether a change in one variable will move the system away from its magic norm, and this, as we have seen, cannot be determined by considering that variable alone. This is clearly relevant when the change is introduced by clinical intervention.

While models of the coronary circulation have yet to grapple with the full intricacies of coronary flow reserve, enough is known to indicate clearly that this aspect of the coronary circulation has evolved to respond to increased demand for coronary blood flow resulting from increased physical activity, rather than to compensate for reduced coronary blood flow in coronary heart disease. As mentioned earlier, the difference between the two situations is one of time scale. Coronary heart disease may reduce blood flow to the heart gradually over a period of months or years, while the increased demand for coronary blood flow following a sudden increase in physical activity may occur within seconds. In both cases the prevailing condition is a deficit in coronary blood flow, but in one case the deficit is small and chronic while in the other it is large and acute. Coronary flow reserve has evolved as a mechanism for dealing with acute deficits in coronary blood flow.

The only mechanism which seems to have evolved for dealing with chronic deficits in coronary blood flow is that of slow "restructuring" of the vasculature to deal with new conditions. Under this general heading may be included some angiogenic responses that are not specific to the heart, such as enlargement of existing vessels to carry higher flow rates, or the development of new vasculature. A response which is more specific to the heart is that of "collateral vasculature" discussed in Section 1.7. It has been adequately demonstrated that in the presence of coronary artery disease the coronary circulation responds by developing new routes for coronary blood flow [15, 14]. It is not clear, however, whether these new routes are based on pre-existing collateral vasculature or are the result of angiogenesis. It is also not clear to what extent this mechanism is effective in dealing with chronic deficits in coronary blood flow since it has evidently not succeeded in preventing heart failures resulting from coronary heart disease [83, 206, 14, 128]. This and many other issues asociated with the efficacy and clinical significance of this mechanism are shrouded with a great deal of controversy. It is widely accepted, however, that the time scale of this mechanism is months or years, not seconds.

We thus have two mechanisms for dealing with coronary blood flow deficits which we shall refer to simply as "fast" and "slow". In addition to the difference in their time scales, another important difference between the two mechanisms is their triggers. It is widely accepted that the trigger for the fast mechanism is an increase in myocardial oxygen consumption. The trigger for the slow mechanism, on the other hand, is repeated episodes of hypoxia and ischemia [83, 128]. The interplay between these two mechanisms in coronary heart disease and in heart failure under different circumstances is discussed in what follows.

We note first that when heart failure results from coronary artery disease, the ultimate cause of the failure is that the heart is not receiving sufficient blood supply for its own metabolic needs. It is thus a failure of the coronary circulation rather than a failure of the heart itself. In the analogy of a car engine, it is a failure of an engine that has run out of fuel rather than an engine that has broken down. While these statements seem to state the obvious, they bear repeating because the terminology used in this subject is somewhat misleading. Thus, what is widely referred to as "coronary heart disease" is a disease not of the heart but of the coronary arteries, and a more accurate term for it is "coronary artery disease". And what is widely referred to as "heart failure" is a failure not of the heart but of sufficient blood supply to reach the heart, and a more accurate term for it would be "heart starvation".

From the point of view of energetics, normal operation of the heart as a pump is based on a very simple equation, whereby under steady state conditions

$$\begin{pmatrix} \text{rate of oxygen consumption} \\ \text{by heart muscle} \end{pmatrix} = \begin{pmatrix} \text{rate of oxygen delivery} \\ \text{to heart muscle} \end{pmatrix} \quad (10.3.1)$$

Coronary blood flow is associated with the right hand side of this equation. More precisely, oxygen delivery to the myocardium depends directly on the rate of coronary blood flow and on "oxygen extraction", which is the percentage of oxygen being extracted from the blood by the heart muscle. While oxygen extraction may vary, under normal modes of operation the dominant factor in the delivery of oxygen to the heart muscle is the rate of coronary blood flow. To simplify matters, therefore, in what follows we shall consider coronary blood flow to be a direct measure of the rate of oxygen delivery to heart muscle.

In what we shall call the "normal" course of events, if the energetics of the heart muscle are in a non-steady state, as, for example, when the level of physical activity of the body increases from rest to a higher level, work of the heart increases appropriately and the rate of oxygen consumption by the heart muscle increases with it. In response to the latter, coronary flow reserve is triggered within seconds to increase coronary blood flow and the rate of oxygen delivery to the heart muscle is thereby increased to a level that would satisfy Eq. 10.3.1. This sequence of events is shown schematically in the form of a flow chart in Fig. 10.3.1. This scenario may be repeated many times, at different levels of activity. If these levels are such that coronary flow reserve is always able to respond and satisfy the oxygen demands of the heart muscle then it is in this sense that we refer to this case as "normal". This term is not intended to imply that the coronary arteries are necessarily free from disease or that the capacity of coronary flow reserve is intact. It only implies that coronary flow reserve is robust to the level of physical activity it is being subjected to.

In what is generally referred to as "coronary heart disease", because of coronary artery disease, the base rate of coronary blood flow has been reduced *and* the capacity of coronary flow reserve has been reduced. A situation may be reached whereby a certain level of physical activity can no longer be supported, and the only way to satisfy Eq. 10.3.1 is to reduce the level of physical activity to a level that can be supported by the impaired coronary flow reserve. This course of events is illustrated schematically in Fig. 10.3.2.

If the conditions of coronary heart disease continue unabated, the capacity of coronary flow reserve will deteriorate to the point where it can no longer augment coronary blood flow in any significant way. And since the base rate of coronary blood flow has also been considerably reduced by the disease, a point is reached where the oxygen demands of the heart muscle can no longer be met even when physical activity has been reduced to a minimum. The heart fails to function as a pump, as illustrated schematically in Fig. 10.3.3.

A course of events that is counter to that of coronary heart disease and heart failure is that of physical "exercise". Here physical activity is taken deliberately to maximal levels so as to create hypoxia and momentary ischemia, and the scenario is repeated so as to presumably trigger the "slow" mechanism of dealing with coronary blood flow deficits, namely the mechanism of vascular restructuring [83, 128]. As mentioned earlier, this may include new angiogenic activity as well as the restructuring (enlarging) of existing normal and any collateral vessels. In the absence of coronary artery disease, this would act to maintain the capacity of coronary vasculature and of coronary flow reserve at sufficiently high levels that they can deal with maximal demands for coronary blood flow, a state of so called cardiovascular "fitness". In the presence of coronary artery disease, or ageing, the restructuring activity triggered by

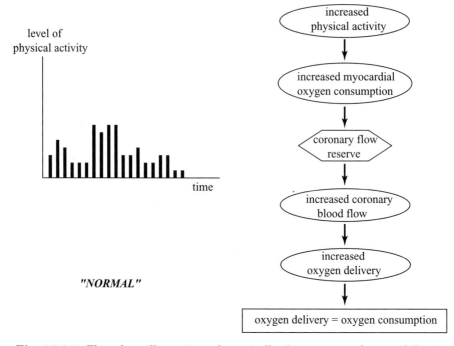

Fig. 10.3.1. Flow chart illustrating schematically the sequence of events following an increase in physical activity, and the role of coronary flow reserve in that sequence under "normal" circumstances, by which is meant that coronary flow reserve is able to provide the necessary rate of coronary blood flow so that the balance of myocardial oxygen consumption and oxygen delivery is satified. The bar chart on the left illustrates, schematically again, the variable level of physical activity as time goes on. Under the "normal" scenario, coronary flow reserve is able to deal with these levels but it is not known if it can handle higher levels or if it is actually compensating for any decline in normal coronary blood flow due to coronary artery disease.

physical exercise may counter and possibly reverse the deteriorating capacity of coronary arteries and of coronary flow reserve, as illustrated schematically in Fig. 10.3.4.

The "bank analogy" mentioned briefly in Section 1.9 can be used to illustrate these scenarios in the context of a worker who earns $1,000 per month and, under normal circumstances, spends as much. Earnings are deposited directly into a "special" bank account and expenses are charged directly to the same account [217]. What is special about the account is that (a) it started out with a reserve of $6,000 and (b) no statements are issued about the balance in the account each month, unless the account is in deficit. Under normal circumstances the worker has no idea whether earnings and expenditures are in balance each month, because any small discrepancies are masked by the large reserve. If such discrepancies continue for a long time, however, a day will

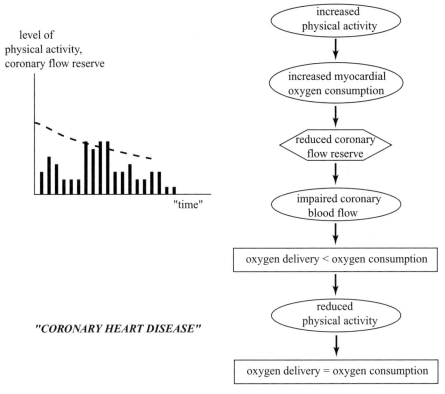

Fig. 10.3.2. Flow chart illustrating schematically the sequence of events following an increase in physical activity, and the role of coronary flow reserve in that sequence under conditions of "coronary heart disease". Normal coronary blood flow and the capacity of coronary flow reserve are both affected negatively by the disease to the extent that a balance between myocardial oxygen consumption and oxygen delivery can only be reached by reducing physical activity. The bar chart on the left illustrates the variable level of physical activity as time goes on, while the dashed line represents the declining capacity of coronary flow reserve. The combination of the two on the time scale is clearly inaccurate because the time scale of variation in physical activity is widely different from the time scale of decline in coronary flow reserve. Nevertheless, it is useful to see the two together, though with the understanding that "time" is different in each case.

be reached when the worker will receive a statement from the bank declaring that his account is in deficit. The worker's options at this time are limited because his reserve is now clearly exhausted. While in the presence of a healthy reserve any monthly discrepancies can be corrected by appropriate changes in spending habits, in the absence of such reserve such measures would be too slow and ineffective. Thus, while the reserve plays a very important role in the worker's financial system, it also inadvertently masks monthly discrepancies which would serve as important warning signs to the worker before the

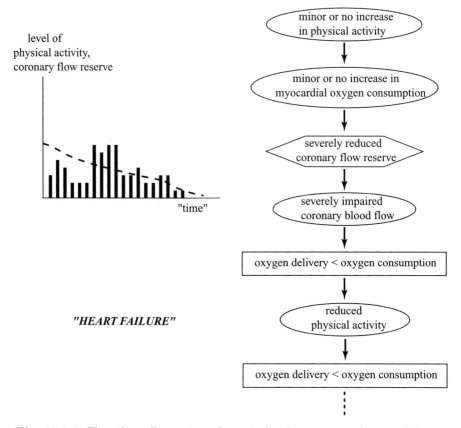

Fig. 10.3.3. Flow chart illustrating schematically the sequence of events following an increase in physical activity, and the role of coronary flow reserve, as in Fig. 10.3.2, but here under conditions of "heart failure". Normal coronary blood flow and the capacity of coronary flow reserve are both severely impaired by the disease to the extent that a balance between myocardial oxygen consumption and oxygen delivery cannot be achieved even at minimal or no physical activity. Remaining caption concerning the chart on the left is as in Fig. 10.3.2.

"terminal" event of a negative statement from the bank. In time, the worker finds that the only way to guard against this terminal event is to devise a method of monitoring the state of his reserve. The method consisted of occasional "spending sprees" that would deliberately challenge the reserve by large conditional expenditures that can easily be reversed if a statement from the bank arrives. As a result of this exercise the worker is constantly aware of the state of the reserve and is able to improve it if necessary by appropriate changes in regular spending habits. While the analogy with the coronary circulation is not accurate in every respect, it serves the purpose of illustrating the conundrum of the coronary flow reserve in coronary heart disease.

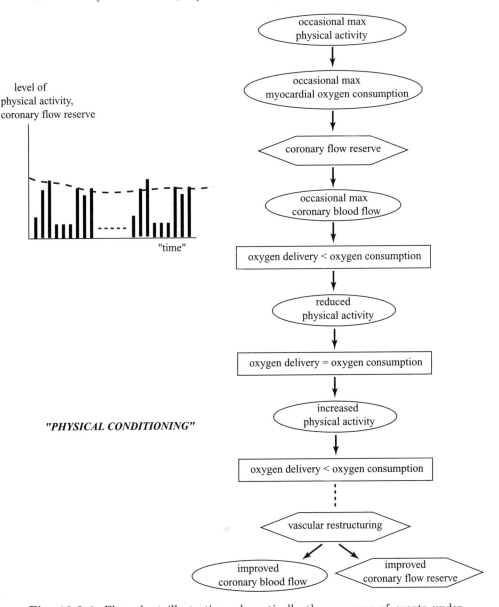

Fig. 10.3.4. Flow chart illustrating schematically the sequence of events under conditions of "physical conditioning" and the roles of coronary flow reserve and of vascular restructuring under these conditions. Only repeated episodes of oxygen imbalance trigger the mechanism of vascular restructuring and the consequent benefits of improved coronary blood flow and improved coronary flow reserve. Remaining caption concerning the chart on the left is as in Fig. 10.3.2.

10.4 Wave Propagation Through a Coronary Bypass

The placement of a vascular graft to bypass an obstruction in a coronary blood vessel is one of the most common surgical cardiovascular procedures in current practice [155, 142, 210, 166]. In its most common form it consists of creating an alternate route for blood flow from the aorta to a point downstream of an obstruction in a main coronary artery as illustrated schematically in Fig. 10.4.1

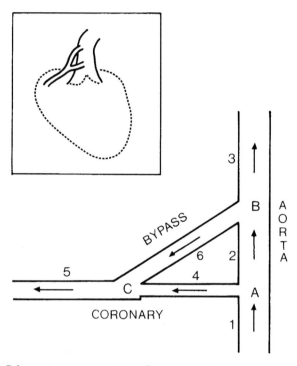

Fig. 10.4.1. Schematic representation of a coronary bypass graft arrangement (inset) and of model used for analysis. The model consists of six vessel segments identified by numbers 1 − 6 and three junctions identified by letters A-C. Segment 4 represents the diseased coronary artery and segment 6 represents the bypass graft, both having variable diameters, to simulate the severity of the disease in one case and the relative size of the graft in the other. From [1].

The clinical basis of coronary bypass procedures is fairly straightforward, namely that of providing coronary blood flow with a new supply route from the aorta. While this has a fairly sound fluid dynamic basis, its tenets lie strictly in the dynamics of steady flow in rigid tubes, thus ignoring the pulsatile nature of coronary blood flow and the elasticity of coronary arteries as well as that of the bypass graft. The dynamics of coronary blood flow, as we have seen, are those of wave propagation and wave reflections in elastic tubes, and the

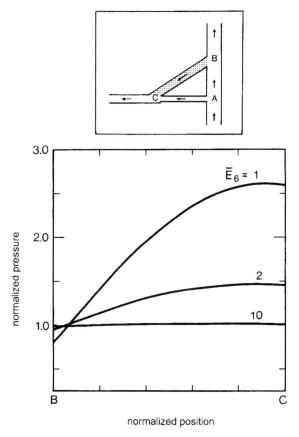

Fig. 10.4.2. Pressure distribution in a coronary bypass graft, created by wave reflections when the Young's modulus of elasticity of the graft is the same as that of the native coronary artery (curve labeled $\overline{E}_6 = 1$), when it is twice as large ($\overline{E}_6 = 2$), and when it is 10 times as large ($\overline{E}_6 = 10$). A stiffer graft produces a pressure difference between the two ends of the graft that is more favorable to flow from the aorta to the coronary artery. From [1].

placement of a coronary bypass graft must correctly be viewed in this broader context.

More specifically, the placement of a graft should be viewed as not only creating a new path for coronary blood flow but also as introducing a new vascular impedance and two new vascular junctions which, as we have seen, can act as wave reflection sites [53, 54]. Thus, the length of the graft, its diameter, and its elasticity are all be important in this context. Analysis of wave propagation and wave reflections in the bypass graft configuration shown in Fig. 10.4.1, taking into account properties of the bypass graft and of the obstructed coronary artery, have been reported elsewhere [1]. We omit the analytical details here and present only a sample of the main findings.

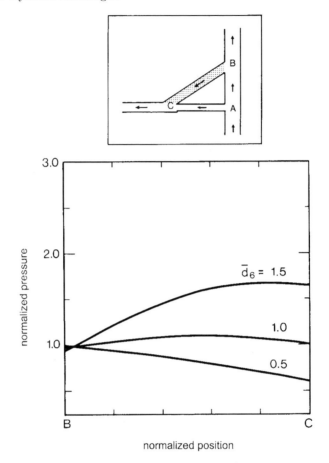

Fig. 10.4.3. Pressure distribution in a coronary bypass graft, created by wave reflections when the diameter of the graft is the same as that of the native coronary artery (curve labeled $\bar{d}_6 = 1$), when it is 50% larger ($\bar{d}_6 = 1.5$), and when it is 50% smaller ($\bar{d}_6 = 0.5$). A graft of *smaller* diameter produces a pressure difference between the two ends of the graft that is more favorable to flow from the aorta to the coronary artery. From [1].

If the elasticity of the bypass graft is the same as that of the native coronary arteries, it is found that wave reflections from the three vascular junctions involved have the net effect of creating a pressure difference within the bypass graft which is positive in the desired flow direction, as shown in Fig. 10.4.2. In other words, as a result of wave reflections the pressure at junction C in that figure is higher than the pressure at junction B. This pressure difference *by itself* would cause the flow to go from the coronary network to the aorta instead of the reverse, for which the bypass is intended. But, of course, this pressure difference is not present by itself, it represents merely a perturbation on the main pressure difference between the aorta and the coronary network. Nevertheless, it is a perturbation with an adverse effect on the main pressure difference driving the flow. The analysis shows that if the bypass graft is chosen to be stiffer (less elastic) than the native coronary artery, this adverse effect can in fact be diminished as shown in Fig. 10.4.2. In the clinical setting the choice of a bypass graft is usually between a saphenous vein graft and a mammary or other arterial graft [57, 82, 30], and it is usually made on surgical grounds. From the perspective of the dynamics of the flow, the results suggest that an arterial or a synthetic graft that is stiffer than a saphenous vein would be preferable. Arteries generally have a higher modulus of elasticity than do veins [34, 70, 71], and arterial grafts are reported to have better performance than vein grafts [30].

Similarly, the pressure difference between the two ends of the bypass graft, created by wave reflections from all junctions, is found to be more favorable to flow from the aorta to the coronary vessel when the diameter of the bypass graft is *smaller* than the native coronary artery, and more adverse when the reverse is true, as shown in Fig. 10.4.3. Common wisdom in coronary bypass surgery is to use a graft of *larger* diameter than that of the native coronary artery [57, 82, 72, 58], clearly, that wisdom is being based on considerations of *steady* flow through the bypass system.

10.5 Wave Propagation Through a Coronary Stent

The placement of a coronary stent within a diseased coronary artery, like the placement of a coronary bypass graft, is a practice based largely on considerations of *steady flow*. The stent functions by scaffolding a diseased coronary artery to compress any lesions protruding into the lumen, thus keeping the vessel open for blood flow as illustrated schematically in Fig. 10.5.1. The basis of this practice is perfectly valid in the context of steady flow where an

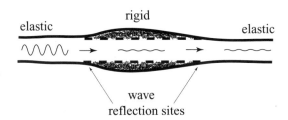

Fig. 10.5.1. From the standpoint of the dynamics of pulsatile flow, a coronary stent represents a rigid tube segment in an otherwise elastic vessel. Because the impedance of the rigid segment is much higher than that of the elastic segments to which it is attached, the propagating wave is blunted within the stent as illustrated schematically, and because of impedance mismatch at the two ends of the stent, two wave reflection sites are created there.

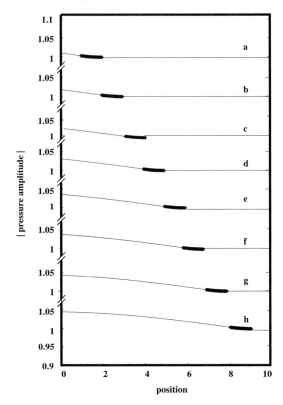

Fig. 10.5.2. Normalized pressure distribution along a coronary artery of length 10 cm, with a stent, shown in bold, of length 1 cm and placed at different position along the vessel, position '0' being the entrance to and '10' being the exit from the artery. When the stent is placed near the entrance of the native vessel it has the effect of producing a pressure distribution (curve a) with a small favorable pressure difference between entrance and exit from the vessel, but this pressure difference becomes increasingly larger as the position of the stent is moved closer to the exit end of the vessel (curves $b - h$). From [2].

open vessel is clearly better than an obstructed one, and several studies have examined various mechanical and other aspects of coronary stents within this context [67, 176, 138, 232, 62, 89, 56, 149, 200, 35, 140].

But in the context of pulsatile flow in elastic tubes a number of additional issues must be addressed. The stent is usually made of a steel mesh with high radial strength [171, 154, 50], and it is found that soon after the stent is implanted the open mesh fills with growing tissue and the stent fuses with the vessel wall [150]. Thus, the combination of stent and vessel wall together become a *rigid* segment within an otherwise elastic vessel, which presents a considerable change in local impedance within the vessel. Because the stented segment is considerably more rigid than the upstream (proximal) and down-

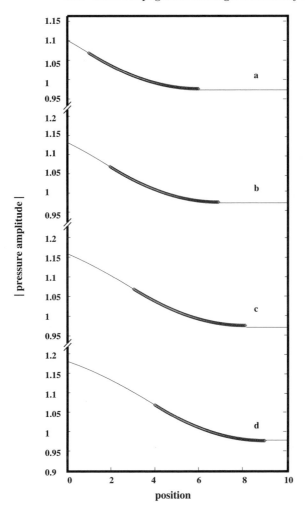

Fig. 10.5.3. Normalized pressure distribution along a coronary artery of length 10 cm, with a stent, shown in bold, of length 5 cm and placed at different position along the vessel, position '0' being the entrance to and '10' being the exit from the artery. When the stent is placed near the entrance of the native vessel it has the effect of producing a pressure distribution (curve *a*) with a significant favorable pressure difference between entrance and exit from the native vessel, and this difference becomes yet larger as the position of the stent is moved closer to the exit end of the vessel (curves *b − d*). From [2].

stream (distal) segments to which it is attached, there is a change in impedance at the two junctions between the stented and non-stented segments and hence the two junctions will act as wave reflection sites (Fig. 10.5.1). These issues have only recently been examined in full elsewhere [2], and the essence of the results is presented below, omitting the analytical details.

Of particular interest is the effect of a coronary stent on the pressure distribution within the vessel in which it is placed. As in the previous section, the pressure distribution in question is only that arising as a result of wave reflections, and it is normalized such that in the absence of wave reflections, which means in the absence of the stent, the pressure distribution would be uniform at a normalized value of 1.0. Any departure from this uniform distribution can thus be attributed entirely to the presence of the stent.

In general it is found that a stent has the effect of producing a pressure distribution that is actually favourable for flow in the native vessel, with the pressure being higher at entrance to the vessel than it is at exit. The extent of this effect depends on two main parameters: the length of the stent in relation to the length of the native vessel, and the position of the stent within the length of the native vessel. The results show that a relatively short stent placed near the entrance of the native vessel has the smallest effect and produces a pressure distribution with the smallest pressure difference between the two ends of the vessel. This effect becomes increasingly more significant, however, as the stent is moved closer to the exit end of the native vessel, as shown in Fig. 10.5.2.

In the case of a relatively long stent it is found that the pressure distribution it produces and the associated pressure difference are fairly significant even when the stent is placed near the entrance of the native vessel, and it becomes considerably more so as the stent is moved closer to the exit end of the vessel, as shown in Fig. 10.5.3.

10.6 Sudden Cardiac Death

A principal theme of this book has been to highlight the dynamics of the coronary circulation and to show the central role that these dynamics can play in coronary blood flow. Nowhere does this theme find a better application than in the phenomenon of sudden cardiac death, and it is therefore appropriate that this subject be discussed in the last chapter of the book. The material to be presented is indeed no more than a brief discussion, based on an extensive review by Osborn [158].

Sudden cardiac death is death due to heart failure in which the timing of the failure is sudden or unexpected [158]. It is a major public health problem, being the leading cause of death in the industrialized world, claiming more than one thousand lives *per day* in the United States alone [158].

The cause of sudden cardiac death in all cases is a precipitous fall in cardiac output to levels that can no longer sustain cerebral or cardiac function. The fall in cardiac output is usually associated with a disruption in heart rhythm in one form or another: ventricular fibrillation, tachycardia, bradycardia, and the like. Common causes of disruption in heart rhythm include myocardial infarction or ischemia and cardiomyopathy, all of which act to disrupt the "electrical uniformity" of the myocardium. Other causes include disruptions in neural activity and environmental or psychological stresses such as startle, fright, anger, and excitement. Coronary artery disease, which is found in most cases of sudden cardiac death in adults, is generally seen to be a precursor to conditions of myocardial ischemia and infarction leading to a disruption in heart rhythm.

Thus, the sequence of events leading to sudden cardiac death according to current theory is as shown schematically in the form of a flow chart in Fig. 10.6.1.

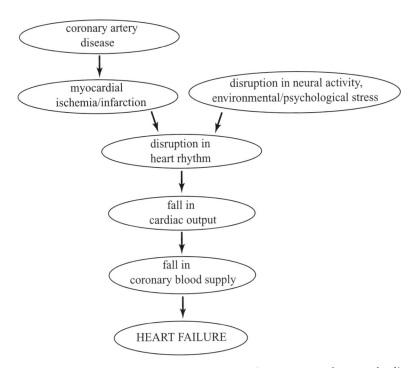

Fig. 10.6.1. A summary of current theory on the sequence of events leading to heart failure in sudden cardiac death.

A difficulty with this sequence of events is that it proposes that a fall in cardiac output is followed by a fall in coronary blood flow. The mechanism for this sequence is not clear, if indeed possible. If the fall in cardiac output leads to a drop in aortic pressure, then it is well known that autoregulation of coronary blood flow ensures that this does not lead to a fall in coronary blood flow. As discussed briefly in Section 1.2, in the absence of this protective regulatory feature of the coronary circulation the system would behave as an unstable "positive feedback" system in which a fall in cardiac output would lead to a fall in coronary blood flow and to a further fall in cardiac output and a further fall in coronary blood flow, etc. It is known that the coronary circulation as a feedback system does not behave in this way, as discussed in Section 1.2.

Thus, a fall or disruption in coronary blood flow is unlikely to be the result of a fall in cardiac output because this is inconsistent with the regulatory mechanisms of the system. Indeed, the *reverse* is more consistent with the regulatory mechanisms of the system and is therefore more likely to be the case. A fall in cardiac output is more likely to be the *consequence* rather than the *cause* of a fall in coronary blood flow in the sequence of events leading to sudden cardiac death.

We have seen throughout this book that a fall in coronary blood flow may result from a disruption in the *dynamics* of the coronary circulation, which in turn can result from coronary artery disease, cardiac damage, or a disruption in neural activity. The fall in coronary blood flow then *leads to* a fall in cardiac output, to possible disruptions in heart rhythm, and to heart failure. This sequence of events is shown schematically in Fig. 10.6.2.

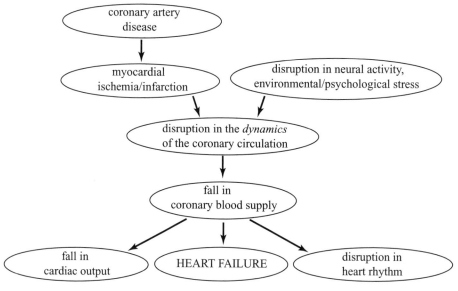

Fig. 10.6.2. A summary of proposed theory on the sequence of events leading to heart failure in sudden cardiac death.

10.7 Broken Heart Syndrome

As this book goes to press, an issue of The New England Journal of Medicine flags one of its leading articles [207] with the banner "Broken Heart Syndrome" and a brief news item to the effect that "Sudden emotional distress ... can sometimes produce severe transient left ventricular dysfunction. This stress-induced cardiomyopathy appears to be a form of myocardial stunning associated with marked sympathetic stimulation." Both the banner and news item provide a particularly fitting epilogue to this chapter, and, indeed, to the principal tenet of this book that dynamic pathologies in the coronary circulation can produce calamities equal to those produced by structural pathologies.

The authors of the article report in more detail that ".. a unique pattern of transient myocardial dysfunction can occur after severe emotional stress. Patients with this syndrome have evidence of exaggerated sympathetic activation, with plasma catecholamine levels several times those in age- and sex-matched patients with Killip class III myocardial infarction. Although our data suggest that catecholamines may be central to the mechanism of stress-related myocardial stunning, a more complete understanding of the pathogenesis of this syndrome awaits further research." Three possible mechanisms are discussed: epicardial coronary arterial spasm, microvascular spasm, and direct myocyte injury. It is concluded that "Emotional stress can precipitate severe, reversible left ventricular dysfunction in patients without coronary disease. Exaggerated sympathetic stimulation is probably central to the cause of this syndrome."

As mentioned in Section 10.6, psychological stresses such as startle, fright, anger and excitement, have been identified to cause disruptions in heart rhythm associated with sudden cardiac death [158]. Emotional stresses that have been identified in conjunction with the broken heart syndrome include death in a family, accident, fear and surprise [207]. These emotional stresses may clearly have some features in common with the above psychological stresses, but the key to the new findings is that the broken heart syndrome is marked by highly elevated plasma levels of catecholamines and "stress-related neuropeptides". These agents are known to control parameters that affect the dynamics of the coronary circulation as they produce changes in vascular calibers and vascular resistance. The broken heart syndrome may therefore be considered another example of a disruption in, indeed an insult to, the dynamics of the coronary circulation leading to a fall in coronary blood supply and to a fall in cardiac output, as outlined schematically in Fig. 10.6.2. Thus, both the sudden cardiac death syndrome and the broken heart syndrome offer examples of dynamic pathologies in the coronary circulation, although an important distinction of the latter is that it is found to be *transient* and *reversible* [207].

One of the most important aspects of dynamic pathologies is that, unlike structural pathologies, they do not leave a "footprint" after they have resolved or produced their damage. Following sudden cardiac death, indeed following any death attributed to heart disease, only structural pathologies can usually be found and they become the focus of attention. Any dynamic pathologies involved are not in evidence because they are no longer at play. The broken heart syndrome on the other hand, with the distinction of being transient, reversible and free from coronary artery disease, offers an opportunity to detect dynamic pathologies while they are at play and in "pure form" because the syndrom is not associated with coronary artery disease.

10.8 Summary

Coronary blood flow can be disrupted by dynamic pathology in the same way that it can be disrupted by vascular pathology. Any disruption in the exquisite harmony between the pulsatile rhythm of the input pressure driving the flow, the pulsatile rhythm of the contracting myocardium muscle and its effect on coronary vessels imbedded within the muscle tissue, and the combination of wave propagation and wave reflections within the coronary vascular tree, can produce dynamic pathology that disrupts coronary blood flow.

Changes in the architecture or mechanical properties of coronary vasculature have significant dynamical consequences and may move the system away from its magic norm or norms, which are modes of operations in which the dynamics of the system appear particularly simplified or optimal in some sense. Every aspect of the distribution and propagation of pressure and flow within the tree model is affected. And while the dynamics of the coronary circulation are far more complicated than those of the 11-level tree model, it is reasonable to speculate that changes in the architecture or mechanical properties of the coronary arteries, whether brought on by neural or humoral agents, under normal or pathological conditions, will have significant effects on the dynamics of coronary blood flow.

Coronary flow reserve and vascular restructuring are two mechanisms by which the coronary circulation can increase coronary blood flow. The trigger for the first is an increase in oxygen consumption by the heart muscle and its response time is of the order of seconds, while the trigger for the second is repeated episodes of hypoxia and momentary ischemia and its response time is at best of the order of weeks but more likely months or years. It would thus seem that coronary flow reserve is designed to deal with acute shortfalls in coronary blood flow, as at the onset of increased physical activity, while vascular restructuring is designed to deal with chronic shortfalls in coronary blood flow, as under the progression of obstructive coronary artery disease. The conundrum of coronary flow reserve is that it may "mask" the gradual shortfall in coronary blood flow at the initial stages of coronary heart disease and thereby prevent the mechanism of vascular restructuring from being triggered. At the latter stages of the disease the capacity of coronary flow reserve is largely depleted and the mechanism of vascular restructuring may not have the time required for its slow course of action. Physical conditioning may forestall this course of events by deliberately and regularly triggering the mechanism of vascular restructuring while coronary flow reserve is still intact.

The placement of a coronary bypass graft is usually viewed in terms of providing a conduit for *steady* flow through the graft. If the pulsatile nature of the flow and the elasticity of the vessels are considered, the effects of wave propagation and wave reflections produce a pressure difference between the two ends of the graft that is more favorable to flow in the desired direction through the graft when the graft is stiffer than the native coronary artery and when the diameter of the graft is smaller than that of the native coronary artery.

Because of impedance mismatch and consequent wave reflections, the placement of a stent within a coronary artery has the effect of producing a pressure distribution with a favorable pressure difference within the native coronary artery such that the pressure at entrance to is higher than that at exit from the artery. This effect is least significant when the stent is small and is placed near the entrance end of the coronary artery, and most significant when the stent is large and is placed near the exit end. From a clinical standpoint the first of these scenarios produces the least disturbance to flow and possibly a reduced chance of restenosis, while the reverse is true in the second scenario.

In the course of events leading to sudden cardiac death a fall in coronary blood supply is more likely to be the *cause* rather than the consequence of a fall in cardiac output. Because of the sudden and unexpected timing of the phenomenon, the fall in coronary blood supply may be caused by a disruption in the *dynamics* of coronary blood flow as discussed in this and previous chapters.

Broken heart syndrome, where transient and reversible ventricular dysfunction is observed in patients without coronary artery disease, may be another example of a disruption in the *dynamics* of the coronary circulation, that is, another example of a dynamic pathology. Furthermore, while in sudden cardiac death the dynamic pathology is usually coupled with structural pathologies, in broken heart syndrome, where the effects are transient, reversible and free from coronary artery disease, the dynamic pathology is in "pure form".

References

1. Alderson H, Zamir M, 2001. Smaller, stiffer coronary bypass can moderate or reverse the adverse effects of wave reflections. J Biomech 34:1455-1462.
2. Alderson H, Zamir M, 2004. Effects of stent stiffness on local haemodynamics with particular reference to wave reflections. J Biomech 37:339-348.
3. Appelbaum E, Nicolson GHB, 1935. Occlusive diseases of the coronary arteries. Am Heart J 10:662-680.
4. Armentano RL, Barra JG, Levenson J, Simon A, Pichel RH, 1995. Arterial wall mechanics in conscious dogs: Assessment of viscous, inertial and elastic moduli to characterize the aortic wall behavior. Circ Res 76:468-478.
5. Arts T, Kruger RTI, Van Gerven W, Lambregts JAC, 1979. Propagation velocity and reflection of pressure waves in the canine coronary artery. Am J Physiol 237:H469-474.
6. Arts T, Reneman RS, 1985. Interaction between intramyocardial pressure (IMP) and myocardial circulation. J Biomech Eng 107:51-56.
7. Atabek HB, Chang CC, 1961. Oscillatory flow near the entry of a circular tube. ZAMP 12:185-201.
8. Atabek HB, Lew HS, 1966. Wave propagation through a viscous incompressible fluid contained in an initially elastic tube. Biophys J 6:481-503.
9. Attinger EO, 1963. *Pulsatile Flow*. McGraw-Hill, New York.
10. Baird RJ, Goldbach MM, de la Rocha A, 1972. Intramyocardial pressure. The persistence of its transmural gradient in the empty heart and its relationship to myocardial oxygen consumption. J Thorac Cardiov Surg 64:635-646.
11. Bassingthwaighte JB, King RB, Roger SA, 1989. Fractal nature of regional myocardial blood flow heterogeneity. Circ Res 65:578-590.
12. Bassingthwaighte JB, Van Beek JHJM, King RB, 1990. Fractal branchings: The basis of myocardial flow heterogenities? Ann NY Acad Sci 591:392-401.
13. Batchelor GK, 1967. *An Introduction to Fluid Dynamics*. Cambridge University Press, Cambridge, UK.
14. Baroldi G, 1983. Diseases of the coronary artery. In: *Cardiovascular Pathology*, Silver MD (ed). University Park Press, Baltimore, MD.
15. Baroldi G, Mantero O, Scomazzoni G, 1956. The collaterals of the coronary arteries in normal and pathologic hearts. Circ Res 4:223-229.
16. Baroldi G, Scomazzoni G, 1967. *Coronary Circulation in the Normal and Pathologic Heart*. Armed Forces Institute of Pathology, Washington, DC.

17. Bean WB, 1937. Infarction of the heart: A morphological and clinical appraisal of three hundred cases, part I. Predisposing and precipitaing conditions. Am Heart J 14:684-702.

18. Beard DA, Bassingthwaighte JB, 2000. The fractal nature of myocardial blood flow emerges from a whole-organ model of arterial network. Vasc Res 37:282-296.

19. Bellamy RF, 1978. Diastolic coronary artery pressure-flow relations in the dog. Circ Res 43:92-101.

20. Benchimol A, Stegall HF, Gartlan JL, 1971. New method to measure phasic coronary blood velocity in man. Am Heart J 81:93-101.

21. Bergman LE, DeWitt KJ, Fernandez RC, Botwin MR, 1971. Effect of non-Newtonian behaviour on volumetric flow rate for pulsatile flow of blood in a rigid tube. J Biomech 4:229-231.

22. Berne RM, Rubio R, 1979. Coronary circulation. In: *Handbook of Physiology. The Cardiovascular System*, Berne RM, Sperelakis N (eds). Sec 2, Vol 1:873-952. American Physiological Society, Washington DC.

23. Berthier B, Bouzerar R, Legallais C, 2002. Blood flow patterns in an anatomically realistic coronary vessel: influence of three different reconstruction methods. J Biomech 35:1347-1356.

24. Beyar R, Sideman S, 1987. Time-dependent coronary blood flow distribution in left ventricular wall. Am J Physiol 252:H417-433.

25. Beyer WH, 1978. *CRC Handbook of Mathematical Sciences*. CRC Press, West Palm Beach, FL.

26. Bloos F, 1998. Coronary circulatory reserve in normotensive hyperdynamic sepsis. PhD Thesis, University of Western Ontario.

27. Blumgart HL, Zoll PM, Freedberg AS, Gilligan DR, 1950. The experimental production of intercoronary arterial anastomoses and their functional significance. Circulation 1:10-27.

28. Brigham EO, 1988. *The Fast Fourier Transform and its Applications*. Prentice Hall, Englewood Cliffs, NJ.

29. Bruinsma P, Arts T, Dankelman J, Spaan JAE, 1988. Model of the coronary circulation based on pressure dependence of coronary resistance and compliance. Basic Res Cardiol 83:510-524.

30. Cameron A, Kemp HG Jr, Green GE, 1985. Internal mammary artery grafts, fifteen years clinical follow-up. Circulation 72:III-293.

31. Camiletti SE, Zamir M, 1984. Entry length and pressure drop for developing Poiseuille flows. Aeronaut J 88:265-269.

32. Canty JM Jr, Klocke FJ, Mates RE, 1985. Pressure and tone dependence of coronary diastolic input impedance and capacitance. Am J Physiol 248:H700-711.

33. Canty JM Jr, Klocke FJ, Mates RE, 1987. Characterization of capacitance-free pressure-flow relations during single diastoles in dogs using an RC model with pressure-dependent parameters. Circ Res 60:273-282.

34. Caro CG, Pedley TJ, Schroter RC, Seed WA, 1978. *The Mechanics of the Circulation*. Oxford University Press, Oxford, UK.

35. Cebral JR, Lohner R, Choyke PL, Yim PJ, 2001. Merging of intersecting triangulations for finite modeling. J Biomech 34:815-819.

36. Chadwick RS, Tedgui A, Michel JB, Ohayon J, Levy BI, 1988. A theoretical model for myocardial blood flow. In: *Cardiovascular Dynamics and Models*, Brun P, Chadwick RS, Levy BI (eds). INSERM, Paris.

37. Chang CC, Atabek HB, 1961. The inlet length for oscillatory flow and its effects on the determination of the rate of flow in arteries. Phys Med Biol 6:303-317.

38. Chang L-J, Tarbell JM, 1988. A numerical study of flow in curved tubes simulating coronary arteries. J Biomech 21:927-937.

39. Changizi MA, Cherniak C, 2000. Modeling the large-scale geometry of human coronary arteries. Can J Physiol Pharmacol 78:603-611.

40. Chilian WM, Marcus ML, 1984. Coronary venous outflow persists after cessation of coronary arterial inflow. Am J Physiol 247:H984-990.

41. Chilian WM, Marcus ML, 1985. Effects of coronary and extravascular pressure on intramyocardial and epicardial blood velocity. Am J Physiol 248:H170-178.

42. Coffman JD, Gregg DE, 1960. Reactive hyperemia characteristics of the myocardium. Am J Physiol 199:H1143-1149.

43. Cogdell JR, 1999. *Foundations of Electric Circuits*. Prentice Hall, Upper Saddle River, NJ.

44. Cohen MV, 1978. The functional value of coronary collaterals in myocardial ischemia and therapeutic approach to enhance collateral flow. Am Heart J 95:396-404.

45. Cokelet GR, 1967. Comments on the Fahraeus-Lidqvist effect. Biorheology 4:123-126.

46. Cox RH, 1969. Comparison of linearized wave propagation models for arterial blood flow analysis. J Biomech 2:251-265.

47. Craiem D, Armentano R, 2003. A new apparent compliance concept as a simple lumped model. Cardiovasc Eng 3:81-83.

48. Crouch JE, 1972. *Functional Human Anatomy*. Lea & Febiger, Philadelphia, PA.

49. Dole WP, Bishop VS, 1982. Influence of autoregulation and capacitance on diastolic coronary artery pressure-flow relationships in the dog. Circ Res 51:261-270.

50. Dolmatch BL, Blum U, 2000. *Stent Grafts. Current Clinical Practice*. Thieme, New York.

51. Downey JM, Kirk ES, 1975. Inhibition of coronary blood flow by a vascular waterfall mechanism. Circ Res 36:753-760.

52. Duan B, Zamir M, 1992. Viscous damping in one-dimensional wave transmission. J Acoust Soc Am 92:3358-3363.

53. Duan B, Zamir M, 1993. Reflection coefficients in pulsatile flow through converging junctions and the pressure distribution in a simple loop. J Biomech 26:1439-1447.

54. Duan B, Zamir M, 1995. Mechanics of wave reflections in a coronary bypass loop model: The possibility of partial flow cut-off. J Biomech 28:567-574.

55. Duan B, Zamir M, 1995. Pressure peaking in pulsatile flow through arterial tree structures. Ann Biomed Eng 23:794-803.

56. Dumoulin C, Cochelin B, 2000. Mechanical behaviour modelling of balloon-expandable stents. J Biomech 33:1461-1470.

57. Edwards WS, Jones WB, Dear HD, Kerr AR, 1970. Direct surgery for coronary artery disease: Technique for left anterior descending coronary artery bypass. JAMA 211:1182-1184.

58. Einav S, Avidor J, Vidne B, 1985. Haemodynamics of coronary artery-saphenous vein bypass. J Biomed Eng 7:305-309.

59. Eng C, Jentzer JH, Kirk ES, 1982. The effects of the coronary capacitance on the interpretation of diastolic pressure-flow relationships. Circ Res 50:334-341.

60. Eng C, Kirk ES, 1984. Flow into ischemic myocardium and across coronary collateral vessels is modulated by a waterfall mechanism. Circ Res 55:10-17.

61. Engel HJ, Torres C, Page HL Jr, 1975. Major variations in anatomical origin of the coronary arteries: Angiographic observations in 4250 patients without associated congenital heart disease. Cathet Cardiovasc Diagn 1:157-169.

62. Fabregues S, Baijens K, Rieu R, Bergeron P, 1998. Hemodynamics of endovascular prostheses. J Biomech 31:45-54.

63. Fahraeus R, Lindqvist T, 1931. The viscosity of the blood in narrow capillary tubes. Am J Physiol 96:562-568.

64. Fedor JM, McIntosh DM, Rembert JC, Greenfield JC Jr, 1978. Coronary and transmural myocardial blood flow responses in awake domestic pigs. Am J Physiol 235:H435-444.

65. Feigl EO, 1983. Coronary physiology. Physiol Rev 63:1-205

66. Folkow B, Neil E, 1971. *Circulation*. Oxford University Press, New York.

67. Fischman DL, Leon MB, Baim DS, Schatz RA, et al, 1994. A randomized comparison of coronary-stent placement and balloon angioplasty in the treatment of coronary artery disease. N Engl J Med 331:496-501.

68. Folts JD, Rowe GG, Kahn DR, Young WP, 1979. Phasic changes in human right coronary blood flow before and after repair of aortic insufficiency. Am Heart J 97:211-215.

69. Fulton WFM, 1965. *The Coronary Arteries*. Thomas, Springfield, IL.

70. Fung YC, 1981. *Biomechanics: Mechanical Properties of Living Tissues*. Springer-Verlag, New York.

71. Fung YC, 1984. *Biodynamics: Circulation*. Springer-Verlag, New York.

72. Furuse A, Klopp EH, Brawley RK, Gott VL, 1972. Haemodynamics of aorta-to-coronary artery bypass. Ann Thorac Surg 14:282-303.

73. Gensini GG, 1975. *Coronary Arteriography*. Futura, Mount Kisco, NY.

74. Gensini GG, Da Costa BCB, 1969. The coronary collateral circulation in living man. Am J Cardiol 24:393-400.

75. Geven MCF, Bohte VN, Aarnoudse WH, van den Berg PMJ, Rutten MCM, Pijls NHJ, van de Vosse FN, 2004. A physiologically representative *in vitro* model of the coronary circulation. Physiol Meas 25:891-904.

76. Ginsberg JH, Genin J, 1984. *Statics and Dynamics*. John Wiley and Sons, New York.

77. Gottwik MG, Puschmann S, Wusten B, Nienaber C, Muller K-D, Hofmann M, Schaper W, 1984. Myocardial protection by collateral vessels during experimental coronary ligation: A prospective study in a canine two-infarction model. Basic Res Cardiol 79:337-343.

78. Gow BS, Schonfeld D, Patel DJ, 1974. The dynamic elastic properties of the canine left circumflex coronary artery. J Biomech 7:389-395.

79. Gow BS, Taylor MG, 1968. Measurement of viscoelastic properties of arteries in the living dog. Circ Res 23:111-122.

80. Gradshteyn IS, Ryzhik IM, 1965. *Table of Integrals, Series, and Products*. Academic Press, New York.

81. Grayson J, 1982. Functional morphology of the coronary circulation. In: Kalsner S (ed), *The Coronary Artery*. Croom-Helm, London.

82. Green GE, Hutchinson JE, McCord C, 1971. Choice of saphenous vein segment for aortocoronary grafts. Surgery 69:924-927.

83. Gregg DE, 1950. *Coronary Circulation in Health and Disease*. Lea & Febiger, Philadelphia, PA.

84. Gregg DE, Fisher LC, 1963. Blood supply to the heart. In: *Handbook of Physiology. Circulation*. Section 2, Vol 2:1517-1584, Am Physiol Soc, Washington, DC.

85. Gregg DE, Green HD, Wiggers CJ, 1935. Phasic variations in peripheral coronary resistance and their determinants. Am J Physiol 130:362-373.

86. Gregg DE, Green HD, 1940. Registration and interpretation of normal phasic inflow into a left coronary artery by an improved differential nanometric method. Am J Physiol 130:114-125.

87. Heineman FW, Grayson J, 1985. Transmural distribution of intramyocardial pressure measured by micropipette technique. Am J Physiol 249:H1216-1223.

88. He X, Ku DN, 1996. Pulsatile flow in the human left coronary artery bifurcation: average conditions. J Biomech Eng 118:74-82.

89. Heuser RR, 1999. *Peripheral Vascular Stenting for Cardiologists*. Martin Dunitz, London.

90. Hoffman JIE, Spaan JAE, 1990. Pressure-flow relations in coronary circulation. Physiol Rev 70:331-390.

91. Holenstein R, Nerem RM, 1990. Parametric analysis of flow in the intramyocardial circulation. Ann Biomed Eng 18:347-365.

92. Horsfield K, Woldenberg MJ, 1989. Diameters and cross-sectional areas of branches in the human pulmonary arterial tree. Anat Rec 223:245-251.

93. Hunter J, 1861. *Essays and Observations on Natural History, Anatomy, Physiology, and Geology*. Van Voorst, London.

94. James TN, 1961. *Anatomy of the Coronary Arteries*. Harper and Brothers, New York.

95. James TN, 1970. The delivery and distribution of coronary collateral circulation. Chest 58:183-203.

96. Judd RM, Levy BI, 1991. Effects of the barium-induced cardiac contraction on large and small vessel intramyocardial blood volume. Circ Res 68:217-225.

97. Judd RM, Mates RE, 1991. Coronary input impedance is constant during systole and diastole. Am J Physiol 260:H1841-1851.

98. Judd RM, Redberg DA, Mates RE, 1991. Diastolic coronary resistance and capacitance are independent of the duration of diastole. Am J Physiol 260:H943-952.

99. Kaazempur-Mofrad MR, Ethier CR, 2001. Mass transport in an anatomically realistic human right coronary artery. Ann Biomed Eng 29:109-120.

100. Kajiya F, Klassen GA, Spaan JAE, Hoffman JIE (eds), 1990. *Coronary Circulation: Basic Mechanism and Clinical Relevance*. Springer-Verlag, Tokyo.

101. Kajiya F, Matsuoko S, Ogasawara Y, Hiramatsu O, Kanazawa S, Wada Y, Tadaoka S, Tsujioka K, Fujiwara T, Zamir M,1993. Velocity profiles and phasic flow patterns in ten patients with non-stenotic left anterior descending arteries during cardiac surgery. Cardiovasc Res 27:845-850.

102. Kajiya F, Tsujioka K, Goto M, Wada Y, Chen XL, Nakai M, Tadaoka S, Hiramatsu O, Ogasawara Y, Mito K, Tomonaga G, 1986. Functional characteristics of intramyocardial capacitance vessels during diastole in the dog. Circ Res 56:310-323.

103. Karch R, Schreiner W, Neumann F, Neumann M, 2000. Modeling of coronary vascular trees: Generating optimized structures. In:*Medical Applications of Computer Modeling: Cardiovascular and Ocular Systems*. Martonen TB (ed), Southampton, Boston: WIT Press, pp:59-74.

104. Kassab GS, Berkley J, Fung Y-CB, 1997. Analysis of pig's coronary arterial blood flow with detailed anatomical data. Ann Biomed Eng 25:204-217.

105. Kassab GS, Le KN, Fung Y-CB, 1999. A hemodynamic analysis of coronary capillary blood flow based on anatomic and distensibility data. Am J Physiol 277:H2158-2166.

106. Kassab GS, Rider CA, Tang NJ, Fung YC, Bloor CM, 1993. Morphometry of pig coronary arterial trees. Am J Physiol 265:H350-365.

107. Katz SA, Feigl EO, 1988. Systole has little effect on diastolic coronary artery blood flow. Circ Res 62:443-451.

108. Khouri EM, Gregg DE, Lowensohn HS, 1968. Flow in the major branches of the left coronary artery during experimental coronary insufficiency in the unanesthetized dog. Circ Res 23:99-109.

109. Kirpalani A, Park H, Butany J, Johnson KW, Ojha M, 1999. Velocity and wall shear stress patterns in the human right coronary artery. J Biomech Eng 121:370-375.

110. Klocke FJ, Mates RE, Canty JM, Ellis AK, 1985. Coronary pressure-flow relationships: controversial issues and probable implications. Circ Res 56:310-323.

111. Klocke FJ, Weinstein IR, Klocke JF, Ellis AK, Kraus DR, Mates RE, Canty JM, Anbar RD, Romanowski RR, 1981. Zero-flow pressures and pressure-flow relationships during single long diastoles in the canine coronary bed before and during maximum vasodilatation: limited influence of capacitive effects. J Clin Invest 68:970-980.

112. Kosa I, Blasini R, Schneider-Eicke J, Dickfield T, Neumann FJ, Ziegler S, Matsunari I, Neverve J, Schomig A, Schwaiger M, 1999. Early recovery of coronary flow reserve after stent implantation as assessed by positron emission tomography. J Am College Cardiol 34:1036-1041.

113. Krams R, 1989. Varying elastance concept may explain coronary systolic flow impediment. Am J Physiol 257:H1471-1479.

114. Krams R, Sipkema P, Zegers J, Westerhof N, 1989. Contractility is the main determinant of coronary systolic flow impediment. Am J Physiol 257:H1936-1944.

115. Kresh JY, Fox M, Brockman SK, Noordergraaf A, 1990. Model-based analysis of transmural vessel impedance and myocardial circulation dynamics. Am J Physiol 258:H262-276.

116. Kreyszig E, 1983. Advanced Engineering Mathematics. Wiley, New York.

117. LaBarbera M, 1990. Principles of design of fluid transport systems in zoology. Science 249:992-1000.

118. LaBarbera M, 1991. Inner currents: How fluid dynamics channels natural selection. Sciences Sept/Oct:30-37.

119. LaBarbera M, Vogel S, 1982. The design of fluid transport systems in organisms. Am Scientist 70:54-60.

120. Larson RG, 1999. The Structure and Rheology of Complex Fluids. Oxford University Press, New York.

121. Lee J, Chambers DE, Akizuki S, Downey JM, 1984. The role of vascular capacitance in the coronary arteries. Circ Res 55:751-762.

122. Levine R, Goldsmith HL, 1977. Particle behaviour in flow through small bifurcations. Microvasc Res 14:319-344.

123. Lew HS, Fung YC, 1970. Entry length into blood vessels at arbitrary Reynolds number. J Biomech 3:23-38.

124. Lighthill M, 1975. *Mathematical Biofluiddynamics*. Society for Industrial and Applied Mathematics, Philadelphia, PA.

125. Ling SC, Atabek HB, 1972. A nonlinear analysis of pulsatile flow in arteries. J Fluid Mech 55:493-511.

126. Lower R, 1669. *Tractatus de corde*. Elsevier, Amsterdam.

127. Luzsa G, 1974. *X-ray Anatomy of the Vascular System*. Lippincott, Philadelphia, PA.

128. Marcus ML, 1983. *The Coronary Circulation in Health and Disease*. McGraw-Hill, New York.

129. Marcus M, Wright C, Doty D, Eastham C, Laughlin D, Krumm P, Fastenow C, Brody M, 1981. Measurements of coronary velocity and hyperemia in the coronary circulation of humans. Circ Res 49:877-891.

130. Mates RE 1987 Coronary blood flow modeling. In: *Systems and Control Encyclopedia*, Singh MG (ed), Pergamon Press, Oxford, UK. Vol.2:850-853.

131. Mates RE, 1993. The coronary circulation. J Biomech Eng 115:558-561.

132. Mayrovitz HN, Roy J, 1983. Microvascular blood flow: evidence indicating a cubic dependence on arteriolar diameter. Am J Physiol 245:H1031-1038.

133. McAlpine WA, 1975. *Heart and Coronary Arteries*. Springer-Verlag, New York.

134. McCormack PD, Crane L, 1973. *Physical Fluid Dynamics*. Academic Press, New York.

135. McDonald DA, 1974. *Blood Flow in Arteries*. Edward Arnold, London.

136. McHale PA, Dube GP, Greenfield JC Jr, 1987. Evidence of myogenic vasomotor activity in the coronary circulation. Prog Cardiovasc Dis 30(2-Sep/Oct):139-146.

137. McLachlan NW, 1955. *Bessel Functions for Engineers*. Clarendon Press, Oxford, UK.

138. Mehan VK, Kaufmann U, Urban P, Chatelain P, Meier B, 1995. Stenting with the half (disarticulated) Palmaz-Schatz stent. Cathet Cardiovasc Diagn 34:122-127.

139. Meriam JL, 1980. *Engineering Mechanics, Statics and Dynamics*. John Wiley and Sons, New York.

140. Migliavacca F, Petrini L, Colombo M, Auricchio F, Pietrabissa R, 2002. Mechanical behavior of coronary stents investigated through the finite method. J Biomech 35:803-811.

141. Milnor WR, 1989. *Hemodynamics*. Williams and Wilkins, Baltimore, MD.

142. Miller DW Jr, 1977. *The Practice of Coronary Artery Bypass Surgery*. Plenum, New York.

143. Moore JE Jr, Guggenheim N, Delfino A, Doriot P-A, Dorsaz P-A, Rutishauser W, Meister J-J, 1994. Preliminary analysis of the effects of blood vessel movement on the blood flow patterns in the coronary arteries. J Biomech Eng 116:302-306.

144. Moore JE, Weydahl ES, Santamarina A, 2001. Frequency dependence of dynamic curvature effects on flow through coronary arteries. J Biomech Eng 123:129-133.

145. Morgan GW, Kiely JP, 1954. Wave propagation in a viscous liquid contained in a flexible tube. J Acoustic Soc Am 26:323-328.

146. Morris JJ, Peter RH, 1971. Coronary circulation. In: Conn HL, Horwitz O (eds), *Cardiac and Vascular Diseases*, Vol 1. Lea & Febiger, Philadelphia, PA.

147. Murray CD, 1926. The physiological principle of minimum work. I. The vascular system and the cost of blood volume. Proc Natl Acad Sci 12:207-214.

148. Myers JG, Moore JA, Ojha M, Johnston KW, Ethier CR, 2001. Factors influencing blood flow patterns in the human right coronary artery. Ann Biomed Eng 29:109-120.

149. Natarajan S, Mokhtarzadeh-Dehghan MR, 2000. A numerical and experimental study of periodic flow in a model of a corrugated vessel with application to stented arteries. Med Eng Phys 22:555-566.

150. Nazarian GK, Bjarnason H, Dietz CA Jr, Bernadas CA, Hunter DW, 1996. Iliofemoral venous stenoses: effectiveness of treatment with metallic stents. Radiology 200:193-199.

151. Nematzadeh D, Rose JC, Schryver Th, Huang HK, Kot PA, 1984. Analysis of methodology for measurement of intramyocardial pressure. Basic Res Cardiol 79:86-97.

152. Neufeld HN, Schneeweiss A, 1983. *Coronary Artery Disease in Infants and Children*. Lea & Febiger, Philadelphia, PA.

153. Nichols WW, O'Rourke WF, 1998. *McDonald's Blood Flow in Arteries*. Edward Arnold, London.

154. Noorani HZ, 1997. *Coronary Stents: Clinical Experience and Cost-effectiveness*. Canadian Coordinating Office for Health Technology Assessment, Ottawa, ON.

155. Norman JC, Lawrence EP, Cooper T (eds), 1975. *Coronary Artery Medicine and Surgery. Concepts and Controversies*. Appleton-Century-Crofts, New York.

156. Nubar Y, 1971. Blood flow, slip, and viscometry. Biophys J 11:252-264.

157. Olsson RA, Bunger R, Spaan JAE, 1986. Coronary circulation. In: *The Heart and Cardiovascular System: Scientific Foundations*, Fozzard HE, Haber E, Jennings RB, Katz AM, Morgan HE (eds), pp:1393-1425. Raven Press, New York.

158. Osborn MJ, 1996. Sudden cardiac death: A. Mechanisms, incidence, and prevention of sudden cardiac death. In: *Mayo Clinic Practice of Cardiology*, pp:862-894. Giuliani ER, Gersh BJ, McGoon, MD, Hayes DL, Schaff HV (eds). Mosby, St. Louis, MO.

159. Perktold K, Hofer M, Rappitsch G, Loew M, Kuban BD, Friedman MH, 1998. Validated computation of physiologic flow in a realistic coronary artery branch. J Biomech 31:217-228.

160. Perktold K, Nerem RM, Peter RO, 1991. A numerical calculation of flow in a curved tube model of the left main coronary artery. J Biomech 24:175-189.

161. Perlmutt S, Riley RL, 1963. Hemodynamics of collapsible vessels with tone: the vascular waterfall. J Appl Physiol 18:924-932.

162. Quick CM, Berger DS, Noordergraaf A, 1998. Apparent arterial compliance. Am J Physiol 274:H1393-1403.

163. Quick CM, Berger DS, Hettrick DA, Noordergraaf A, 2000. True arterial system compliance estimated from apparent arterial compliance. Ann Biomed Eng 28:291-301.

164. Quick CM, Berger DS, Noordergraaf A, 2002. Arterial pulse wave reflection as feedback. IEEE Trans Biomed Eng 49:440-445.

165. Rabbany SY, Kresh JY, Noordergraaf A, 1989. Intramyocardial pressure: Interaction of myocardial fluid pressure and fiber stress. Am J Physiol 257:H357-364.

166. Rahimtoola SH, 1982. Coronary bypass surgery for chronic angina–1981. A perspective. Circulation 65:225-241.

167. Rodbard S, 1975. Vascular caliber. Cardiology 60:4-49.

168. Rouse H, Ince S, 1957. *History of Hydraulics*. Dover Publications, New York.

169. Rumberger JA, Nerem, RM, 1977. A method-of-characteristics calculation of coronary blood flow. J Fluid Mech 82:429-448.

170. Santamarina A, Weydahl E, Siegel JM Jr, Moore JE Jr, 1998. Computational analysis of flow in a curved tube model of the coronary arteries: effects of time-varying curvature. Ann Biomed Eng 26:944-954.

171. Savoie I, Sheps S, 1996. *Coronary Stents: An Appraisal of Controlled Clinical Studies*. British Columbia Office of Health Technology Assessment, Discussion Paper Series, The University of British Columbia.

172. Schaper W (1971). *The Collateral Circulation of the Heart*. North-Holland, Amsterdam.

173. Schlesinger MJ, 1938. An injection plus dissection study of coronary artery occlusions and anastomoses. Am Heart J 15:528-568.

174. Schlichting H, 1979. *Boundary Layer Theory*. McGraw-Hill, New York.

175. Schreiner W, Neumann F, Neumann M, Karch R, End A, Roedler SM, 1997. Limited bifurcation asymmetry in coronary arterial tree models generated by constrained constructive optimization. J Gen Physiol 109:129-140.

176. Serruys PW, De Jaegere P, Kiemeneij F, Macaya C, et al, 1994. A comparison of balloon-expandable-stent implantation with balloon angioplasty in patients with coronary artery disease. N Engl J Med 331:489-495.

177. Sexl T, 1930. Über den von E.G. Richardson entdeckten "Annulareffekt." Zeit Physik 61:349-362.

178. Sherman TF, 1981. On connecting large vessels to small. The meaning of Murray's Law. J Gen Physiol 78:431-453.

179. Silver MD (ed), 1983. *Cardiovascular Pathology*. University Park Press, Baltimore, MD.

180. Sjoquist P-O, Duker G, Almgren O, 1984. Distribution of the collateral blood flow at the lateral border of the ischemic myocardium after acute coronary occlusion in the pig and the dog. Basic Res Cardiol 79:164-175.

181. Smith NP, Pullan AJ, Hunter PJ, 2000. Generation of an anatomically based geometric coronary model. Ann Biomed Eng 28:14-25.

182. Soto B, Russel RO, Moraski RE, 1976. *Radiographic Anatomy of the Coronary Arteries: An Atlas*. Futura, New York.

183. Spaan JAE 1991. *Coronary Blood Flow*. Kluwer Academic Publishers, Dordrecht, The Netherlands.

184. Spaan JAE, Breuls NPW, Laird JD 1981. Diastolic-systolic coronary flow differences are caused by intramyocardial pump action in the anesthesized dog. Circ Res 49:584-593.

185. Spalteholz W 1907. Die Coronararterien des Herzen. Verhandlungen der Anatomischen Gesellschaft (Supplement to Anatomischer Anzeiger) 21st Meeting at Wurzburg, pp 141-153.

186. Spiegel MR, 1968. *Mathematical Handbook of Formulas and Tables*. McGraw-Hill, New York.

187. Spiller P, Schmiel FK, Politz B, Block M, Fermor U, Hackbarth W, Jehle J, Korfer R, Pannek H, 1983. Measurement of systolic and diastolic flow rates in the coronary artery system by x-ray densitometry. Circulation 68:337-347.

188. Stergiopulos N, Westerhof BE, Westerhof N, 1999. Total arterial inertance as the fourth element of the windkessel model. Am J Physiol 276:H81-88.

189. Suga H, Sagawa K, Shoukas AA, 1973. Load independence of the instantaneous pressure-volume ratio of the canine left ventricle and effects of epinephrine and heart rate on the ratio. Circ Res 32:314-322.

190. Thompson D'AW, 1942. *On Growth and Form.* Cambridge University Press, Cambridge, UK.

191. Tillmanns H, Ikeda S, Hansen H, Sarma JSM, Fauvel JM, Bing RJ, 1974. Microcirculation in the ventricle of the dog and turtle. Circ Res 34:561-569.

192. Tokaty GA, 1971. *A History and Philosophy of Fluidmechanics.* Foulis, Henley-on-Thames, Oxfordshire.

193. Tritton DJ, 1988. *Physical Fluid Dynamics.* Clarendon Press, Oxford.

194. Uchida S, 1956. The pulsating viscous flow superimposed on the steady laminar motion of incompressible fluid in a circular pipe. ZAMP 7:403-422.

195. Van Huis GA, Sipkema P, Westerhof N, 1987. Coronary input impedance during the cardiac cycle as determined by impulse response method. Am J Physiol 253:H317-324.

196. Vlodaver Z, Neufeld HN, Edwards JE, 1975. *Coronary Arterial Variations in the Normal Heart and in Congenital Heart Disease.* Academic Press, New York.

197. Walker JS, 1988. *Fourier Analysis.* Oxford University Press, New York.

198. Watson GN, 1958. *A Treatise on the Theory of Bessel Functions.* Cambridge University Press, Cambridge, UK.

199. Weibel ER, 1984. *The Pathways to Oxygen: Structure and Function in the Mammalian Respiratory System.* Harvard University Press, Cambridge, MA.

200. Wentzel JJ, Whelan DM, van der Giessen WJ, van Beusekom HMM, et al, 2000. Coronary stent implantation changes 3-D vessel geometry and 3-D shear stress distribution. J Biomech 33:1287-1295.

201. West GB, Brown JH, Enquist BJ, 1997. A general model for the origin of allometric scaling laws in biology. Science 276:122-126.

202. Westerhof N, 1990. Physiological hypothesis– Intramyocardial pressure. A new concept, suggestions for measurement. Basic Res Cardiol 85:105-119.

203. Westerhof N, Elzinga G, Sipkema P, 1971. An artificial arterial system for pumping hearts. J Appl Physiol 31:776-781.

204. Westerhof N, Noordergraaf A, 1970. Arterial viscoelasticity: a generalized model-effect on input impedance and wave travel in the systemic tree. J Biomech 3:357-379.

205. Westerhof N, Sipkema P, VanHuis GA, 1983. Coronary pressure-flow relations and the vascular waterfall. Cardiovasc Res 17:162-169.

206. Willerson JT, Hillis LD, Buja LM, 1982. *Ischemic Heart Disease. Clinical and Pathological Aspects.* Raven Press, New York.

207. Wittstein IS, Thiemann DR, Lima JAC, Baughman KL, Schulman SP, Gerstenblith G, Wu KC, Rade JJ, Bivalacqua TJ, Champion HC, 2005. Neurohumoral features of myocardial stunning due to sudden emotional stress. N Engl J Med 352:539-548.

208. Womersley JR, 1955. Oscillatory motion of a viscous liquid in a thin-walled elastic tube-I: The linear approximation for long waves. Phil Mag 46:199-221.

209. Yater WM, Traum AH, Brown WG, Fitzgerald RP, Geisler MA, Wilcox BB, 1948. Coronary artery disease in men eighteen to thirty-nine years of age. Am Heart J 36:683-722.

210. Yu PN, Goodwin JF, (eds) 1977. *Progress in Cardiology.* Lea & Febiger, Philadelphia, PA.

211. Zamir M, 1972. Blood flow, slip, and viscometry. Biophys J 12:703-704.
212. Zamir M, 1976. Optimality principles in arterial branching. J Theor Biol 62:227-251.
213. Zamir M, 1978. Nonsymmetrical bifurcations in arterial branching. J Gen Physiol 72:837-845.
214. Zamir M, 1988. Distributing and delivering vessels of the human heart. J Gen Physiol 91:725-735.
215. Zamir M, 1988. The branching structure of arterial trees. Comments Theor Biol 1:15-37.
216. Zamir M, 1990. Flow strategy and functional design of the coronary network. In: *Coronary Circulation: Basic Mechanism and Clinical Relevance*, Kajiya F, Klassen GA, Spaan JAE, Hoffman JIE, (eds), Springer Verlag, Tokyo.
217. Zamir M, 1996. Secrets of the heart. Sciences 36(5):26-31.
218. Zamir M, 1996. Tree structure and branching characteristics of the right coronary artery in a right-dominant heart. Can J Cardiol 12:593-599.
219. Zamir M, 1998. Mechanics of blood supply to the heart: Wave reflection effects in a right coronary artery. Proc R Soc Lond B 265:439-444.
220. Zamir M, 1999. On fractal properties of arterial trees. J Theor Biol 197:517-526.
221. Zamir M, 2000. *The Physics of Pulsatile Flow*. Springer-Verlag, New York.
222. Zamir M, Brown N, 1982. Arterial branching in various parts of the cardiovascular system. Am J Anat 163:295-307.
223. Zamir M, Chee H, 1986. Branching characteristics of human coronary arteries. Can J Physiol Pharmacol 64:661-668.
224. Zamir M, Chee H, 1987. Segment analysis of human coronary arteries. Blood Vessels 24:76-84.
225. Zamir M, Medeiros JA, Cunningham TK, 1979. Arterial bifurcations in the human retina. J Gen Physiol 74:537-548.
226. Zamir M, Phipps S, 1987. Morphometric analysis of the distributing vessels of the kidney. Can J Physiol Pharmacol 65:2433-2440.
227. Zamir M, Phipps S, Langille BL, Wonnacott TH, 1984. Branching characteristics of coronary arteries in rats. Can J Physiol Pharmacol 62:1453-1459.
228. Zamir M, Silver MD, 1985. Morpho-functional anatomy of the human coronary arteries with reference to myocardial ischemia. Can J Cardiol 1:363-372.
229. Zamir M, Sinclair P, 1988. Roots and calibers of the human coronary arteries. Am J Anat 183:226-234.
230. Zamir M, Sinclair P, Wonnacott TH, 1992. Relation between diameter and flow in major branches of the arch of the aorta. J Biomech 25:1303-1310.
231. Zeng D, Ding Z, Friedman MH, Ethier CR, 2003. Effects of cardiac motion on right coronary artery hemodynamics. Ann Biomed Eng 31:420-429.
232. Zhao SZ, Collins MW, 1996. Flow patterns in stented arteries. Biomed Pharmacother 50:387 (abstract).
233. Zhou Y, Kassab GS, Molloi S, 1999. On the design of the coronary arterial tree: A generalization of Murray's law. Phys Med Biol 44:2929-2945.

Index

Volumes Published in This Series: